John M. McIntosh
Organic Chemistry
De Gruyter STEM

Also of Interest

Industrial Organic Chemistry.
Benvenuto, 2017
ISBN 978-3-11-049446-4, e-ISBN 978-3-11-049447-1

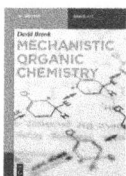

Mechanistic Organic Chemistry
David Brook, 2022
ISBN 978-3-11-056461-7, e-ISBN 978-3-11-056466-2

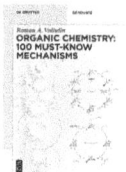

Organic Chemistry: 100 Must-Know Mechanisms
Roman Valiulin, 2020
ISBN 978-3-11-060830-4, e-ISBN 978-3-11-060837-3

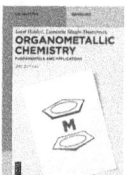

Organometallic Chemistry
Ionel Haiduc, 2022
ISBN 978-3-11-069526-7, e-ISBN 978-3-11-069527-4

John M. McIntosh

Organic Chemistry

—

Fundamentals and Concepts

2nd Edition

DE GRUYTER

Author
Prof. Dr. em. John M. McIntosh
University of Windsor
Department of Chemistry & Biochemistry
401 Sunset Ave.
Windsor ON N9B 3P4
Canada
jandj.mcintosh@gmail.com

ISBN 978-3-11-077820-5
e-ISBN (PDF) 978-3-11-077831-1
e-ISBN (EPUB) 978-3-11-077844-1

Library of Congress Control Number: 2022935913

Bibliographic information published by the Deutsche Nationalbibliothek
The Deutsche Nationalbibliothek lists this publication in the Deutsche Nationalbibliografie;
detailed bibliographic data are available on the Internet at http://dnb.dnb.de.

Preface 1st Edition

This book, which was originally entitled "The ABC's of Organic Chemistry," was derived from a set of lecture notes that had been developed over a decade of teaching introductory organic chemistry. One might ask if another text in this area was *really* necessary. In the past few years, many good texts, with superb graphics and multi-colored drawings have appeared. However, the amount of material presented in such texts is far in excess of the requirements for a one- or even two-semester introductory course and the size and cost of such commercial texts led to a significant number of hernias, both physical and financial!

In order to incorporate the desired amount of material in a reasonably sized volume, several arbitrary decisions were required. Spectroscopy is an integral part of today's science of organic chemistry and has been included as a chapter since 2005. The chemistry of carbanions formed by acid-base reactions, carbenes, rearrangements, heterocycles, and numerous other important topics are still absent due to limitations of space and time. The material and its arrangement reflect the authors prejudices. Reactions are considered by mechanistic type beginning with electrophilic additions. I have found this to be the best vehicle for introducing mechanistic concepts including resonance and inductive effects and carbocations. In general, the approach taken was to provide a concise, readable introduction to the subject *without* becoming encyclopedic. Early introduction to stereochemistry is included as it provides the necessary background for understanding the nuances of reaction mechanisms.

The text has been designed to be used in conjunction with the included workbook, which introduces and teaches organic nomenclature. Therefore, very little of this subject, which usually occupies a large part of introductory texts is included. Other, much more comprehensive texts are available in the library which can be consulted when more detailed explanations are desired. All have the title "Organic Chemistry" or some variation of this.

Problems are included at the end of each chapter. In some of these sets, problems specifically dealing with material from *previous* chapters have been included without warning. Many of the problems require application of the principles learned to new situations and therefore are an integral part of the learning process.

The author would like to thank Dr. John Hayward for generating the computed NMR spectra in Chapter 11. Also infrared spectra used in this book are reproduced from the National Institute of Standards and Technology (NIST) and are used with their permission[1].

John McIntosh, Windsor, Ontario, 2017

---

1 S.E. Stein, "Infrared Spectra" in **NIST Chemistry WebBook, NIST Standard Reference Database Number 69**, Eds. P.J. Linstrom and W.G. Mallard, National Institute of Standards and Technology, Gaithersburg MD, 20899, doi:10.18434/T4D303, (retrieved October 3, 2017).

https://doi.org/10.1515/9783110778311-201

Preface 2nd Edition

The second edition has removed some errors found in the original text, added a few new problems at the chapter ends and, most importantly, added Appendix A which contains the answers to the problems in Part I. An introduction to some of the most important molecules of nature has also been added. I hope these make the book more useful and adds to its acceptance.

John McIntosh, Windsor, Ontario, 2022

Organic Chemistry: The Fundamental Truths

1. Carbon atoms have no more than four bonds connected to them. Never five. Double bonds count as two, triple as three.
2. Carbon has no more than four bonds. Never five. Count double bonds as two. This is the same as the first rule. It is so important that it is restated.
3. If carbon has three bonds it must have a charge or be a free radical. So it must have a "+", "-", or ".".
4. The arrowhead of curved arrows points at an electrophilic center. The tail of the arrow begins at the nucleophile. The nucleophile is a Lewis base. The electrophile is a Lewis acid.
5. E1 and Sn1 reaction mechanisms involve intermediates, E2 and Sn2 do not.
6. Use dotted lines to imply delocalized charges only with great care. It is safer to avoid them and use resonance structures.
7. There are three bonds to nitrogen, and two to oxygen. You can have one more but then there must be a positive charge on the atom.
8. Hydrogen has one bond.
9. Carbon has no more than four bonds. Never five. See Rule 1.
10. Always wear safety glasses in the lab.
11. An acid transfers a proton to a base. A base removes a proton from an acid. Water is a weak acid and a weak base.
12. You cannot depend on good luck in proposing a synthesis that can yield many products besides the one you want.
13. Molecular models are good for you.
14. Chirality applies to whole objects or molecules.
15. Neither incantations nor witchcraft will cause a reaction to occur where you want it to if another, more reactive site is present in the molecule.
16. Organic chemistry still must obey the physical laws of nature. The principles learned in general chemistry regarding equilibria and Le Chatlier's Principle, acids and bases, solubilities, valence, etc., all are still important here. (See also Rules 7, 8, 9, 11 above.)
17. *[Perhaps the most important] People who teach organic chemistry give better marks to those who know these truths!!*

Contents

XIV —— Contents

Part I: **Fundamentals**

1 Introduction

1.1 What Is Organic Chemistry?

Perhaps the best operating definition of organic chemistry is "the chemistry of compounds containing carbon atoms, but excluding the carbon oxides and their metallic salts." Although CO_2, CO, HCO_3^-, and CO_3^{-2} do contain carbon, they are usually not considered to be "organic" in nature. In one sense the chemistry of organic compounds is much like that of inorganic materials, but in other ways, the chemistry differs quite considerably. It requires a more pictorial approach and usually requires some time to become familiar with its language.

Organic chemistry has the reputation among students of being a "memory work course." While it is true that a certain amount of memory work is required – as it is in any course – every effort will be made to minimize this by considering groups of reactions and the principles that guide them. This mechanistic approach will allow you to apply basic principles to new situations and arrive at sound conclusions.

In this course, we will begin where all chemistry courses begin – with a consideration of what holds atoms together in molecules – i.e., bonding principles. It is expected that most of this will be a review and therefore it will be covered rather quickly. If you encounter difficulties, please consult your freshman chemistry text.

1.2 Bonding

1.2.1 Ionic Bonds

Bonds are the forces that hold atoms together in larger arrays called molecules. These can be either *ionic* or *covalent* in nature.

It will be very helpful if you commit to memory the first ten elements of the periodic table and their atomic numbers and weights.

Recall two basic properties of atoms: i.e., the concepts of electronegativity and the electronic configuration of atoms and the stability associated with the "stable octet" of electrons corresponding to the rare gas configuration. Consider the reaction of sodium and chlorine atoms. Because sodium can achieve the neon electronic configuration by loss of one electron and chlorine can achieve the argon configuration by gaining one electron, a great deal of energy will be evolved in this electron transfer. Another way of stating the same thing is that by the transfer of one electron from the electropositive atom (Na) to the electronegative atom (Cl), the reacting system becomes *less energetic or more stable*.

https://doi.org/10.1515/9783110778311-001

$$\text{Na·} + \text{:}\overset{..}{\underset{..}{\text{Cl}}}\text{·} \longrightarrow \overset{\oplus}{\text{Na}} + \text{:}\overset{..}{\underset{..}{\text{Cl}}}\text{:}^{\ominus}$$

Because of the transfer of the electron, the sodium atom becomes positively charged and the chlorine atom becomes negatively charged – i.e., they become ions. Since opposite charges attract, these ions can be held together by electrostatic forces called *ionic bonds*.

> An ionic bond forms due to the electrostatic forces between oppositely charged ions. It is spherically symmetrical: that is it has no directional properties.

This latter point is very important. In the isolated molecule Na^+Cl^-, the chloride ion is *not* in any specific position in space relative to the sodium ion. This is *not true* in the crystal lattice of course.

Molecules held together by purely ionic bonds have several physical characteristics in common. They are generally quite soluble in water and other solvents of high dielectric constant. They are also usually crystalline materials with high melting points. Typical examples of ionic compounds are the alkali metal halides and hydroxides (e.g., LiOH, NaOH, KCl, etc.). In general, the larger the difference in the electronegativities of the atoms, the more likely a compound is to be ionic.

> The importance of the general physical properties of various classes of compounds may not, at first sight, seem obvious. However, as we progress, this importance will become clear. We will discuss these at various points and considerable attention should be paid to them.

1.2.2 Covalent Bonds

Consider the carbon atom which has the electronic configuration $1s^2\ 2s^2\ 2p^2$. In order for carbon to achieve a rare gas electron arrangement, it would need to either gain four electrons or lose four.

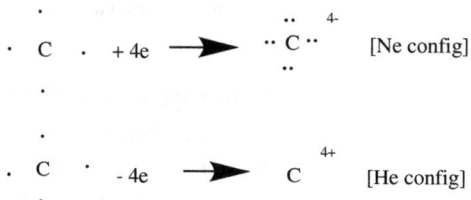

$$\cdot\ \overset{\cdot}{\underset{\cdot}{\text{C}}}\ \cdot\ +\ 4e \longrightarrow \text{:}\overset{..}{\underset{..}{\text{C}}}\text{:}^{4-} \quad \text{[Ne config]}$$

$$\cdot\ \overset{\cdot}{\underset{\cdot}{\text{C}}}\ \cdot\ -\ 4e \longrightarrow \text{C}^{4+} \quad \text{[He config]}$$

Since like charges repel each other and carbon is a small atom, these quadruply charged ions would have a very high charge density and would therefore be very unstable and so it is unlikely that either of these processes will occur. Another type of bonding must be present in molecules containing atoms like carbon.

Lewis proposed that outer-shell (valence) electrons of reacting atoms could be either transferred (leading to ionic bonds) *or shared*. Two atoms might both reach the energetically desirable stable rare gas configuration by mutual sharing of electrons. Such bonds are called *covalent bonds*. As we will see, *a covalent bond always involves a pair of electrons*. This should always be kept in mind. Since the bond is usually represented by a dash (−), it sometimes leads the beginning student to think of it as one electron. This leads to many difficulties. The concept of covalent bonds will be developed more fully in succeeding sections.

It is important to recognize that there are *two* requirements for bond formation: electrons and a place to put them (orbitals). Having said this, all bonds are not created equal. Some are stronger than others and it is important to be able to evaluate the bond strengths. A good measure of the strength of a bond is in its *length*. The stronger a bond is, the shorter it will be.

Some situations we will encounter frequently in the future involve the use of electrons on atoms such as oxygen. Consider how the nonbonding electrons on such an atom might be used. Water is a good example. The oxygen atom in water has two unshared electron pairs and therefore it is a base. When it accepts a proton to form a hydronium ion, it uses one of these electron pairs to form the new O–H bond. Therefore, it becomes positively charged.

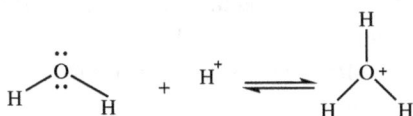

Since oxygen is an electronegative atom, this is a higher energy situation than the neutral case and therefore water is a weak base. It is important to note that the oxygen atom still has its stable octet of electrons.

Compare this to the following.

Breaking the bond with both electrons remaining associated with X (ionization) again affords a positively charged oxygen atom. However, this now has only *six* valence electrons. This is *much* more unfavorable (higher energy, less stable) than the previous case. Obviously, more than just the charge must be considered when determining the stability of an ion.

1.2.3 Molecular Orbitals

Electrons in atoms are restricted to certain areas of space around the nucleus. These areas are called orbitals. Each orbital can accommodate a maximum of *two electrons*.

Consider two hydrogen atoms. Each has one electron situated in a 1s orbital. When they combine to form molecular hydrogen (H_2), the two electrons are shared by both atoms leading to a situation where both atoms have the Helium configuration. It is important to note that both atoms are electrically neutral. Each has one-half share of a pair of electrons to balance the single positive charge on the nucleus. *In order for the electrons to be shared, the orbitals on each atom must overlap in space.* This can be represented as shown in Fig. 1.1. While these pictures are convenient, they are misleading in one important respect. Rather than being the "meat in the sandwich," the electrons are actually the "string around the package."

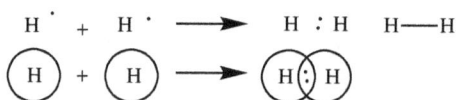

Fig. 1.1

When two atoms interact, their *atomic orbitals* (AO's) interact to form a new set of orbitals that encompass both of the nuclei. These new orbitals are called *molecular orbitals* (MO's). If n AO's are combined, quantum mechanics tells us that n MO's must be formed. According to the *Linear Combination of Atomic Orbitals* (LCAO) method, these are combined according to equations of the form

$$MO_1 = AO_1 - AO_2$$
$$MO_2 = AO_1 + AO_2$$

where AO_1 and AO_2 are the atomic orbitals on the reacting atoms. Energetically this can be represented as shown in Fig. 1.2. Note that one of the new MO's has a lower energy than the AO's and one has a higher energy. These are called the bonding and antibonding orbitals, respectively. The antibonding orbital is signified by a * symbol. Now if each AO contains only one electron and each MO can contain a maximum of two electrons, both electrons will go into the lower energy MO and a bond will be formed.

What would happen with helium which has two electrons in each atom? We still have only two AO's and therefore the diagram in Fig. 1.2 applies, but there are now four electrons so that two must go into the antibonding (higher energy) orbital. This cancels out the advantage gained by allowing two electrons to enter the bonding MO and no advantage (energetically) is gained by forming He_2. Helium is a monoatomic gas.

The molecular orbitals formed encircle both of the reacting atoms and are like the "string on the package." Nevertheless, the *highest density* of electrons lies between the two atoms and so bonds are usually written as shown in Fig. 1.1. Overlap of two s-orbitals produces a covalent bond with cylindrical symmetry. Such bonds are called σ (Greek sigma) bonds.

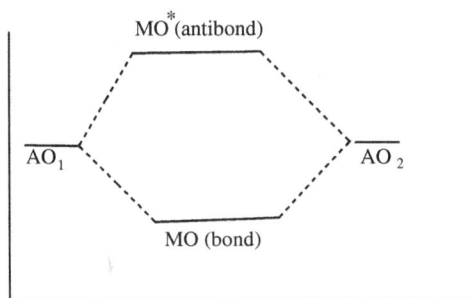

Fig. 1.2

The same concepts apply to the use of p-orbitals in forming molecular orbitals. If a p-orbital overlaps with an s-orbital, a σ bond is formed. However, if two p-orbitals are used, a slightly different situation arises due to their particular shape. Recall that p-orbitals have a nodal plane and the electron density is concentrated above and below this plane. When two p-orbitals interact, they must do so in such a way that the areas of electron density in the two atoms overlap. For this to occur, the two p-orbitals must be *coplanar*. Bonds formed in this way have electron density above and below the plane of the nuclei only and are called π (Greek pi) bonds.

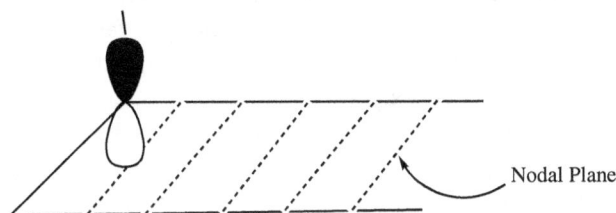

Nodal Plane

1.2.4 Hybridization

Let us return to our main topic of interest – carbon. This atom has the ground-state electronic configuration $1s^2 2s^2 2p_x^1 2p_y^1$. This might lead one to predict that, since there are only two unpaired electrons on the atom, only two bonds could be formed. For example, the reaction of a carbon with two hydrogen atoms might be schematically represented as shown below. This would satisfy the rare gas configuration of hydrogen but note that carbon still has only six valence electrons rather than the eight of the neon configuration. In fact, the simplest stable compound containing only carbon and hydrogen has the molecular formula CH_4.

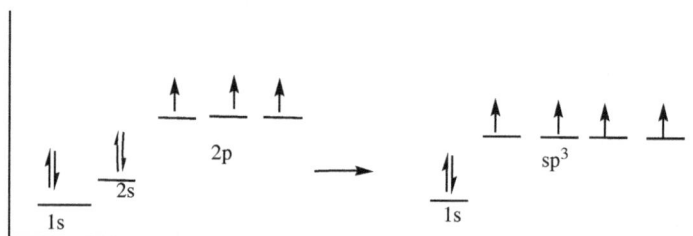

Fig. 1.3

In order to accommodate this finding we must generate a carbon atom capable of forming four bonds. To do this, energy must be put into the atom – i.e., it is no longer in the ground state. This process is called hybridization and can be represented in the following way (Fig. 1.3).

The 2s and three 2p orbitals are "mixed" (a mathematical operation) to form four new orbitals (quantum mechanics demands that the number of new orbitals be equal to the number of original orbitals). These four new orbitals are said to be degenerate because they are all of equal energy. Because they are formed from one s- and three p-orbitals they are called sp^3-hybrid orbitals. Since they are all of equal energy, the four electrons in the original orbitals will distribute themselves one in each of the sp^3-hybrid orbitals (Hund's Rule). Each hybrid now has one electron available to form a bond by overlap with an orbital from another atom – i.e., four bonds can be formed. A covalent bond resulting from the formation of a molecular orbital by the end-to-end overlap of atomic orbitals like sp^3 or s is called a sigma (σ) bond.

One might ask why this would be a favorable process. After all, it involves an input of energy and the formation of a less-stable species. The answer is complex and involves VSEPR (valence shell electron pair repulsion) theory, but the main point can be summarized by pointing out that the energy released in the formation of the two additional bonds more than compensates for the energy required for hybridization.

Hybridization can also involve only two of the three p-orbitals. In this case, the hybrid is called sp^2 and the hybridized atom has three equivalent hybrid orbitals and one electron remaining in a p-orbital. Similarly, sp-hybrids have two equivalent hybrid orbitals and two electrons in p-orbitals (Fig. 1.4). The importance of these hybridized states of carbon is enormous and will be of constant concern to us throughout this course.

1.3 Three-Dimensional Properties of Carbon

A major factor in the chemistry of organic compounds is the three-dimensional shape of the various kinds of hybridized carbon atoms. These concepts will be introduced here and will be developed more fully as we proceed.

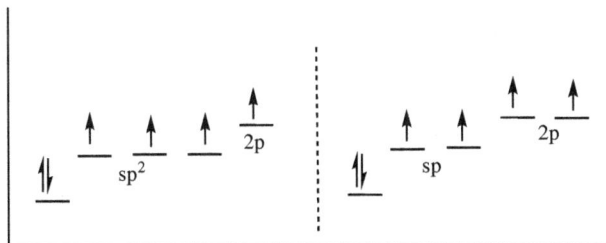

Fig. 1.4

The shape of a carbon atom including the atoms bonded to it depends on the hybridization state of the carbon atom. VSEPR theory tells us that sp^3-hybridized carbon, which has four degenerate (i.e., of equal energy) orbitals, will have these orbitals arranged in a tetrahedral shape (Fig. 1.5). This arrangement puts the atoms bonded to the carbon (a, b, c, d) at the corners of a tetrahedron with carbon at the center. The special property associated with this shape is that each of the bonded atoms is equidistant from all the others. The distances ad = bc = cd, etc. This means that a molecule such as CH_2Cl_2 can have only one arrangement provided that the carbon is sp^3-hybridized. If the four bonds were arranged in (for example) a planar fashion, two forms would be possible (Fig. 1.6). In the figure, the H–H distance in (a) is different to that in (b). However, it is known that only one molecule with this formula can be formed and the tetrahedral model accounts for this fact very nicely.

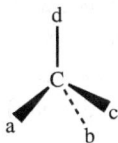

Fig. 1.5

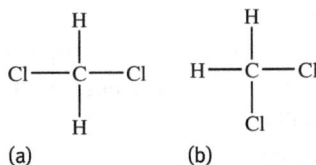

Fig. 1.6

For carbon in the sp^2-hybridized state, a different geometry exists. In this case, the three degenerate sp^2 orbitals lie in *one plane* with an angle of 120° between them. The remaining p-orbital lies perpendicular to this plane; this is depicted in Fig. 1.7. The three sp^2-hybrid orbitals can overlap effectively (i.e., form bonds) with other s- or hybrid-orbitals but the p-orbital, due to its shape, can only overlap effectively with another p-orbital leading to the pi (π) bond type and so sp^2 carbon atoms will form three σ bonds and one π bond which is usually to an atom already attached with the σ bond. Therefore, although sp^2 carbons still form four bonds, these are to only *three* atoms.

For sp-hybridized carbon, the two sp-hybrid orbitals are *colinear*. The remaining two p-orbitals are oriented as shown in Fig. 1.7. For the reasons discussed above, sp-hybrid carbon will form two σ bonds and two π bonds. In the same way as for the sp^2 carbon, the sp atom will still form four bonds, but only to *two* other atoms.

It is important to note that all three hybridized states of carbon form *four* bonds. In the sp^3 state, these are bonded to four *different* atoms, while in the sp^2 and sp states, this is not so.

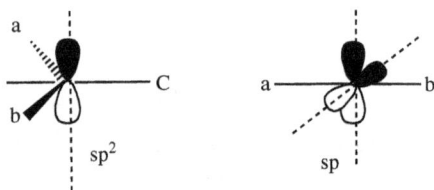

Fig. 1.7

1.4 Polar Covalent Bonds

When a covalent bond forms between two atoms of different electronegativities, the electrons in the bond will, on a time average, be closer to the more electronegative atom. (If complete transfer occurred, the bond would, of course, be ionic). This will cause the more electronegative element to become partially negatively charged, with a corresponding partial positive charge on the other atom. The bond is called a *polar covalent* bond. This is an extremely important concept in organic chemistry as it controls the reactivities of the molecules to a very large extent. The Greek symbol δ (delta) is used to indicate a partial charge on an atom (Fig. 1.8).

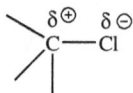

Fig. 1.8

1.5 Bonding Between Carbon Atoms

One of the most striking contrasts between inorganic and organic chemistry is shown by the apparent variable valence of carbon. For example, most metallic halides have only a few discrete formulae known (e.g., Hg_2Cl_2 and $HgCl_2$; Cu_2Cl_2 and $CuCl_2$, etc.). In contrast, organic compounds with the formulae CH_4, C_2H_6, C_3H_8, C_4H_{10}, and

C_5H_{12} are all well known. In these compounds the "apparent valence" of carbon is 4, 3, 8/3, 10/4, and 12/5.

The explanation for this lies in the ability of carbon to bond to other carbon atoms in chains. One sp^3-hybridized carbon may bond with four hydrogen atoms to give the molecule CH_4. However, it may also bond with another carbon atom and six hydrogen atoms to give C_2H_6. Some examples of this are shown in Fig. 1.9.

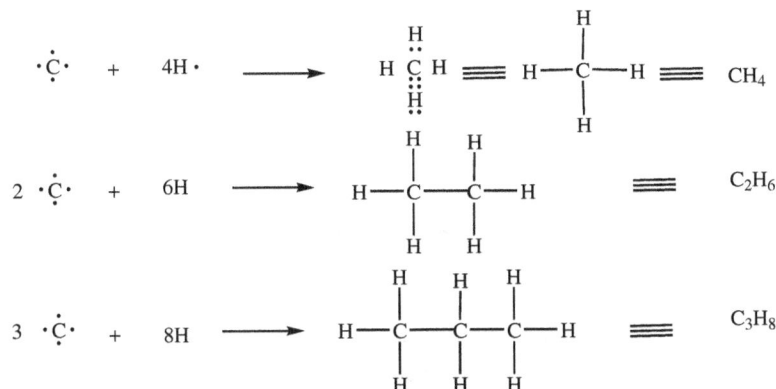

$$\cdot \overset{\cdot}{\underset{\cdot}{C}} \cdot \quad + \quad 4H \cdot \quad \longrightarrow \quad H\,\overset{\overset{\displaystyle H}{\cdots}}{\underset{\underset{\displaystyle H}{\cdots}}{C}}\,H \;\equiv\; H\!-\!\overset{\overset{\displaystyle H}{|}}{\underset{\underset{\displaystyle H}{|}}{C}}\!-\!H \;\equiv\; CH_4$$

$$2\,\cdot \overset{\cdot}{\underset{\cdot}{C}} \cdot \quad + \quad 6H \quad \longrightarrow \quad H\!-\!\overset{\overset{\displaystyle H}{|}}{\underset{\underset{\displaystyle H}{|}}{C}}\!-\!\overset{\overset{\displaystyle H}{|}}{\underset{\underset{\displaystyle H}{|}}{C}}\!-\!H \;\equiv\; C_2H_6$$

$$3\,\cdot \overset{\cdot}{\underset{\cdot}{C}} \cdot \quad + \quad 8H \quad \longrightarrow \quad H\!-\!\overset{\overset{\displaystyle H}{|}}{\underset{\underset{\displaystyle H}{|}}{C}}\!-\!\overset{\overset{\displaystyle H}{|}}{\underset{\underset{\displaystyle H}{|}}{C}}\!-\!\overset{\overset{\displaystyle H}{|}}{\underset{\underset{\displaystyle H}{|}}{C}}\!-\!H \;\equiv\; C_3H_8$$

Fig. 1.9

If sp^2-hybridized carbons are used, the situation alters somewhat because p-orbitals can only bond with other p-orbitals (do you remember why?). Such molecules contain a carbon–carbon double bond consisting of one σ bond and one π bond. Similarly, sp-hybridized carbon will form carbon–carbon triple bonds consisting of one σ and two π bonds. Because two bonds are stronger than one, a C=C bond will be *shorter* than a C–C (single) bond and a C≡C will be shorter than a C=C bond.

It must be noted that any hybridized form of carbon can form σ bonds with carbon in any of the three hybridized states. Thus, bonds like sp-sp^3, sp^2-sp^3 and sp-sp^2 are all well-known.

We are now ready to proceed to look at some simple organic molecules.

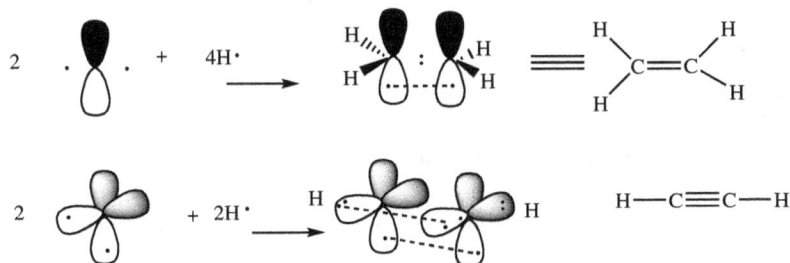

1.6 Summary

The concepts presented in this chapter are expected to be quite familiar to you and they have been presented as an overview and review. The material is of crucial importance to future discussions. *Organic chemistry is cumulative!* You must master the material presented in each chapter before proceeding to the next. Keep up with your studying and do the assigned questions. These will help you to determine what you do know and what needs more work.

1.7 Problems

1-1. Ionic compounds are generally higher melting than covalent compounds. What forces determine the melting point of a compound and why are ionic compounds higher melting?

1-2. When crystalline compounds are dissolved in a solvent, energy in the form of heat is usually absorbed: i.e., the temperature of the solution goes down. Why?

1-3. What is wrong with a structure like C–H–C? Explain your answer in terms of electron counts.

1-4. Show that in the molecule C_2H_6, each atom has a rare-gas electronic configuration and is electrically neutral.

1-5. Although CCl_4 has polar bonds, it does *not* have a dipole moment. Explain this fact.

1-6. In the molecule shown below, what atomic orbitals on each atom are used to form the numbered bonds

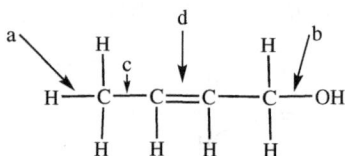

Label the numbered bonds as being pi or sigma.

1-7. Indicate, using the δ symbol, the atoms that have a partial positive or negative charge.
   a) C–N,   b) C–O,   c) N–O,   d) N–H,   e) H–O,   f) C–H

1-8. Complete the following table for the three hybridized states of carbon.

	sp^3	sp^2	sp
shape			
Number of bonds to carbon atom			
Number of sigma bonds			
Number of pi bonds			
Number of atoms bonded			

2 The Simplest Organic Molecules – The Hydrocarbons

2.1 Definitions

To begin this chapter, we need some working definitions:

1. *The hydrocarbons* are molecules that contain *only* carbon and hydrogen atoms.
2. *Saturated hydrocarbons* or *Alkanes* are molecules that contain only hydrogen and sp^3-hybridized carbon atoms. This means that there are no double or triple bonds in the molecule.
3. *Unsaturated hydrocarbons* are hydrocarbons that contain one or more double or triple bonds: i.e., they contain sp^2 and/or sp-hybridized carbon atoms. (Recall that if an sp^2-carbon is bonded to only other carbon or hydrogen atoms, one of the bonded carbons must also be either sp^2- or sp-hybridized.) Compounds that contain carbon–carbon *double* bonds are called *Alkenes* (the older name is olefins). Those that contain carbon–carbon *triple* bonds are called *Alkynes* (the older name is acetylenes).
4. *Molecular formula* is the formula for a molecule that gives the ratio of the contained elements (e.g., CH_4, C_2H_6). It tells us nothing about which atoms are directly bonded.
5. *Structural formula* is the formula that indicates specifically to which atoms each atom is bonded.
6. *Isomers* are molecules with the *same molecular* formula, but *different structural* formula. There are many subclasses of isomers and we will see examples of these as we proceed.

2.2 Positional (or Constitutional) Isomers

To illustrate the most basic type of isomerism, we will consider how the structure with the molecular formula C_4H_{10} can be built. There are two compounds known with this formula. They can be constructed in the following way (Fig. 2.1). When two sp^3-hybridized carbons join together they can do so in only one way. The two carbons in C:C are indistinguishable since simply turning the molecule end-for-end exchanges them. Therefore, adding a third carbon (or any other atom or group for that matter) on one end of the C_2 unit generates exactly the same structure as adding it to the other end. The two carbons are said to be *equivalent*. Also remember that, because sp^3 carbons are tetrahedral and sp^3 orbitals are equivalent, there is no difference between

C:C:C and C:C
         ..
         C.

https://doi.org/10.1515/9783110778311-002

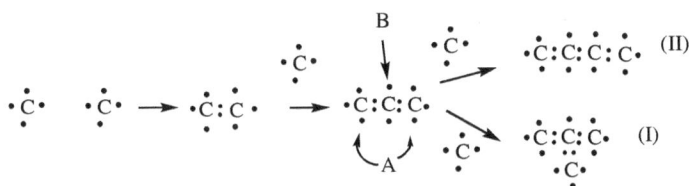

Fig. 2.1

If we now consider the C_3 unit we see that all carbons are *not* equivalent and there are two different places (A and B, Fig. 2.1) where the fourth carbon can be attached. The two atoms labeled A *are* equivalent, but they are different to B. Therefore, two isomeric products (I and II) can be formed. Completing the molecules by pairing the remaining electrons with those from eight hydrogen atoms, we get

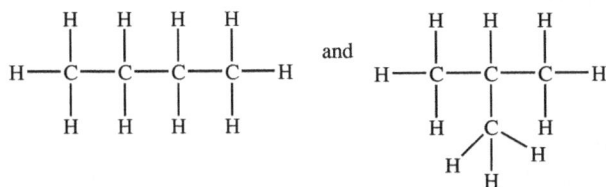

both of which have the formula C_4H_{10} but which are clearly different in terms of their structure. Such compounds are *isomers* because one atom (in this case carbon) occupies a different position on the chain.

> It is important to remember that positional isomers are different compounds. They have different physical and chemical properties. These differences may be very small or very large depending on the particular case. For example, compound I (Fig. 2.1) has a boiling point (bp) of $-12\,°C$, while that for II is $-0.5\,°C$. Conversely, ethanol and dimethyl ether, which both have the molecular formula C_2H_6O, differ in their boiling points by $103\,°C$.

2.3 Writing Formulae

Organic chemists, in common with most of humanity, are basically lazy. This manifests itself in many ways, but one aspect of this is the attempt to convey a specific piece of information using as few pen strokes as possible. For example, writing structural formulae showing every hydrogen atom and carbon–hydrogen bond is very time consuming (try writing out the full structure of one isomer of $C_{15}H_{32}$!) and several "short forms" are commonly used.

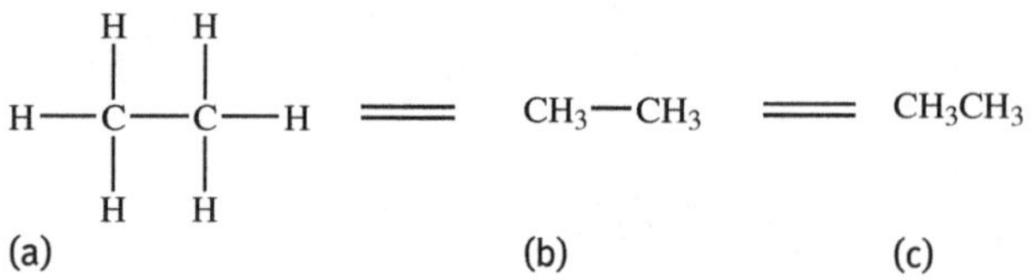

$$
\begin{array}{ccc}
\underset{\substack{|\\ H}}{\overset{\substack{H\\ |}}{H-C}}-\underset{\substack{|\\ H}}{\overset{\substack{H\\ |}}{C}}-H & = & CH_3-CH_3 & = & CH_3CH_3 \\
(a) & & (b) & & (c)
\end{array}
$$

Fig. 2.2

The first simplification is to omit all the C–H bonds. The molecule C_2H_6, rather than being drawn as in Fig. 2.2(a), can be abbreviated CH_3-CH_3 [Fig. 2.2(b)]. Note that the bond shown is still meant to indicate a C–C bond. A further contraction is made by eliminating all bonds and simply writing the formula as in (c) (Fig. 2.2).

2.4 Index of Hydrogen Deficiency

One can calculate the number of hydrogen atoms required to form a saturated hydrocarbon from any given number of carbon atoms. For any acyclic (this term will be introduced shortly) saturated hydrocarbon, the molecular formula will correspond to C_nH_{2n+2} where n is an integer. An alkane with 12 carbon atoms will require $(12\times2)+2 = 26$ hydrogen atoms.

Introduction of one double bond (called a center of unsaturation) will decrease this number of hydrogens by two. The presence of a ring in the molecule will have the same effect. A triple bond is equivalent to two double bonds. The number of centers of unsaturation is called the Index of Hydrogen Deficiency. It is equal to the number of hydrogens missing relative to the saturated material – *divided by two*.

Example: A hydrocarbon has the molecular formula C_7H_{10}. What can be said about its structure?

Answer: A saturated hydrocarbon with seven carbon atoms would have $(7\times2)+2 = 16$ hydrogen atoms. For C_7H_{10}, there are six hydrogens missing. Therefore, there are $6/2 = 3$ units of unsaturation. These might be any combination of the following:
- three double bonds
- one double bond and one triple bond
- three rings
- two rings and one double bond
- one ring and two double bonds
- one ring and one triple bond

Information about the structure of a molecule can be derived just from a knowledge of its molecular formula.

It should be noted that the presence of oxygen in the molecule does not change these calculations. Also, the presence of halogen (Cl, Br, I) will cause the formula to have one less hydrogen than expected since each halogen will replace one hydrogen atom.

Another example. Calculate the IHD (Index of Hydrogen Deficiency) for the molecule with the molecular formula $C_{10}H_{15}O_2Cl$. The saturated hydrocarbon with 10 carbon atoms would have $(2 \times 10 + 2) = 22$ hydrogens. There are 15 H's PLUS ONE Cl (which replaces H). Therefore, there are the equivalent of 16 hydrogens, 6 less than the saturated compound. The IHD is $6 \div 2 = 3$.

2.5 Naming Organic Compounds

In order to convey information about structures of organic molecules one must be able to talk about them using names that are unambiguous. In the early days of organic chemistry, when a chemist discovered a new compound, he gave it a name that usually indicated its source, e.g., caproic acid (isolated from goat's milk), butyric acid (from the German for butter as it is isolated from rancid butter). As the number of known compounds increased, this system became unworkable and a systematic method of naming compounds was devised. This system is called the IUPAC system (because it was devised by the International Union of Pure and Applied Chemistry). This method of nomenclature (naming) permits the assignment of a name to each organic molecule; this will allow anyone else who knows the system to accurately determine which molecular structure you are referring to. Unfortunately some of the older, nonsystematic names still persist and some of these have to be learned.

This text is designed to be used in conjunction with a self-teaching style of workbook which is solely concerned with organic nomenclature. Therefore, only a very short idea of the system will be presented here. We will, however, start using the names immediately. *It is imperative that you keep up with the assignments from the nomenclature book!*

The systematic (IUPAC) method starts by dividing organic compounds into classes that are defined by the presence of specific groups of atoms. A name is then constructed in the following way:

$$(prefix) + (root) + (suffix) = name.$$

The root indicates how many carbon atoms are in the longest contiguous chain of carbon atoms, the suffix indicates to which class the molecule belongs, and the prefixes gather up all the information not given by the root and the suffix.

As mentioned, the root word indicates the number of carbon atoms in the longest chain. These roots are shown below. The suffix indicates what *kind* of molecule it is. A saturated hydrocarbon has the suffix ane. (Note the general class name is alkane).

Root	# of Carbons	Root	# of Carbons
meth	1	hex	6
eth	2	hept	7
prop	3	oct	8
but	4	non	9
pent	5	dec	10

CH_4 is methane, C_2H_6 is ethane, and C_3H_8 is propane. Note that these names give all the information required to write a complete molecular formula.

The suffix for a compound that contains a double bond is *ene*. (Note the general class name is *alkene*.) Some simple compounds of this type with their names are shown. Some older nonsystematic names are also given in parenthesis. What happens when we have a structure like

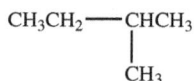

$$CH_3CH_2-\underset{\underset{\displaystyle CH_3}{|}}{C}HCH_3$$

Here not all the carbon atoms can be incorporated into one contiguous chain. To handle this situation, the longest such chain is located and named. The carbons are then numbered starting at one end of the chain, in such a way that the first encountered carbon atom which is attached to the group that is not included in the chain is given the lowest possible number. (The groups not included in the main chain are called *substituents*). In the example above, numbering the main chain from left to right assigns the number 3 to the substituted carbon whereas numbering the chain from right to left assigns it the number 2 and so the latter is correct.

To name the substituent, we note that it contains one carbon and therefore the root will be meth. To this we add the suffix -yl. One carbon with three hydrogens is called a methyl group. The full IUPAC name of the compound is then 2-methylbutane.

$H_2C{=\!=}CH_2$ $\qquad$ $H_2C{=\!=}CHCH_3$ $\qquad$ $H_3C-\underset{H}{C}{=\!=}\underset{H}{C}-CH_3$

ethene $\qquad\qquad$ propene $\qquad\qquad$ 2-butene
[ethylene] $\qquad\qquad$ [propylene]

Some alkenes

The *punctuation* in IUPAC names is important. In general, numbers are separated from numbers by commas and numbers are separated from words by hyphens. Words with no numbers between them are generally run together (e.g., 2-methylpropane) but some exceptions to this occur.

Besides simple alkyl (a general word to describe a group consisting of only sp^3-carbon and hydrogen atoms – derived from alk(ane) + "yl" for substituent) substituents, we will frequently use halogen as the substituent. Since fluorine, chlorine, bromine, and iodine are similar to hydrogen in that they require only one electron to reach the nearest rare gas configuration, any C–H bond can be replaced with a C–X bond where X = F, Cl, Br, or I. When this occurs the halogen is treated as a substituent called, respectively, a fluoro, chloro, bromo, or iodo group. (Learn the spellings of these! Note the "h" in chloro!). The molecule CH_3CH_2-Cl is 1-chloroethane. Some other examples are given in Fig. 2.3.

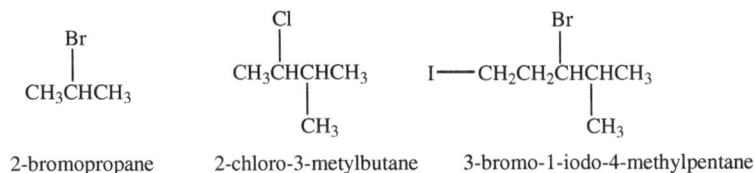

Br
|
CH_3CHCH_3

Cl
|
$CH_3CHCHCH_3$
|
CH_3

Br
|
$I—CH_2CH_2CHCHCH_3$
|
CH_3

2-bromopropane   2-chloro-3-metylbutane   3-bromo-1-iodo-4-methylpentane

Fig. 2.3

To allow more time for discussion of the important aspects of structure and reactions, a more explicit description of organic nomenclature will not be given in this book. Do the assigned work from the nomenclature text. We will, however, begin immediately to use the nomenclature here and some practice problems are included on this topic at the end of the chapter.

2.6 Functional Group Concept

If each organic molecule underwent reactions that were different to all the others, the situation would be impossible to handle. Since there are literally an infinite number of possible organic compounds, there would be an infinite number of reactions to remember. Fortunately, this situation does not arise. Most molecules have a grouping of atoms contained in them which is the point at which most reactions occur. This atom or group of atoms is called a *functional group*. There are about 20 relatively important functional groups, but only a very few of these will concern us in this course.

Furthermore, within certain limits, the *reactions of a particular functional group are independent of the nature of the rest of the molecule*. For example, the group C–O–H (an alcohol) reacts in the same way under a given set of conditions whether the actual molecule is CH_3OH (methanol), CH_3CH_2OH (ethanol), or $CH_3CH_2CH_2CH_2-OH$ (1-butanol). Very frequently, the structures are generalized as, for example, R–OH where the symbol R represents any hydrocarbon (alkyl) group.

Table 2.1 lists some of the most important organic functional groups, their structures, and the suffixes for the IUPAC name. A further method of identifying some specific types of carbon atoms can be quite useful.

Tab. 2.1: List of functional group classes.

Class name	Structure	IUPAC ending
Alkane	Only C–H and C–C	ane
Alkene	C=C	ene
Alkyne	C≡C	yne
Alcohol	C–OH	ol
Aldehyde	$\overset{\displaystyle O}{\overset{\|}{C—C}}—H$	al
Ketone	$\overset{\displaystyle O}{\overset{\|}{C—C}}—C$	one
Carboxylic acid	$\overset{\displaystyle O}{\overset{\|}{C—C}}—OH$	oic acid
Ester	$\overset{\displaystyle O}{\overset{\|}{C—C}}—OR$	oate
Ether	C–O–C	
Amine	$C–NH_2$	
Alkyl halide	C–Cl, C–Br, C–I	
Acyl halide	$\overset{\displaystyle O}{\overset{\|}{C—C}}—Cl$	
amide	$\overset{\displaystyle O}{\overset{\|}{C—C}}—NH_2$	

A *primary* (1°) carbon is a carbon that is bonded to only one other carbon atom. A *secondary* (2°) carbon is one that is bonded to two others, while *tertiary* (3°) and *quaternary* (4°) carbon atoms are bonded to three and four others, respectively.

Similarly a hydrogen atom bonded to a 1° carbon is called a primary hydrogen. Some examples of these designations are shown Fig. 2.4.

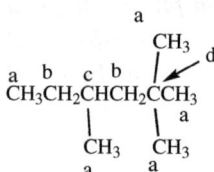

Fig. 2.4

The carbons labeled *a* are primary, those labeled *b* are secondary, those labeled *c* are tertiary, and those labeled *d* are quaternary.

2.7 Additional Definitions

We started this chapter by introducing some new terms. We will conclude it in the same way.

We have seen that the group CH_3 is called a methyl group. Frequently we want to refer to a $-CH_2-$ group as in, for example, the middle carbon atom of propane. CH_2 groups are called *methylene* groups. Similarly, a CH group is a *methine* group.

A series of similar compounds, each of which differs from the previous members of the series by the addition of one methylene group is called a *homologous* series. So, for example, methane, ethane, propane, and butane are a part of the homologous series of alkanes.

We are now ready for the next two chapters, which will focus more attention on the detailed structures of some simple compounds. This is a necessary prelude to the discussion of the chemical reactions of organic compounds which is, of course, our main interest.

2.8 Problems

2-1. Give the IUPAC name for each of the following compounds.

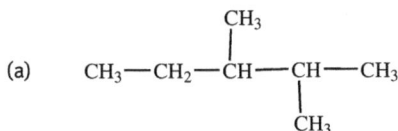

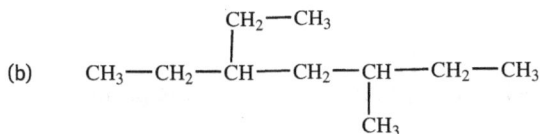

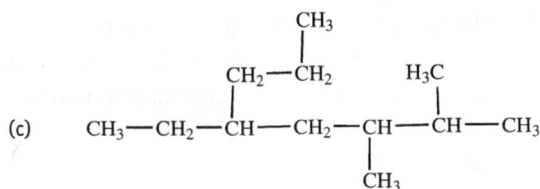

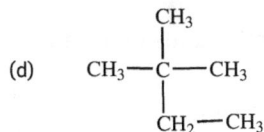

(a)
$$CH_3-CH_2-\underset{\underset{CH_3}{|}}{CH}-\underset{\overset{|}{CH_3}}{CH}-CH_3$$

(b)
$$CH_3-CH_2-\underset{\overset{CH_2-CH_3}{|}}{CH}-CH_2-\underset{\overset{|}{CH_3}}{CH}-CH_2-CH_3$$

(c)
$$CH_3-CH_2-\underset{\underset{CH_2-CH_2}{|}\;\overset{CH_3}{|}}{CH}-CH_2-\underset{\overset{|}{CH_3}}{CH}-\underset{\overset{H_3C}{|}}{CH}-CH_3$$

(d)
$$CH_3-\underset{\underset{CH_2-CH_3}{|}}{\overset{\overset{CH_3}{|}}{C}}-CH_3$$

(e)
$$CH_3—CH_2—\overset{\overset{\displaystyle Br}{|}}{CH}—\overset{\overset{\displaystyle CH_3}{|}}{CH}—\overset{\overset{\displaystyle H}{|}}{\underset{\underset{\displaystyle Cl}{|}}{C}}—\overset{\overset{\displaystyle Br}{|}}{CH}—\overset{}{\underset{\underset{\displaystyle CH_3}{|}}{CH}}—CH_3$$

(f) $Br—H_2C—HC{=\!=\!=}CH—CH_2—CH_3$

(g)
$$H_3C—\overset{\overset{\displaystyle I}{|}}{CH}—\overset{\overset{\displaystyle H}{|}}{C}{=\!=\!=}CH—\underset{\underset{\displaystyle CH_3}{|}}{C}{=\!=\!=}CH_2$$

(h)
$$H_3C—CH_2—\overset{\overset{\displaystyle HC{\diagup\!}^{CH—CH_3}}{|}}{CH}—CH_2—\underset{\underset{\displaystyle Cl}{|}}{CH}—CH_3$$

(i)
$$H_3C—CH_2—\overset{\overset{\displaystyle Br}{|}}{CH}—\underset{\underset{\displaystyle CH_3}{|}}{CH}—C{≡}CH$$

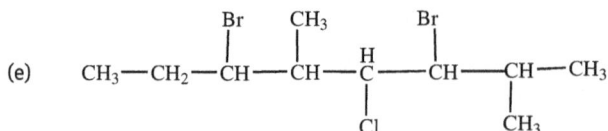

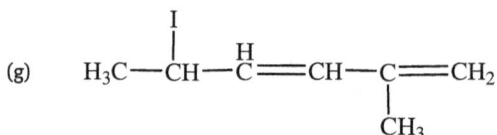

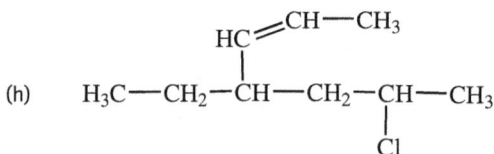

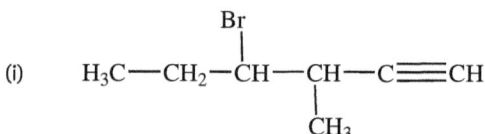

2-2. Write the structural formula that corresponds to each of the following names:
(a) 3-methylpentane
(b) 4-chloro-3-methyloctane
(c) 2-bromo-1-chloropropene
(d) 3,4,5-tribromooctane
(e) 5-isopropyl-2,3-dimethyldecane
(f) 6-chloro-3,9-dimethyl-2,6-decadiene

2-3. Identify each of the carbon atoms in question 2.1 a, b, c, d as being primary, secondary, tertiary, or quaternary.

2-4. Calculate a *molecular* formula for a hydrocarbon on the basis of the following information. One gram of the compound is burned in oxygen. The carbon dioxide is collected and found to weigh 3.2 g. The water formed weighs 1.1 g. The equation for the process is

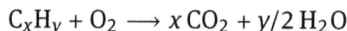

$$C_xH_y + O_2 \longrightarrow x\,CO_2 + y/2\,H_2O$$

2-5. For the following molecular formulae, determine how many "units of unsaturation" are present. (i.e., the IHD).
(a) C_5H_{12}, (b) $C_{10}H_{20}$, (c) $C_{10}H_8$, (d) $C_{19}H_{36}O$, (e) $C_{14}H_{24}Cl_2$

2-6. Draw and name at least three positional isomers of chlorohexane.

2-7. How many positional isomers of the molecular formula $C_6H_{13}Cl$ can you draw? (Be careful of redundancies!).

2-8. Draw and give the IUPAC names for the following:
(a) Four isomeric molecules with the formula C_4H_9Cl
(b) Eight isomeric molecules with the formula $C_5H_{11}Br$

2-9. Which of the following pairs of compounds represent identical structures and which are positional isomers?

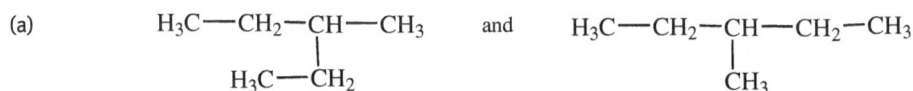

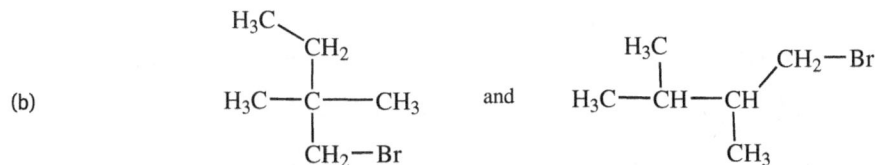

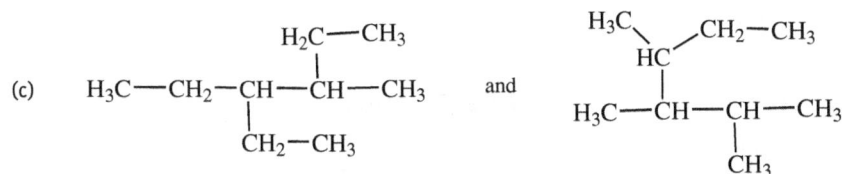

(a)
$$H_3C—CH_2—CH—CH_3$$
$$|$$
$$H_3C—CH_2$$
and
$$H_3C—CH_2—CH—CH_2—CH_3$$
$$|$$
$$CH_3$$

(b)
$$H_3C{\diagdown}CH_2$$
$$H_3C—\underset{\underset{CH_2-Br}{|}}{C}—CH_3$$
and
$$H_3C$$
$$H_3C—\underset{\underset{CH_3}{|}}{CH}—\underset{\diagup CH_2-Br}{CH}$$

(c)
$$H_3C—CH_2—\underset{\underset{CH_2-CH_3}{|}}{CH}—\underset{\underset{H_2C-CH_3}{|}}{CH}—CH_3$$
and
$$H_3C{\diagdown}HC{\diagup CH_2-CH_3}$$
$$H_3C—\underset{\underset{CH_3}{|}}{CH}—\underset{}{CH}—CH_3$$

2-10. Determine the hybridization state of the indicated carbon atoms.

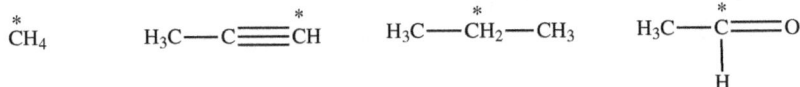

$$\overset{*}{C}H_4 \qquad H_3C—\overset{*}{C}{\equiv}CH \qquad H_3C—\overset{*}{C}H_2—CH_3 \qquad H_3C—\underset{\underset{H}{|}}{\overset{*}{C}}{=}O$$

3 The Shapes of Organic Molecules – Stereochemistry 1

3.1 Introduction

Many of the physical and chemical characteristics of molecules are strongly influenced by their three-dimensional shape. The introduction of the various hybridized states of carbon gave us an inkling of why this may be so. In this and the next chapter we will consider these aspects in much greater detail in order to set the stage for understanding how they influence the reactions of organic molecules.

In this regard we must practice thinking in three dimensions. This is an art which, like most, is developed through use and practice. It is useful to obtain an inexpensive set of molecular models so that you can demonstrate to yourself the various points to be discussed. Extensive use of molecular models will be made in lecture demonstrations and you will have an opportunity to obtain some experience using them in a laboratory session. We will also be making frequent use of 3-D drawings.

> It is important that you do these drawings carefully. A sloppy drawing will not illustrate the points clearly and will usually cause more confusion than no drawing at all. So prepare your drawings carefully and practice using them.

3.2 The Structure of Some Simple Alkanes

Remember that alkanes = saturated hydrocarbons = all carbons in the sp^3-hybridized state. Also recall that the shape of sp^3-hybrid carbon is tetrahedral, which means that the four attached atoms are all equidistant from each other. Although the structural formula of methane for example is written as if it were flat (Fig. 3.1a), this is just a convenient shorthand for the actual, tetrahedral structure (Fig. 3.1b).

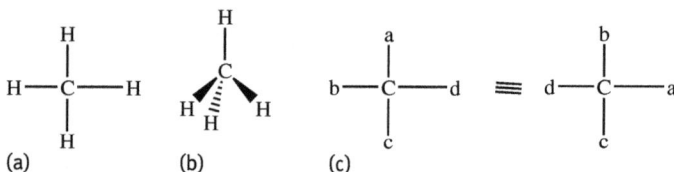

(a)   (b)   (c)   **Fig. 3.1**

Note again that there is no difference between the drawings shown in Fig. 3.1c because of the actual shape of the carbon atom. In Fig. 3.1b, the wedged lines are intended to indicate that they are protruding in front of the plane of the paper, the dashed lines are behind the paper, and the simple lines are within the plane of the paper.

https://doi.org/10.1515/9783110778311-003

Now consider the molecule ethane (C_2H_6). Each carbon atom is sp³-hybridized and we can draw the structure shown below. Note that, because of the tetrahedral shape of the carbon atoms, each carbon is equidistant from the three hydrogens on the attached carbon atom.

When carbon atoms are attached to each other by only a single bond, it is found that one carbon atom can rotate relative to the next. The drawings in Fig. 3.2 all represent the same molecule, but with different relative positions of the hydrogen atoms. Each of these structures can be converted into any of the others merely by rotation about the C–C bond. Since thermal energy is large at room temperature, the groups are in constant motion and all of the below shapes are possible and present at any given time.

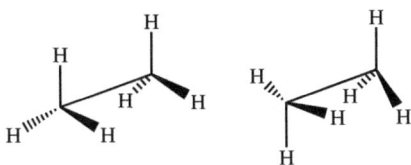

Fig. 3.2

3.3 Newman Projections and Conformations

The drawings shown in Fig. 3.2 can become very complex when more than two or three carbon atoms are involved. To simplify this situation, a new type of drawing must be introduced: the so-called *projection drawings*. We will see different applications of these, but in general, they serve to convert three-dimensional drawings into two-dimensional ones. To allow this, certain restrictions must be applied, the violation of which may lead to erroneous conclusions. It is very important to fully understand what is involved in making a projection drawing and what the restrictions are for each type.

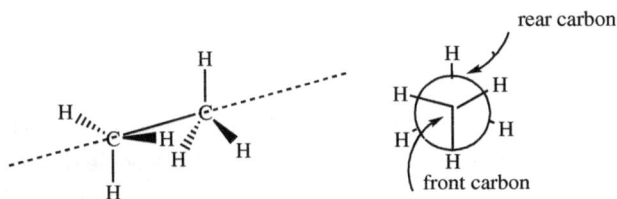

Fig. 3.3

If a two-carbon fragment is turned so that the two carbons under consideration are colinear with the line of sight and then a light beam in projected down this axis, the molecule will cast a shadow which is the two-dimensional projection of the actual molecule (Fig. 3.3). This shadow shows the front carbon atom as a dot in the middle and three of the four bonds attached to it are now oriented at 120° to each other. (The fourth C–C bond is behind the front atom and therefore not visible.) The rear carbon is represented by the larger circle. Of the four bonds attached to it, three are oriented at 120° to each other. Since these are behind the rear carbon atom, they cannot be seen until they extend beyond the radius of the carbon. Consequently, these bonds begin at the circumference of the circle representing the carbon. This type of projection drawing is called a *Newman projection*. This type of drawing can be used for larger molecules. However, *only one* C–C bond can be shown at a time. The Newman projections of some simple alkanes are shown in Fig. 3.4.

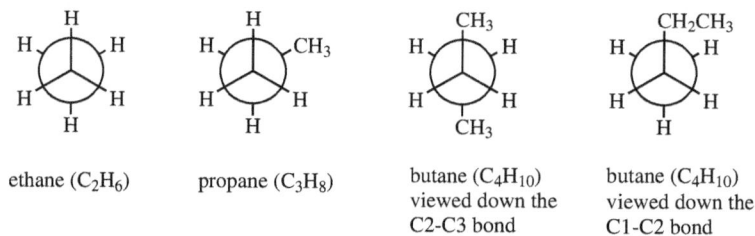

ethane (C_2H_6)   propane (C_3H_8)   butane (C_4H_{10}) viewed down the C2-C3 bond   butane (C_4H_{10}) viewed down the C1-C2 bond

Fig. 3.4

Recall that carbons joined by only single bonds can rotate with respect to one another. Let us look at the Newman projections for ethane undergoing this rotational process. We will start with the structure where all six hydrogen atoms are spaced equally around the drawing (A). Holding the front carbon still and rotating the back 60° brings us to B, in which each of the hydrogen atoms on the back carbon is lined up or "eclipsed" with one on the front carbon. A further 60° rotation brings us back to A.

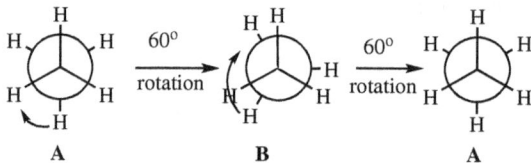

Such rotational forms or isomers are called *conformations*. It is important to note that there are an infinite number of conformations of ethane or any other molecule of this type, each different from the others by a small rotation about a carbon–carbon bond. The two limiting cases for ethane are called the *staggered* form (A) and *eclipsed* form (B). These are not of equal energy.

3.4 Stabilities of Conformational Isomers

It is a common and very useful practice to keep track of energy changes that occur during a chemical or physical change. These "potential energy diagrams," when used properly, present a lot of information in a simple manner. Think of the potential energy associated with a soccer ball being kicked toward the top of a hill with a deep valley on the other side.

Two possibilities, with some subsets exist.

1. Insufficient energy is given to the ball to allow it to reach the top of the hill. In this case, even though C has lower energy than A, the ball cannot get there.

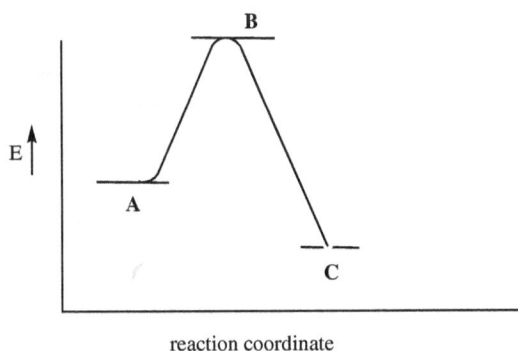

reaction coordinate

2. If sufficient energy is given to the ball (i.e., if it is kicked hard enough), the ball which initially has potential energy A will climb the hill to B. Kinetic energy has been transformed into potential energy. Since B is of higher potential energy than either C or A, the ball does not stop there. If possible, *natural processes always prefer to end in the position of lowest potential energy* (C in the case shown in the Figure). Potential energy is reconverted to an equal amount of kinetic energy in this process. The lower potential energy of the ball is reflected in a greater (higher) stability.

   (a) If insufficient kinetic energy can be obtained by the ball to get back to B, the ball is effectively trapped at C. This will result in an "irreversible process."

   (b) If the ball can acquire sufficient kinetic energy from the environment to get back to B, it can then roll down the hill in the opposite direction – i.e., to A. Since both A and C can be accessed, the process is "reversible" and the two states can be said to be in *equilibrium*. The difference in energy between A and C in this situation will determine the amount of time the ball will spend in each place. The bigger the ΔE between A and C is, the more time the ball will spend in the lower energy state.

When talking about *molecules* instead of soccer balls, we are rarely thinking about *one* molecule. Rather, a sample of millions of molecules is usually under consideration. Energy diagrams such as that shown are then best interpreted in terms of the *population* of any given energy level. If a stop-action photo of a large number of molecules could be taken, and the number in each of the conformations counted, then there would be more in the lower energy state than in the upper one and the difference in population is directly related to the size of the energy difference between the two states. The relationship is given by:

$$\Delta E = -RT \ln K_{eq}$$

Atoms are three-dimensional. They require space and attempts to force two atoms to occupy the same space will cause such forms to be of higher energy (i.e., lower stability) than those forms in which the atoms are well separated. This applies particularly to nonbonded atoms such as the hydrogen atoms in ethane. Note we are referring to the interaction of hydrogen atoms on C-1 of ethane with those on C-2 of ethane. The distance between two hydrogens bonded to the *same* carbon is fixed by the tetrahedral geometry of the carbon. In the staggered form of ethane, stretching and bending motions of the hydrogen–carbon bonds do not lead to serious interactions between the hydrogens on the carbons because of the distance between them. However, in the eclipsed form, these atoms are much closer together and when they undergo stretching or bending motions, they can interfere with each other. This raises the energy of these forms relative to the staggered forms. This can be drawn as an "energy profile" (Fig. 3.5). The maxima represent the eclipsed conformations and the minima the staggered ones.

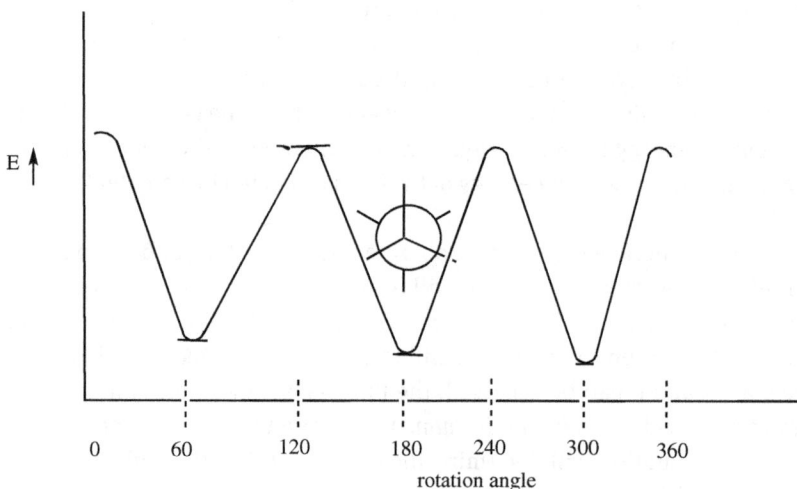

Fig. 3.5

Consider the Newman projection of butane viewed down the C_2–C_3 bond. The situation is now somewhat more complex than in ethane (Fig. 3.6a). In conformation A, the two large methyl groups are as far apart as possible. A 60° rotation again brings us to an eclipsed form (B), the bulky methyl groups now being eclipsed with a small hydrogen atom. Continuing the rotation, the next staggered form (C) puts the two large groups much closer together than in A. While this is not as unstable as an eclipsed form, it clearly is less stable than A.

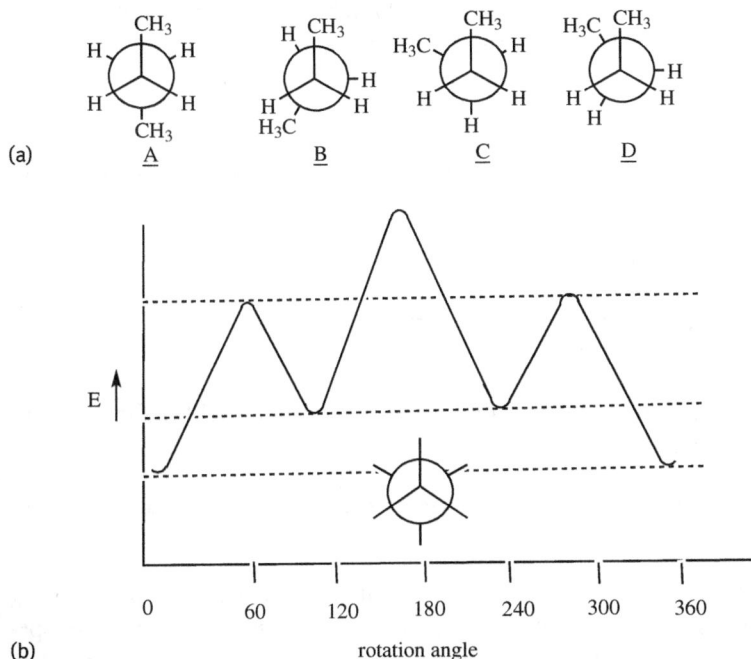

(a)

(b)   rotation angle

Fig. 3.6

Finally, conformation D has the two largest groups eclipsed – the least stable situation of all. The energy profile for this sequence is shown in Fig. 3.6.b.

Names are associated with the forms A, B, C, and D. Conformation A, where the largest groups are on opposite sides is called the *trans* or *antiperiplanar* conformation; B, where a large and small group are eclipsed is called the *anticlinal* conformation; C, in which the two largest groups are beside one another is called the *synclinal*, *gauche*, or *skew* (all names are used) conformation, while D is called the *cis* or *synperiplanar* conformation. These concepts will have great significance when we come to discuss cyclic compounds. Since thermal energy is large at room temperature, all atoms can spin relatively freely on their bonds. Therefore, at any given instant some molecules will be in all of the staggered or eclipsed conformations shown in Fig. 3.6. As noted earlier, the number in each (i.e., the population) will be determined by its energy and so more molecules will be in A than B, C, or D while, of these four, C will have the next highest population (Fig. 3.6). As the temperature is raised, the higher energy states will increase their population because the molecules have absorbed thermal energy.

3.5 Stereoisomers

Recall the definition of the term isomers (Sect. 2.1). When this term was introduced, we saw one particular kind of isomer called the positional isomer and at that time it was mentioned that other types would follow soon. In this section we will learn about some of these.

> It is important to learn the definitions that follow as one builds on the previous one and failure to keep up will cause considerable difficulty!

We start by defining the term *stereoisomer*.

> *Stereoisomers* are isomers that differ *only* with respect to how atoms are arranged in space and *not* with respect to which atoms are joined together.

Stereoisomers are a large class that contains several easily identifiable subgroups, all of which are still stereoisomers.

> *Geometrical isomers* are stereoisomers with restricted rotation about the carbon–carbon double bonds.

Recall the shape of sp²-hybridized carbon: three bonds oriented in one plane and a p-orbital containing one electron oriented perpendicular to that plane. When two sp²-carbons are joined, the two p-orbitals must be coplanar for the second (π) bond to form between them (Fig. 3.7). In order for rotation of one carbon relative to the other to occur, the π bond must be broken. This requires more energy than is usually available and therefore rotation of the type discussed in the previous section cannot occur. Note that all six atoms of the system – the two carbons and the four hydrogens (or whatever might be attached) *must* be in one plane: i.e., the molecule is flat.

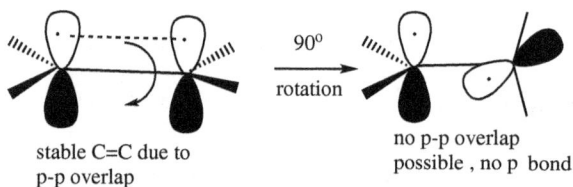

stable C=C due to
p-p overlap

90°
rotation

no p-p overlap
possible , no p bond

Fig. 3.7

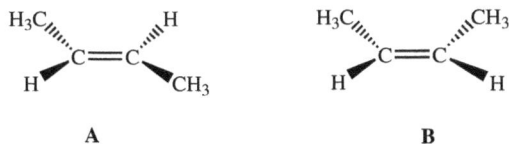

Fig. 3.8

Consider the molecule 2-butene (Fig. 3.8). It is possible to write *two* structures of the type shown above. In one case, the two methyl groups are on *opposite* sides of the molecule and in the other, they are on the *same* side. Note that in molecule A, every atom is bonded to the same atom to which it is bonded in molecule B. In order to convert A into B, a rotation about the C=C bond is necessary. This is not possible and A and B fit the definition given for stereoisomers and for geometrical isomers. The names given to these two stereoisomers are *trans* (A) (substituted on opposite sides) and *cis* (B) (substituted on the same side). When naming geometrical isomers the terms *cis* and *trans* refer to the *configuration* of the double bond in the main chain. Some examples are shown in Fig. 3.9. In this Figure, note that the third compound cannot be named either as *cis* or *trans* since both chain lengths are the same.
    Note the new term ("Configuration").

The *configuration* of a molecule is its three-dimensional arrangement of atoms, which cannot be changed without making and breaking bonds.

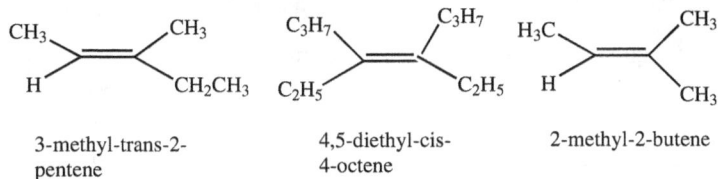

3-methyl-trans-2-
pentene

4,5-diethyl-cis-
4-octene

2-methyl-2-butene

Fig. 3.9

The distinction between the terms *conformation* and *configuration* is an important one which is not always immediately obvious to the beginning student. Remember that *conformations* can be changed by rotational processes, but bonds must be broken to interconvert *configurations*. In geometric isomers, interconversion of *cis* and *trans* isomers requires breaking of the π bond as noted above.

    A more modern method of stereochemical nomenclature is somewhat more complex, but is more generally applicable and solves some perplexing problems that occur with the older *cis*, *trans* system. The system is called the Cahn–Ingold–Prelog (CIP) system. It will be introduced here and developed more fully in future sections. To see why it is necessary, consider the two simple molecules shown in Fig. 3.10.

Trying to name these as *cis* or *trans* leads to the question does *cis* mean the Br and Cl or Br and methyl groups? In other words, the system is ambiguous.

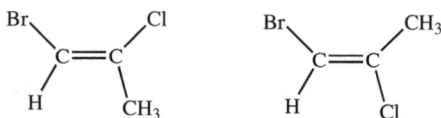

Fig. 3.10

3.6 Stereochemical Designations – The Cahn–Ingold–Prelog System

The general approach of the systematic method of designating stereochemical configuration is as follows.

1. "Priorities" are assigned to the two groups directly attached to each of the sp^2-hybridized carbon atoms.
2. If the higher priority group on both sp^2-hybridized carbons are on the same side of the molecule it is called Z (for *zusammen* – German for "together").
3. If the higher priority group on both sp^2-hybridized carbons are on opposite sides of the molecule it is called E (for *entgegen* – German for "apart").

(A cynic whose mother tongue is English might note that the symbol Z appears to represent groups on opposite sides, and E on the same side – the *exact opposite* of what is true!).

This system looks very similar on the surface to the older *cis, trans* system, but as we will see, it does eliminate the ambiguities found in the older system and is capable of extension to other systems that are harder to handle.

In order to use the system, we must learn how to assign the "priorities" (step 1. above). These are assigned by guidelines called "sequence rules." A full description of these will be found in the organic nomenclature book you are using, but the main points are outlined below.

The rules are called "sequence rules" for a reason. They are applied in sequence: that is rule #1 is applied and only if this fails to give an unambiguous result, is rule #2 used and you must remember not only the rules, but also the order in which they come!

3.7 The Sequence Rules

Rule #1. Divide the molecule through the center of, and perpendicular to, the plane of the double bond. Now consider the two atoms directly bonded to each of the sp²-hybridized carbons. If they have different atomic numbers (Z), the atom with the higher atomic number has the higher priority.

For example, consider *cis*-2-butene (Fig. 3.8b) where the atoms directly attached are one hydrogen (Z = 1) and one carbon (Z = 6) in each case. Since carbon has the higher atomic number, it has the higher priority. Since the higher priority groups on both sp²-carbons are on the same side, this molecule has the Z-configuration.

In the drawing Fig. 3.10b, the atoms attached to the left end of the double bond are Br (Z = 35) and H (Z = 1). On the right end, the atoms are C (Z = 6) and Cl (Z = 17). In this case, the higher priority groups on each end of the double bonds are on opposite sides of the molecule and so it has the E-configuration. Note that you really do not have to know the actual atomic number, only which is higher. Therefore, a knowledge of *where on the periodic table a particular element is will usually be sufficient to assign the priorities.*

Rule #2. If the atoms directly attached to the sp²-carbons are the same, move down the two chains, atom by atom, until a difference is found in atomic numbers. Once again, the atom with the higher atomic number has the higher priority.

Consider the molecular fragment shown.

Both atoms attached to the sp²-carbon are carbon and rule #1 does not distinguish between them. Rule #2 says in this situation, to move one atom down each chain. In the case of CH_3, the next atom is a hydrogen (Z = 1), but in the ethyl group, the next atom is carbon (Z = 6). The ethyl group is of higher priority.

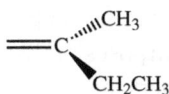

$$=\!\!\!C\!\!\overset{\text{\tiny{···}}\!\!\!\text{III}CH_3}{\underset{CH_2CH_3}{\diagdown}}$$

You may find it useful to use the following notation to define the priority of carbon atoms in terms of their substituents. The carbon is written with the three atoms attached to it in order of their atomic numbers: e.g., C(O,N,H) for an atom in a chain that is attached to one oxygen, one nitrogen, and one hydrogen atom. When priority assignments are made between two carbons noted in this fashion, the process is very simple. The atomic numbers of the substituents are compared one by one until a difference is found and the assignment is made on the basis of that difference. Some examples will be shown in the succeeding sections.

Rule #3. If a branch point or substituent is encountered, the path is chosen such that the earliest differentiation on the basis of atomic numbers can be made. Also, the atom with the *larger* number of substituents of higher atomic number has the higher priority.

In structure 1 in Fig. 3.11, on the left sp^2-carbon, the lower substituent has a carbon with two carbon substituents. The same carbon in the upper substituent has only one. The first carbon of the *upper* substituent is attached to two hydrogen and one carbon atom. It can therefore be abbreviated as C(C,H,H) (note that the substituents are entered in order of priority). In the same way, the first carbon in the lower *substituent* on the left could be abbreviated C(C,C,H). The lower substituent has the higher priority set of substituents and therefore is of higher priority. For the same reason, the priorities of the substituents on the right-hand end of the double bond are as shown and structure 1 shows the molecule in the E-configuration. Structure 2 (Fig. 3.11) is also shown in the E-configuration. Make sure you can work this out. Your nomenclature book has many other examples.

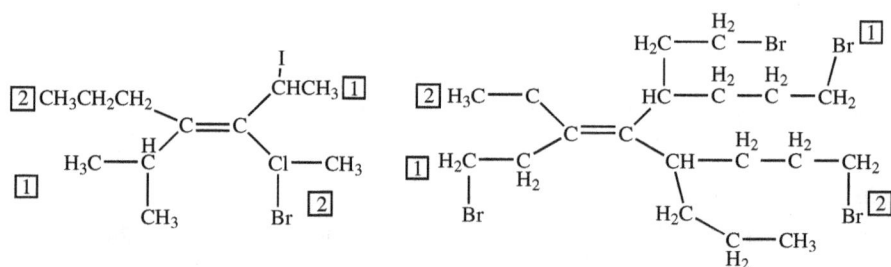

Fig. 3.11

You will undoubtedly be able to visualize complex situations that these three simple sequence rules will not answer. However, they are sufficient for our purposes at this point. We will add further refinements to them in Chapter 4. Your nomenclature text will also expand them.

The next sections will introduce some related, but new applications of the ideas just considered. However, make sure you master the applicable problems at the end of the chapter before you proceed.

In the forgoing sections, two extremely important terms were introduced – namely *conformation* and *configuration*. If you are not completely sure of the differences between these, review the appropriate pages before proceeding. Note that the important distinction between them concerns the ability to interconvert conformations *without* making or breaking bonds, which cannot be done with configurations.

3.8 Alicyclic Compounds

We have previously referred briefly to organic compounds that contain *rings*: i.e., chains of carbon atoms whose ends are joined. Some examples are shown in Fig. 3.12. Note that the nomenclature uses the prefix *cyclo* to indicate a ring. The full nomenclature of these compounds is treated in your nomenclature work-book. As shown in the Figure, it is customary to abbreviate these structures by omitting all the carbons and hydrogens. In the "line-drawings," each corner of the figure represents a carbon and enough hydrogen atoms to fulfil the required tetravalency of each carbon are understood to be present. The general name for these compounds is *alicyclic* compounds.

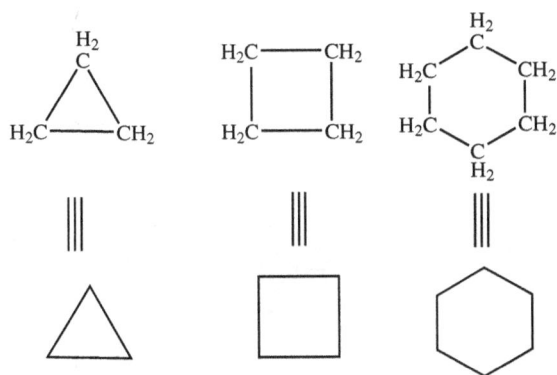

Fig. 3.12

Since all the carbons in the molecules shown are sp³-hybridized – this does not have to be so as we shall see – they are all tetrahedral. This leads to some interesting conclusions regarding their three-dimensional shapes. Since it is always possible to draw a plane through any three points in space, cyclopropane must have all three carbon atoms in one plane. The associated six hydrogen atoms lie in two areas: three above the plane of the ring and three below the plane (Fig. 3.13). It is not possible to distinguish between these hydrogen atoms. (You should prove this to yourself.)

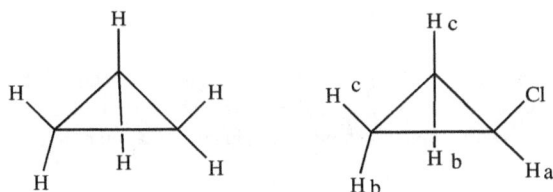

Fig. 3.13

Make a model of cyclopropane, mark one of the hydrogen atoms, turn the molecule so that you can't see the mark, and see if you can identify the marked atom. These hydrogen atoms are said to be *equivalent*. (Equivalent atoms are those that lie in exactly the same spatial relationship with every other atom in the molecule – see Sect. 2.2.) If any of these hydrogens are replaced by a substituent (e.g., Cl), only one isomer is possible. In the substituted compound, there are now three different sets of equivalent hydrogens (labeled a, b, and c in Fig. 3.14). (The two hydrogens labeled b can be exchanged and they remain in exactly the same relationship with the Cl atoms. Therefore, they are equivalent.) Thus, introduction of a second substituent (X) can lead to three different isomers and these are shown in Fig. 3.14. Isomers B and C, and B and D are positional isomers (naming these will prove this). However, isomers C and D are stereoisomers – that is they differ only in the way Cl and X are oriented in space. In a very real sense, rings can be considered as special "double bonds." The isomer C is referred to as the *cis* isomer (substituents on the same side of the ring) and D is the *trans* isomer. (The E,Z system can also be used in the same way as was done for alkenes).

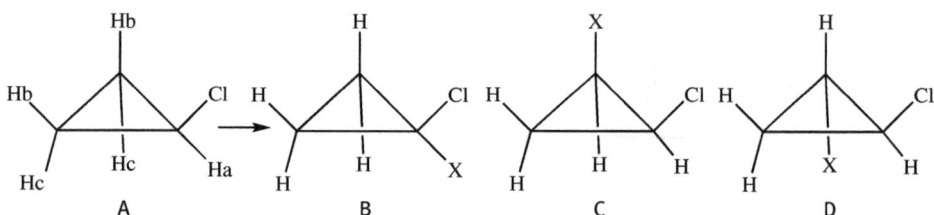

Fig. 3.14

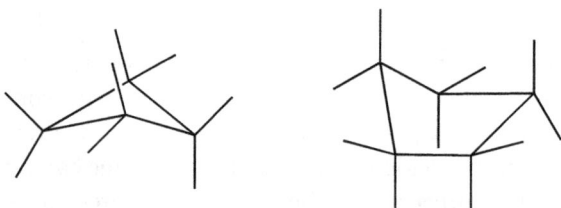

Fig. 3.15

Both cyclobutane and cyclopentane are more complex. We will skip over these except to note that the ring carbons do not lie in the same plane due to the tetrahedral nature of the carbon atoms. The actual shapes of these molecules are shown in Fig. 3.15.

It is important to realize that the chair and boat forms are conformations of the same molecule: that is they can be interconverted by rotation about all six C–C bonds simultaneously. This is difficult to show on paper but lecture demonstrations with molecular models should convince you of this.

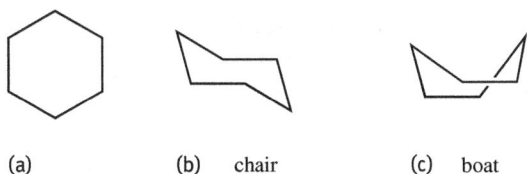

(a)   (b)   chair   (c)   boat   **Fig. 3.16**

3.9 Cyclohexane

By far the most interesting and important case for us is the molecule cyclohexane (C_6H_{12}). This is usually drawn as shown in Fig. 3.16(a) but we must remember that this ring cannot be flat. The normal bond angles of sp^3-hybridized carbon are approx. 109°. The internal angles of a regular hexagon are 120° and to avoid the strain involved in stretching the angles to this size the molecule adopts one of a number of possible three-dimensional shapes. If models are used, it becomes apparent that there are two forms that will allow the normal bond angles to be present. These are called the chair and the boat forms and are shown in Fig. 3.16 (b and c). In order to consider the detailed structure and stability concepts for cyclohexane, the use of Newman projections becomes necessary. If we view the molecule along the axis of the C1–C2 and C4–C5 bonds, the chair and boat forms appear as shown in Fig. 3.17. Applying the criteria and nomenclature developed in Chapter 3, it should be clear that the chair form represents a staggered conformation whereas the boat is an eclipsed form and so we should expect the chair form to be more stable. This is found to be the case and we will focus most of our attention on the chair form.

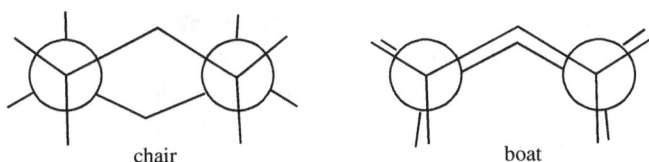

chair   boat

Fig. 3.17

If we look more closely at the Newman projections of the chair form of cyclohexane (Fig. 3.18) we will see that two different kinds of hydrogen atoms are present. *Each carbon atom* has one hydrogen which is oriented perpendicular to the plane defined by C1, C2, C4, C5 and one hydrogen which is not perpendicular to this plane. These are called, respectively, the *axial* (a) and *equatorial* (e) positions. Carbons 3 and 6 also have one of each of these. A second fact that can be seen is that *if the axial hydrogen is oriented above the plane of the ring, the equatorial one on the same carbon is below the plane*. Finally, it must be noted that as you proceed around the ring, the axial bonds *alternate* above and below the plane, and the equatorial bonds do the same.

It must also be understood that these concepts apply regardless of what is actually attached to the bonds. In the case of cyclohexane, all the bonds are joined to hydrogen but any other atoms or groups can be substituted without altering the concepts. The chair form in a non-projection drawing is also shown in Fig. 3.18. This is done in three stages. The first two show only the axial and equatorial bonds respectively, while the third assembles the complete molecule.

Practice is required to draw these molecules correctly. When preparing your drawings take great care to orient the bonds properly. Failure to do this will cause great difficulties in succeeding sections.

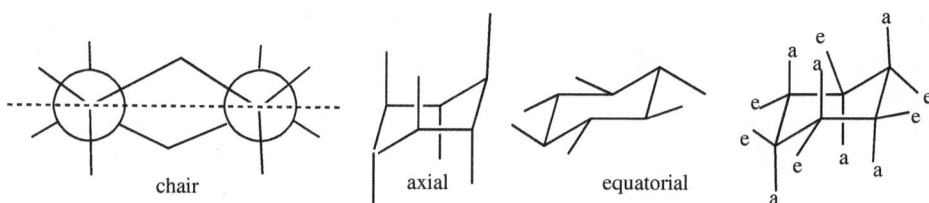

Fig. 3.18

Going back to the Newman projection, the sequence shown in Fig. 3.19 can be written. The chair form can convert into the boat form by a conformational process, and this can revert to the chair in one of two ways depending on which atom, C3 or C6, moves below the plane. This change does not appear to be very exciting but it has profound consequences. If the axial and equatorial bonds are followed from the left to the right drawing, it will be seen that they reverse positions! This same change is shown in Fig. 3.19 using non-projection drawings. Note that these are *conformational* changes and in cyclohexane itself they occur with great rapidity at room temperature.

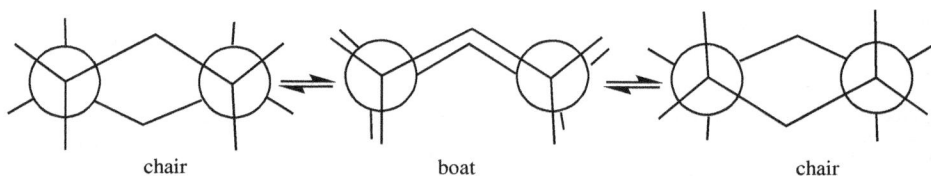

Fig. 3.19

3.10 Monosubstituted Cyclohexanes

Let us consider the situation when one of the hydrogen atoms of cyclohexane is replaced by a substituent – e.g., CH_3. Two positions appear to be available – the axial
and equatorial ones. Drawing these two (Fig. 3.20) shows that the axial substituent is
in a synclinal (skew or gauche) position relative to the ring C–C bond, whereas in the
equatorial position, this is not so. From our knowledge of the energies of the various
conformations of butane (Chapter 3) we should conclude that the axially substituted
compound should be less stable than the equatorially substituted one. Since these
two forms can interconvert by conformational change, they are always in equilibrium
and the equatorially substituted isomer will be the largest component of the conformational mixture.

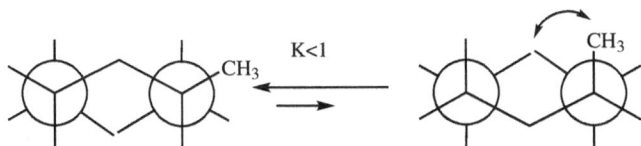

Fig. 3.20

In simple molecules containing substituted cyclohexane rings, the conformer with the larger substituents in the equatorial position is the more stable.

Another reason for the decreased stability of the axially substituted cyclohexanes lies
in a phenomenon called 1,3-diaxial interactions (Fig. 3.21). In the axial position, a
substituent is in close proximity to the atom on the axial bond two atoms away (H
in Fig. 3.21). This causes an interaction that destabilizes this relative to the equatorially substituted conformer. The larger the substituents are, the greater this 1,3-diaxial
interaction is.

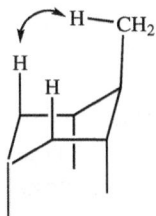

Fig. 3.21

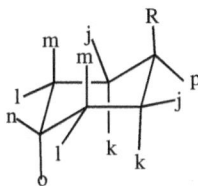

Fig. 3.22

3.11 Disubstituted Cyclohexanes

It has been shown in this chapter that monosubstituted cyclopropane has three sets of equivalent hydrogen atoms. In exactly the same way monosubstituted *cyclohexane* has seven sets (Fig. 3.22, j ⟶ p). Therefore, there must be seven isomers which can be generated when a second substituent is added. Four of these are positional isomers – the 1,1-, 1,2-, 1,3-, and 1,4-isomers (Fig. 3.23).

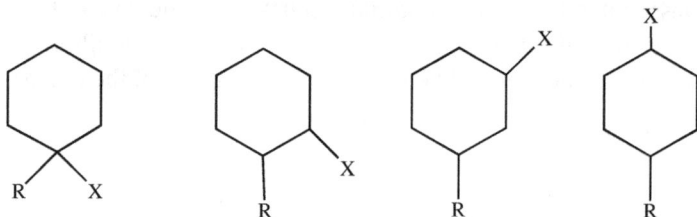

Fig. 3.23

However, each of the 1,2-, 1,3-, and 1,4-isomers can exist in *cis*- and *trans*-forms (i.e., stereoisomers) (Fig. 3.24).

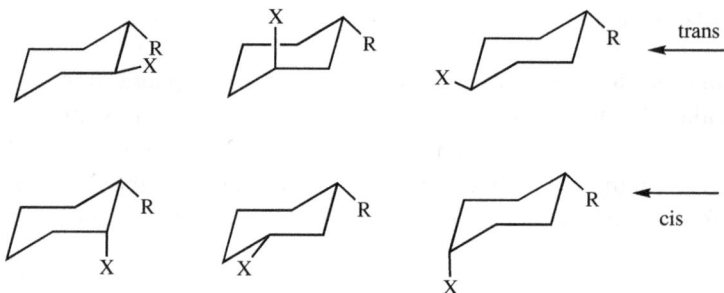

Fig. 3.24

Furthermore, each of these seven can exist as two possible chair conformations. An examination of models will reveal the following facts. Because of the alternating up and down orientations of the axial and equatorial bonds, the positional isomer designated as 1,2-disubstituted will have both substituents *on the same side of the ring* only when one of the substituents is axially oriented and the other is equatorially disposed. The arrangement is designated the a,e (axial, equatorial) conformation. Conversely, the two substituents can be on *opposite* sides of the ring if both are axially oriented (a,a conformation) *or* both are equatorially oriented (e,e conformation). Since inverting the conformation of the ring changes all axial bonds to equatorial ones, and since equatorially substituted rings are more stable than axially substituted ones, the *trans* configuration of the 1,2-disubstituted isomer will be more stable than the *cis* configuration which always *must* have one substituent axial.

When the same arguments are applied to the 1,3-disubstituted isomers, it will be found that the situation is exactly reversed. That is – it is the *cis* configuration that can have both its substituents equatorial and which is therefore more stable. The situation for the 1,4-disubstituted isomer reverts back to being the same as for the 1,2-isomer; *trans* is the more stable configuration (Fig. 3.25).

Note that you cannot interconvert *cis*- and *trans*-isomers by rotational (conformational) motions. *Cis*- and *trans*-isomers are different *configurations*.

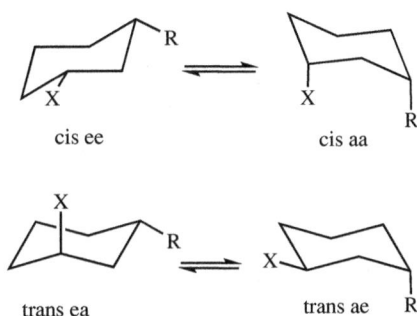

Fig. 3.25

Notice that the foregoing allows you to predict on the basis of very simple ideas which *configuration* is preferred. Another application of the same type of ideas will determine which *conformation* of any given configuration will be the predominant one in the conformational equilibrium.

Which conformation of a pair, one of which must be a,a and the other of which must be e,e, is more stable is clear, the diequatorial form will predominate. Which conformation of a pair, one of which will have the a,e conformation and the other of which will have the e,a conformation, will be more stable depends on which of R or X is larger, *the larger substituent on the equatorial bond is favored*.

Note that the last sentence above refers to the physical size of the substituents and *has no relationship* to their CIP priorities! Bromine (an atom) is smaller than CH_3 (a group of four atoms).

When using Newman projections to examine conformations of disubstituted cyclohexanes, it is necessary to number the carbons so that *both* substituents are located on the carbons represented by the circles and dots. This can always be done. 1,2-Disubstituted isomers use both of one set of circle and dot, 1,3-disubstituted isomers use both circles or both dots, and the 1,4-isomers use the circle on one side and the dot on the other. These are shown in Fig. 3.26.

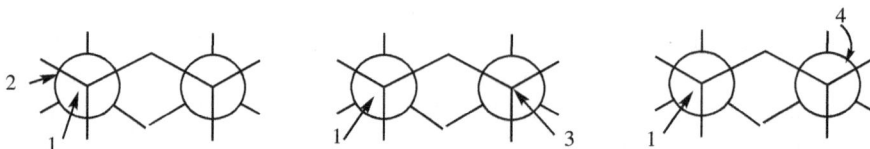

Fig. 3.26

The type of stereochemistry we have discussed so far is all related to the definition of geometric isomers (stereoisomers due to restricted rotation about bonds because of the presence of double bonds or rings). We will now go on to look at other aspects of reactions and structure, including some new kinds of stereochemistry.

3.12 Problems

3-1. Draw the energy diagram for the various eclipsed and staggered conformations of 2-methylbutane as viewed down the C2–C3 bond.

3-2. Bonds between carbon and chlorine atoms are polar. The *molecule* 1,2-dichloro-ethane has a dipole moment that increases as the temperature increases. Why is this so? (Hint: consider conformational factors).

3-3. Give the correct stereochemical designation (E or Z) for the double bonds in the following structures.

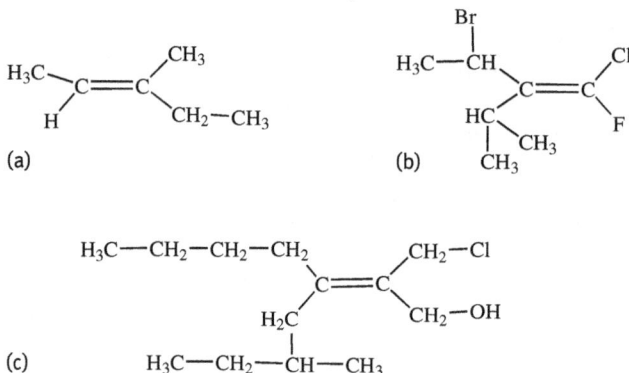

3-4. Give the complete IUPAC names (including stereochemical descriptors) for the first two compounds shown in question 3.3.

3-5. Draw the Newman projections of the following:
a) 2-Methylpentane in any skew conformation viewed down the C3–C4 bond.
b) An eclipsed conformation of 1,2-dibromobutane viewed down the C1–C2 bond.
c) A synperiplanar conformation of 3-methylhexane viewed down the C2–C3 bond.
d) The *most stable* conformation of 2,3-dichloro-2,3-dimethylbutane viewed down the C2-C3 bond.
e) The *least stable* conformation of 2-bromo-3-ethylheptane viewed down the C2–C3 bond.

3-6. Explain why *cis* and *trans* 2-butene do not interconvert.

3-7. Draw the structural formula for the possible geometrical isomers of each of the following and label each as Z or E.
a) 4-methyl-2-hexene
b) 1,2-dibromo-1-butene
c) 4-methyl-3-hexene-1-yne
d) 1-methylcyclohexene

3-8. Arrange the following groups in the order their priorities would be if they were attached to an sp^2 carbon of an alkene. Arrange the list in order of *decreasing priority*.

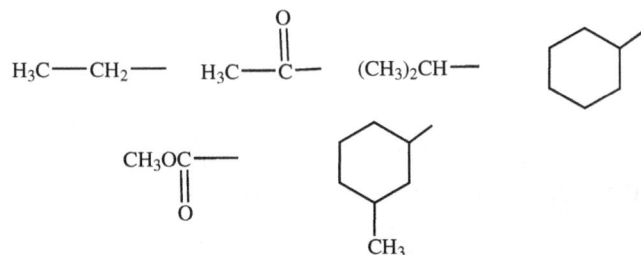

H_3C—CH_2— H_3C—$\overset{\overset{O}{\|}}{C}$— $(CH_3)_2CH$—

CH_3OC—

$\overset{\|}{O}$

CH_3

3-9. Draw a structural formula for each of the following and predict whether it is the more stable of the two possible geometric isomers.
a) (E) 2-pentene
b) (Z) 3-isopropyl-2-hexene
c) (E) 2,3-dibromo-2-pentene
d) (Z) 1-chloro-3-(chloromethyl)-2,4-dimethyl-2-hexene (Note – Chloromethyl is a –CH_2Cl substituent)

3-10. How many isomers are there of a molecule containing a cyclobutane ring, a methyl group, and a chlorine atom (C_5H_9Cl)? Draw each of these and provide a full IUPAC name that includes stereochemistry where appropriate.

3-11. How many isomers (including stereoisomers are there of the molecular formula $C_5H_{11}Br$? Draw them.

3-12. Construct a chart of orientation vs. stability for the dimethylcyclohexanes by filling out the following:

Isomer	Conformations (ae, ea or aa, ee)	Configuration (*cis* or *trans*)	Most stable?
1,2	—— ——	—— ——	( )
	—— ——	—— ——	( )
1,3	—— ——	—— ——	( )
	—— ——	—— ——	( )
1,4	—— ——	—— ——	( )
	—— ——	—— ——	( )

3-13. Draw the chair form of the *less* stable *conformation* of the *most* stable *configuration* of the following compounds. Label the substituents as axial or equatorial and indicate whether the compound is *cis* or *trans*.
- 1-bromo-3-methylcyclohexane
- 1-bromo-4-isopropyl-3-methylcyclohexane
- 1-cyclopentyl-3-chlorocyclohexane

3-14. Draw the Newman projection of *cis* 4-methyl-1-ethylcyclohexane in its most stable conformation and label the substituents as being axial or equatorial.

3-15. Draw the chair form of a cyclohexane ring that has an axial chlorine atom on C-1 and an equatorial ethyl group on C-3. Is this the most stable conformation? Is this the *cis* or *trans* isomer?

3-16. The tertiary butyl group [$(CH_3)_3C-$] has a much larger preference for occupying the equatorial position than either a propyl or butyl group. Why?

4 Reactions – Basic Principles (or Where, Why, What, How Fast, and How Far)

4.1 Introduction

In this chapter, we will learn a new term which will occupy us continuously as we study chemical reactions of organic compounds. This term is *mechanism*.

> A mechanism is a step-by-step description of how reactants interact to produce a product. A *reaction* is a process by which one molecule is changed into another. It involves the making and breaking of bonds.

Reactions can be considered as a series of simple steps, occurring consecutively, which lead to the final product. Basic principles already discussed, in combination with others that we will soon learn will allow us to *predict* what product will be obtained from a given reaction *before the reaction is finished* with a reasonable degree of certainty. This is most important and is the major objective of this chapter. However, before we can start this we must take a brief look at what general factors are important in reactions.

4.2 Acids, Bases, and Moving Electrons

We have to begin with a review of acid-base chemistry. This will be very brief and is intended to remind you of what you already "know."

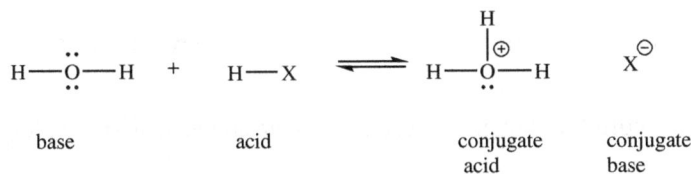

| base | acid | conjugate acid | conjugate base |

Fig. 4.1

The Brønsted–Lowry theory of acids and bases defines an acid as a proton donor and a base as a proton acceptor. A reaction between an acid and a base produces a conjugate acid and a conjugate base (Fig. 4.1).

Several generalizations can be mentioned here.
1. The weaker the acid, the stronger the conjugate base.
2. The weaker the base, the stronger the conjugate acid.
3. The relative acidity of a series of compounds H–X, where the X's are related in a vertical column of the Periodic Table, is an inverse function of the bond strength

https://doi.org/10.1515/9783110778311-004

between H and X. In general, this bond strength *decreases*, and therefore the acidity *increases* as we proceed down the column. The acidity increases in the series HF, HCl, HBr, and HI.

4. As a consequence of #3, the *conjugate base strength decreases* F^-, Cl^-, Br^-, I^-.

for example

$$H_2O \rightleftharpoons H^+ + OH^- \qquad K=10^{-14}$$

$$CH_3OH \rightleftharpoons H^+ + CH_3O^- \qquad K=10^{-17}$$

Thus water is a stonger acid than methanol, but OH^- is a weaker base than CH_3O^-

The more general Lewis theory defines acids as electron-pair acceptors and bases as electron-pair donors. This removes the restriction inherent in the Brønsted–Lowry theory that a proton must be involved. The following reaction then qualifies as an acid-base reaction.

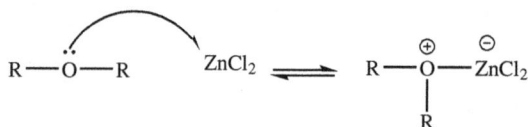

Two important points are quietly introduced in this equation. The oxygen atom has lost a 1/2 share of a pair of electrons in forming the new Zn–O bond. It therefore must be *positively charged*!

The ability to count electrons and assign charges to atoms is fundamental to organic chemistry. Practice it!!

Secondly, the "curved" or "curly" arrow, so beloved by organic chemists is shown for the first time.

The curved arrow denotes the flow of electron pairs in a reaction. It *never* shows the movement of atoms. The *direction* the arrow points is important. It always points toward a positive charge or at least an atom with an orbital capable of accepting the electrons.

Curved arrows usually start with an unshared pair or bonding pair of electrons. If the moving electrons are an unshared pair, *the atom they are leaving will gain a positive charge*. If the moving electrons are from a bonding pair, *the bond will be broken!* The moving electron pairs must also have some place to go – i.e., there must be an orbital available to accept them. (Remember that an orbital can contain a maximum of two electrons.) The following equations represent some correct usages of this helpful notation (Fig. 4.2).

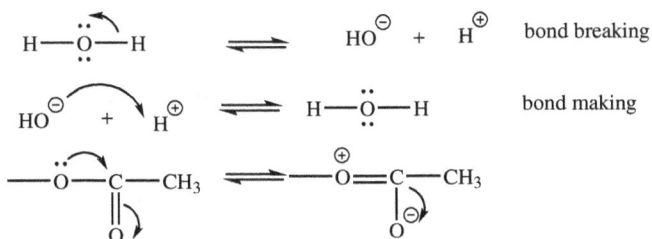

Fig. 4.2

4.3 Reversibility

You should recall from earlier courses that all reactions are, in principle, reversible *provided enough energy is available!* Thus, we can for example write,

$$A \rightleftharpoons B$$

and the equilibium constant is given by the expression $K_{eq} = [B]/[A]$. Two fundamental questions can be asked about this reaction: viz. what is the size of K_{eq} and how fast is equilibrium reached? The first of these two questions is a thermodynamic one, while the second is kinetic.

4.4 Chemical Thermodynamics (How Large is K_{eq})

Thermodynamics is a science concerned with systems at equilibrium. The fundamental equation that mathematically describes a system at equilibrium is

$$\Delta G = \Delta H - T\Delta S$$

where G is the "free energy," H is the enthalpy, and S is the entropy of the system.

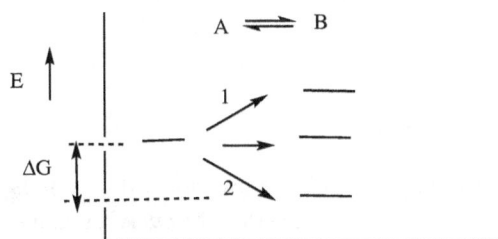

Fig. 4.3

In most cases, ΔS is quite small relative to ΔH and therefore it is convenient to use the more easily determined ΔH as a rough measure of ΔG. In the following pages we will use this approximation freely. A nonmathematical picture of the effects of thermodynamics is suggested by Fig. 4.3. The vertical axis is in arbitrary energy units and the horizontal axis is in units of a loosely defined term called the "reaction coordinate." The energy of A and that of B are indicated. Thermodynamics tells us that the equilibrium constant for this reaction is given by the expression

$$\Delta G = -RT \ln K_{eq}$$

If $\Delta G = 0$ (i.e., A and B have the same energy), $K_{eq} = 1$. This means that, at equilibrium, equal amounts of A and B will be present.

If $\Delta G > 0$ (i.e., B is of higher energy than A), $K_{eq} < 1$. This means that, at equilibrium, there will be more of A present than B. The larger the energy difference, the more of A will be present.

where ΔG = (energy of B) – (energy of A). From this expression you will see that thermodynamics says nothing about how A and B interconvert. Nevertheless, if they do, and if B is of significantly higher energy than A, it is useless to try and convert A into B. A more picturesque way of saying this is that nature, like a ball on a hill, always wants to go downhill. It follows from this that if you want to obtain B in the most efficient manner, you should choose a reaction that has the largest K_{eq} possible.

If $\Delta G < 0$ (i.e., B is of lower energy than A; this is the case shown in Fig. 4.3), $K_{eq} > 1$. The more negative ΔG is, the more of B will be present at equilibrium.

Figure 4.3 shows two pathways labeled 1 and 2. The first involves conversion of A into a more energetic form. This type of transformation is termed *endothermic* since energy (usually in the form of heat) must be added to the system. The second pathway (#2) involves an *exothermic* transformation as energy is liberated. Exothermic reactions have large K_{eq}, while those for endothermic reactions have small ones ($K_{eq} < 1$).

4.5 Kinetics (How Fast)

Consider the reaction

$$C_2H_4 + H_2 \longrightarrow C_2H_6$$

This reaction is highly exothermic – i.e., the K_{eq} is very large. However, if ethene is mixed with hydrogen – *nothing happens*. The reason for this is that the *speed* at which equilibrium is reached is infinitely slow. Questions concerning how fast a reaction proceeds (i.e., the rate of reaction) are the concern of kinetics. This subject is largely concerned with how A is transformed into B: i.e., the pathway or mechanism followed.

When molecules react it is frequently found that some energy must be put into the system in order to get the reaction going. This energy, called the *activation energy*, can be likened to the kick a ball is given to get it over a hill into a deep valley. In Fig. 4.4, the diagram used in Fig. 4.3 is repeated, but an additional feature is added. It can be seen that initially molecule A must absorb energy to allow it to reach the top of the "energy hill." After it passes this point, all the energy put in *plus* the energy equivalent to ΔG is released. Therefore, the overall reaction is exothermic by the amount of ΔG.

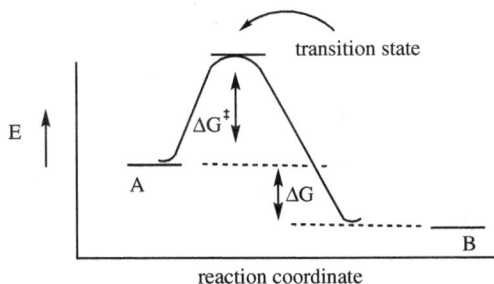

Fig. 4.4

The amount of energy required to allow the reacting molecule A to reach the top of the highest energy barrier is called the *activation energy* ($\Delta G^{\ddagger}$).

The speed at which a reaction proceeds depends on the size of the activation energy and not on the difference in energy between starting material and product. Referring again to Fig. 4.4, it should be obvious that the activation energy for B $\longrightarrow$ A is much larger than for A $\longrightarrow$ B. Therefore, if the energy available to the system is sufficient to allow A $\longrightarrow$ B, and the excess energy developed as A goes to B (i.e., ΔG) is removed (by cooling for example) then there will be insufficient energy to allow B $\longrightarrow$ A: i.e., the reaction becomes *irreversible*.

The maximum on the energy curve of the type shown in Fig. 4.4 is called a *transition state*. It is the point at which the reacting molecules start to spontaneously form product.

If a reaction occurs via several steps (e.g., A $\longrightarrow$ B $\longrightarrow$ C $\longrightarrow$ D $\longrightarrow$ P) where P is the final product, the energy diagram might look as shown in Fig. 4.5. Note that there are four transition states, one for each step in the sequence, but the transition state between C and D is the highest barrier and this represents the transition state for the whole reaction.

At this point we need to define another term:

An *intermediate* is an energy *minimum* lying between starting material and product. (Note that a transition state is an energy maximum). In Fig. 4.5, B, C, and D are all intermediates.

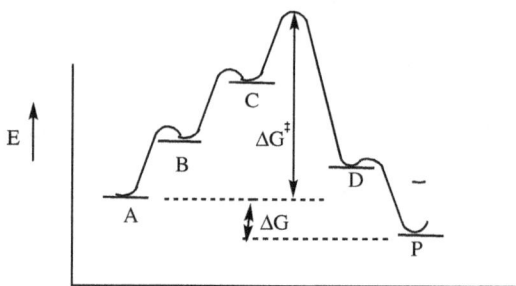

Fig. 4.5

4.6 Kinetic Versus Thermodynamic Control

Suppose two molecules A and B can react via two different pathways to give two different products, C and D. One situation that may arise is illustrated in Fig. 4.6.

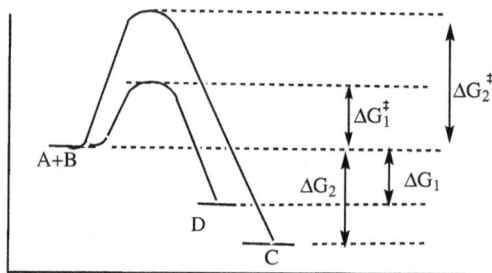

Fig. 4.6

Look first at the reaction giving D. If there is enough energy available for A + B to overcome the activation energy ($\Delta G_1^{\ddagger}$) but insufficient for C to overcome the activation energy for formation of A + B ($\Delta G_1 + \Delta G_2^{\ddagger}$) then, once formed, D cannot revert to A + B and the reaction becomes *irreversible*. (Since ΔG_1 is a common component of the energy requirements of both the forward and reverse reactions, a large negative ΔG_1 will favor irreversible actions.)

Now compare the two reactions that form C and D. For the reaction A + B $\longrightarrow$ D, the activation energy ($\Delta G_1^{\ddagger}$) is small and so D can be formed when the energy input into the system is small. Under these conditions, insufficient energy to surmount the activation barrier leading to C ($\Delta G_2^{\ddagger}$) is available and D will be formed, *even though C is more stable*! As the temperature is increased (i.e., more energy is put into the system), there will come a point where the path leading to C becomes possible. At this point, there is also enough energy to allow D to revert to A + B (i.e., $\Delta G_1 + \Delta G_1^{\ddagger} < \Delta G_2$); i.e., the reaction becomes reversible. Since C is more stable than D, D will be converted, via A + B, into C. You should convince yourself that, if the conversion of A + B > C is irreversible, *all* of D will be transformed into C, while if C is formed reversibly, the equilibrium expression

$$\Delta G_{(D \longrightarrow C)} = -RT \ln K_{eq(D \longrightarrow C)}$$

applies. The reaction occurring to give D is described as being under *kinetic control* since the size of the activation energies controls which reaction occurs. The reaction occurring to give C is described as being under *thermodynamic control* since product stabilities determine which reaction occurs.

4.7 Molecularity of Reactions

Consider a reaction A + B ⟶ C. If we ask how this reaction proceeds, we might come up with several answers but in general, these can be simplified to two possibilities.

1. A and B might interact directly and be converted directly to C.
2. A might undergo a reaction independent of B to form a new species I which then reacts with B to form C.

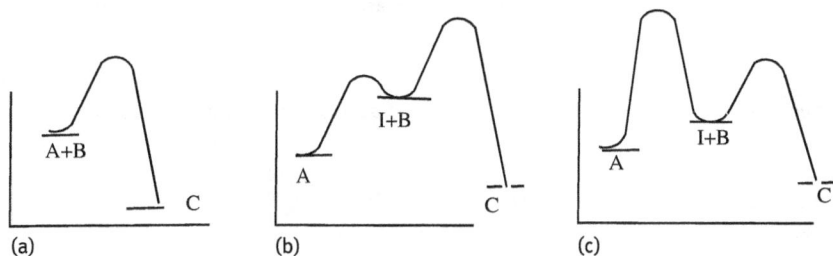

Fig. 4.7

The energy diagrams for these are shown in Fig. 4.7. In Fig. 4.7a, we see the diagram corresponding to the direct reaction. There are no intermediates. Since A and B must interact in order to undergo reaction and the frequency of their interaction will depend on how many molecules of each are present in a *unit volume*, the rate of reaction depends on the concentration of both molecules, and the reaction is said to be *bi-molecular*. The equation that describes the rate at which this reaction proceeds will have the form

$$\text{rate } \alpha [A][B]$$

In Fig. 4.7b and c two cases are shown, both corresponding to the conversion of A into I which then reacts with B to give C. Note that both of these have an intermediate. The difference between the two cases lies in which step, A ⟶ I or I + B ⟶ C is faster. In case b, the second step is slower because it has the higher activation energy. In such a case, the rate expression will be identical to that shown above: i.e., the rate will be dependent on the product of two concentration terms.

In Fig. 4.7c, the second step is faster than the first. In this case, the rate at which C is formed depends *only* on how fast I is formed (i.e., the rate of the slow step). Since B is not involved in this step, the rate expression will look like:

$$\text{rate } \alpha [A]$$

i.e., the reaction is *unimolecular*. An excellent generalization of these points is

Only those reactants that are involved in or before the slowest step in a reaction mechanism will appear in the rate expression.

4.8 Nucleophiles and Electrophiles

Two new terms will now be introduced that will be used constantly as long as you study organic chemistry.

A *nucleophile* is an atom or group of atoms that can act as an electron-pair donor to an atom other than hydrogen (usually carbon). It may or may not be negatively charged.

An *electrophile* is an atom or group of atoms that can act as an electron-pair acceptor. It may or may not be positively charged.

Some common nucleophilic species are OH^-, Cl^-, Br^-, CN^-, $H_2O:$, $:NH_3$

The most common electrophile is the proton (H^+) although many others exist.

While the proton (H^+) is an electrophile, and water is a nucleophile (because of the lone pairs of electrons on the oxygen atom), the hydronium ion (H_3O^+) is *neither*! Because it is positively charged, it cannot be a nucleophile. Also, because neither the hydrogen atoms nor the oxygen atom can accept another pair of electrons, it cannot function as an electrophile *in spite of the positive charge*.

At this point you should note the similarity of these definitions to those of Lewis bases and Lewis acids, respectively. Clearly, any nucleophile is capable of acting as a base and any electrophile is capable of acting as an acid. One of our main tasks will be to sort out when each of these types of properties will be important. In organic chemistry, a large number of reaction types can be considered as reactions between nucleophiles and electrophiles.

4.9 Inductive and Steric Effects

Consider the reaction shown below, i.e., the ionization of ethanoic (acetic) acid. The reaction involved is breaking of the O–H bond. Any effect in the molecule that serves to decrease the electron density in this bond should weaken it and make its breakage easier. This in turn should be reflected in a larger equilibrium constant: i.e., the acid will be made stronger.

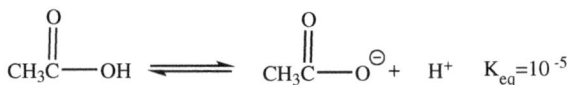

$$CH_3C{-}OH \rightleftharpoons CH_3C{-}O^{\ominus} + H^+ \quad K_{eq}=10^{-5}$$

What kind of effect might achieve this? Since atoms like Cl are more electronegative than carbon, if a hydrogen is replaced by Cl, this should pull electrons away from carbon. The positive charge on carbon resulting from this polar covalent bond should, in turn, attract electrons from the O–H bond (Fig. 4.8). Such an effect is called an *inductive effect*. Inductive effects operate *through bonds* and are caused by the presence of polar covalent bonds.

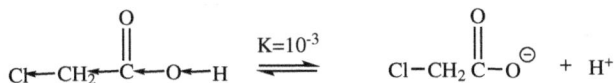

$$Cl{\leftarrow}CH_2{-}C{\leftarrow}O{\leftarrow}H \quad \overset{K=10^{-3}}{\rightleftharpoons} \quad Cl{-}CH_2\,C{-}O^{\ominus} + H^+$$

Fig. 4.8

Groups or atoms that attract electrons to themselves by an inductive effect are termed (-I) groups. Those that release (donate) electrons are termed (+I) groups.

Inductive effects fall off rapidly as the number of bonds through which they operate increases. Generally, they are effective through four or fewer bonds.

In general, (-I) groups increase the acidity of a compound while (+I) groups decrease it. It follows that (-I) groups decrease the basicity of a compound whereas (+I) groups increase it.

Table 4.1 gives the acidities (pK_as) of some organic compounds that illustrate these points. You should take a few minutes to look at the structures of these compounds and try to qualitatively relate these to their respective pK_as.

Another effect that has great influence on reactions is called a *steric effect*.

A steric effect is an effect that arises by interaction of atoms through space.

We have already seen an example of this effect when we considered the conformations of ethane (Sect. 3.4). The reason that the staggered conformation is more stable than the eclipsed one is that the atoms on the two carbons in the staggered form are less likely to try to occupy the same spatial volume at the same time: i.e., they are farther apart and the *steric* effect is lower! In general, a steric effect will be important whenever a relatively large number of atoms are confined to a small volume.

4.10 Resonance Effects

At this point, we must introduce one of the most powerful and important effects to be found in organic chemistry. Our treatment of it here will be confined to the basic principles and these will be developed more fully as we encounter its application at various points. The concept is very subtle but its importance cannot be overemphasized.

The basic principle involved can be derived from quantum mechanics. The so-called "electron-in-a-box" model of molecules predicts that

If an electron in a molecule can become associated with more than two atoms, the stability of that molecule will be increased relative to the situation where the electron is restricted to only two atoms. Furthermore, the more atoms with which it can become associated, the more stability will be gained.

If we consider the nature of sigma and pi bonds, we will see that, since the electrons in π bonds are less-strongly attracted to atoms (they are farther away), and since atoms that are joined by π bonds are also joined by σ bonds, we might well conclude that electrons in π bonds should be more able to move around molecules and become associated with more atoms. This is certainly a valid prediction.

For example, consider the ion produced by the ionization of an organic acid (Fig. 4.9). This ion is drawn in two ways in this figure. (The p-orbitals that are not involved are omitted for simplicity.) From the picture which shows the orbitals we can see that the pair of electrons on the charged oxygen atom might be used to form a new π bond to the carbon atom. However, since an orbital can hold only two electrons, if this were to happen, the electrons in the existing bond would have to "move" out onto the other oxygen atom. This is depicted in Fig. 4.9. (Again, note the use of the "curly arrows" to denote the movement of electron pairs.)

The two structures shown in Fig. 4.9 are called resonance or canonical forms. It is important to realize that the symbol ↔ does *not* mean that the molecule is in one form at one instant and in the other form the next instant: i.e., the two forms are *not* in equilibrium. (This would be indicated by the symbol ⇌). Rather the single molecule is a species that has some of the characteristics of both. It is said to be a *resonance hybrid*.

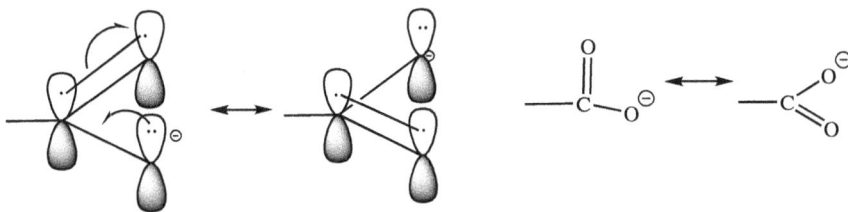

Fig. 4.9

Perhaps the best way of illustrating what is meant by this terminology is through analogy. A mule is a hybrid of a donkey and a horse, two biologically distinct species. This does not mean that a mule is, at one instant a horse and in the next instant a donkey. Rather a mule is a distinct entity that has some of the characteristics of both a donkey and a horse. The anion RCO_2^- is a discrete molecule that has some of the characteristics of both the structures in Fig. 4.9. In this case, both of the structures are the same, but this need not always be so.

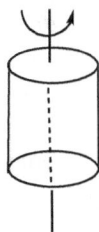

Fig. 4.10

Another useful analogy is as follows. Suppose you have a can of blue paint and a can of yellow paint. You paint one half of a cylinder with each color as indicated in Fig. 4.10. When the cylinder is rotated rapidly around the longitudinal axis, it will appear to be green since your eye cannot keep track of the individual colors. However, if you mix the two colors together and paint the cylinder, it will appear green at all times whether spinning or not. The green paint is a "hybrid" of the blue and yellow: it has some of the characteristics of both but is a discrete entity unto itself. This method of drawing structures is really an attempt to convey information about a molecule that cannot be put in one picture.

A classic case occurs in the molecule called benzene (Fig. 4.11). The six sp^2-hybridized carbon atoms each have one electron that can be paired up to form three π bonds. This could be done in two different ways and it appears there might be two isomers of benzene in which there would be alternating double and single bonds. In fact, there is only one isomer with this constitution and *all* the bonds are of the same

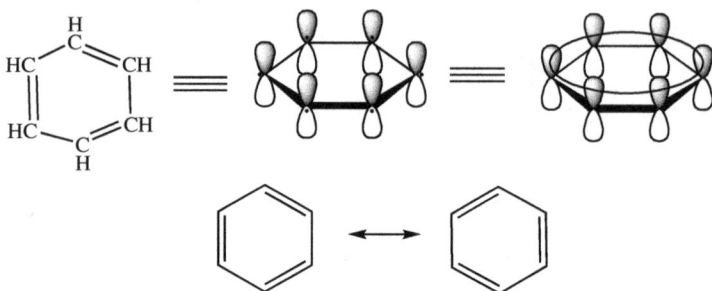

Fig. 4.11

length, and are intermediate in length between typical values for single and double bonds. The structure of the molecule is such that all the C–C bonds are the same (or equivalent). In fact, in a system like this, the six electrons can move around the ring in a continuous circle, much like a closed electrical circuit. The actual structure is a *weighted average* of all possible canonical forms. When the possible forms are identical in energy (Fig. 4.9) the resonance stabilization is at a maximum. As one or more of the forms becomes less likely (due to separation of unlike charges, high charge density on one atom, positive charge on an electronegative atom, etc.) their contribution to the *total* structure decreases and the resonance stabilization energy is lowered. We will consider this case in much greater detail later (see Chapter 9).

When π electrons can become associated with more than two atoms, the molecule gains stability. The energy difference between the actual molecule and the molecule which it would be if resonance did not occur is called the resonance stabilization energy.

It is important to note that the resonance stabilization energy is an energy difference between the actual molecule and what it would be if resonance was not present (a theoretical value). It is *not correct* to say that one molecule is more stable than another because it can be resonance-stabilized. This is similar to comparing apples and oranges by saying, "We know that orange-colored oranges are riper than green-colored ones so therefore orange-colored oranges must be riper than green apples as well," a statement that is not necessarily true.

The terms *reactivity* and *stability* should not be confused. A compound's reactivity is measured by the activation energy (Sect. 4.3) required to undergo reaction: i.e., it is a kinetic parameter. This, in turn depends on the particular reaction being considered. A compound's stability is a thermodynamic parameter related to the energy required to break it into its constituent atoms. It is entirely possible for the more stable of two compounds to also be the more reactive!

4.11 How Important is Resonance?

In the parlance of 2018, resonance is the "Connor McDavid" or "Neymar" of organic chemistry. That is – when it is in the game it usually is the controlling factor. Consider a simple example.

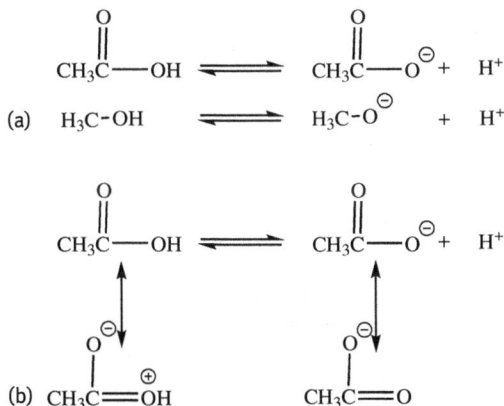

(a)

(b)

Fig. 4.12

Why is acetic acid ($pK_a = 5$) a stronger acid than methanol ($pK_a = 17$)? To answer this, we look at the equations for the two reactions (Fig. 4.12a). In the methanol case, no possibility for resonance exists on either side of the equation. Therefore, there is no gain in stability from (or driving force for) ionization. In the acetic acid case, resonance forms can be written for both starting material and products as shown in Figure 4.12b. The resonance hybrid on the left involves a separation of unlike charges (you should prove to yourself that the charges shown are real). Since it requires work to separate unlike charges (i.e., the charged form is less stable than the uncharged one), this delocalization of electrons is not favorable and does not allow resonance to be a significant factor in the stability of the *un-ionized* acid. However, in the product, this unfavorable situation is eliminated and a large increase in stability is gained due to the delocalization of the negative charge. This provides a driving force for ionization. The resonance effect changes the ionization constant by a factor of 10^{12}! We will encounter this effect many times in future chapters and come to understand it better through frequent exposure to it.

4.12 Reaction Types

Many types of reactions are known. However, our main concern in this course will be with four basic types. To complete this chapter, we will look at these in a very elementary way.

To start with some definitions are required.

A substitution reaction is a reaction in which one group of atoms replaces another. The main characteristic of this type of reaction is that the carbon atom undergoing substitution does *not* change its hybridization state. For example:

$$CH_3I + OH^- \longrightarrow CH_3OH + I^-$$

In this case, the hydroxide ion replaces the iodide ion and the carbon remains in the sp^3-hybridized state.

An *addition* reaction is one in which one molecule adds to another. In organic chemistry, this requires that one molecule has at least one double or triple bond present. If all the carbon atoms are sp^3-hybridized, no further atoms can be added without something else leaving (thus it is *saturated*). The main characteristic of addition reactions is that the hybridization of the reacting carbon changes either from sp^2 to sp^3 or sp to sp^2. In the case shown below, the bromine adds to the ethene molecule.

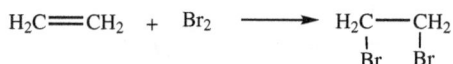

$$H_2C{=\!=}CH_2 \ + \ Br_2 \longrightarrow \underset{\underset{Br}{|}}{H_2C}{-}\underset{\underset{Br}{|}}{CH_2}$$

Elimination is the reverse of addition. It involves loss of a molecule from a larger one and the formation of a new molecule that has one additional unit of unsaturation. For an example, see below.

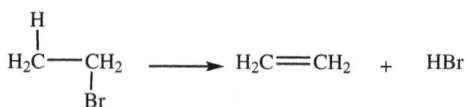

$$\overset{\overset{H}{|}}{H_2C}{-}\overset{}{\underset{\underset{Br}{|}}{CH_2}} \longrightarrow H_2C{=\!=}CH_2 \ + \ HBr$$

Oxidation and *reduction* processes in organic chemistry are less easily defined than in classical inorganic chemistry where changes in the amount of charge on ions are easily monitored. A good operating definition of *oxidation* is – the removal of hydrogen atoms from, and/or the addition of oxygen atoms to, a molecule. *Reduction* is the reverse process: i.e., the addition of hydrogen atoms to, and/or the removal of oxygen atoms from, a molecule.

4.13 Summing Up

In this chapter we have been introduced to many new terms for the succeeding chapters. As usual, you are encouraged to master this material NOW and prove your mastery by doing the problems at the end of the chapter.

Tab. 4.1: Acidities of Some Selected Organic Compounds.

Acid	Conjugate Base	pK_a
CH_3CH_3	$CH_3CH_2^-$	>40
NH_3	NH_2^-	36
$CH_2=CHCH_3$	$CH_2=CHCH_2^-$	35
$HC≡CH$	$HC≡C^-$	25
$CH_3C(O)CH_3$	$CH_3C(O)CH_2^-$	20
CH_3OH	CH_3O^-	18
H_2O	HO^-	14
CH_3CH_2SH	$CH_3CH_2S^-$	11
R_3NH^+	R_3N	9.5
HCO_3^-	$CO_3^=$	10
C_6H_5OH	$C_6H_5O^-$	10
HCN	CN^-	10
H_2S	HS^-	7
CH_3COOH	CH_3COO^-	5
$ClCH_2COOH$	$ClCH_2COO^-$	3
$Cl_2CHCOOH$	Cl_2CHCOO^-	1
Cl_3CCOOH	Cl_3CCOO^-	0.7
F_3CCOOH	F_3CCOO^-	0.2
HNO_3	NO_3^-	-1.4
H_3O^+	H_2O	-1.7
HCl	Cl^-	-7

4.14 Problems

4-1. Classify each of the following as a nucleophile, electrophile, or neither.
CN^-, CH_3OH, HO^-, H^+, Br^-, BF_3, H_2.

4-2. Classify each of the following substituents as being either +I or -I. Use electronegativities and consider its effect on acidity.

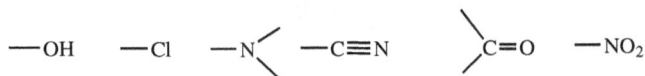

$—OH$ $—Cl$ $—N\big\langle$ $—C≡N$ $\big\rangle C=O$ $—NO_2$

4-3. Classify each of the following as a substitution, addition, elimination, or oxidation reaction.

4-4. Define the terms transition state and intermediate.

4-5. If a reaction of A and B proceeds via a three-step mechanism to give C and B is involved in the second step, draw the reaction energy profile and write the form of the rate equation if:
(a) The first step is the slowest step.
(b) The first and third transition states are lower in energy than the second.

4-6. Considering *only* steric effects, predict which of the following would be more stable.
(a) The synclinal or synperiplanar conformations (around C2–C3) of butane.
(b) *cis* or *trans* 1,3-dimethylcyclohexane.
(c)

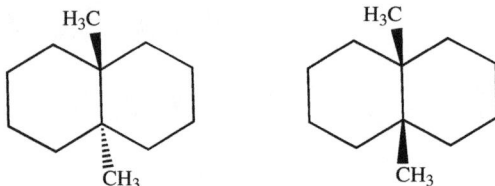

4-7. Identify which of the systems below are capable of being stabilized by resonance. Draw as many resonance forms as possible for those structures that can undergo this process.

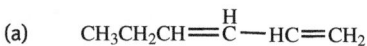

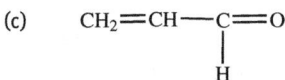

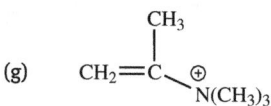

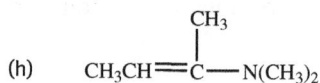

(a) $CH_3CH_2CH=\overset{\text{H}}{\underset{}{C}}-HC=CH_2$

(b) $H_2C=CH-CH_2OH$

(c) $CH_2=CH-\underset{\text{H}}{\overset{}{C}}=O$

(d) $CH_2=CH-OCH_3$

(e) $CH_2=CH-CH_2-CH=CH_2$

(f) $CH_3CH=CH-Cl$

(g) $CH_2=\overset{\overset{\text{CH}_3}{|}}{C}\diagdown\underset{N(CH_3)_3}{\overset{\oplus}{}}$

(h) $CH_3CH=\overset{\overset{\text{CH}_3}{|}}{C}-N(CH_3)_2$

5 Reactions of Alkanes, Alkenes, and Alkynes

5.1 Introduction

In the preceding four chapters you have been introduced to a wide range of basic principles that govern the structure, shape, and reactivity of organic molecules. We are finally ready to start applying these principles to actual molecules and their reactions. In this chapter, we will look at some reactions of hydrocarbons, and particular attention will be paid to *alkenes*. Some of the reactions we will see do not fit the general mechanistic types we will be developing and therefore must be learned separately. However, most will be considered from the viewpoint of what is actually happening as the molecules react: i.e., the mechanism.

5.2 Reactions of Alkanes

This section will be quite brief simply because, on the usual scale of reactivities, alkanes (saturated hydrocarbons) are quite unreactive. Furthermore, those reactions they do undergo do not fit the type of mechanistic pathways we will be considering.

5.2.1 Oxidation

The most general reaction undergone by alkanes is combustion: i.e., their oxidation in air. For example

$$CH_4 + 2O_2 \longrightarrow CO_2 + 2H_2O + \text{heat}$$

This, of course is the reaction that heats houses and powers internal combustion engines. It is also useful for determining the molecular formula of organic molecules (see Problem 2.4).

> The ultimate goal of any organic chemistry course is to be able to predict, from a knowledge of mechanism and/or by analogy with similar molecules, how a particular molecule will react under given conditions. It is strongly suggested that you start a list of the reactions we have discussed and keep it up-to-date, lecture by lecture. This will greatly simplify review. It is also important to realize that the reactions must be learned frontwards and backwards. That is – we will see a reaction where A gives B under certain conditions. You should remember this in terms of *how* A reacts *and* also how to prepare B.

https://doi.org/10.1515/9783110778311-005

5.2.2 Halogenation

The replacement of hydrogen atoms by halogen (usually chlorine) is another common reaction of alkanes. The products are called *alkyl halides*. (Alkyl is the term used to describe a general structure of the type C_nH_{2n+1}). The reaction is used frequently in industrial processes

$$CH_3CH_3 + Cl_2 \longrightarrow CH_3CH_2Cl + HCl$$

The products, particularly if they are polyhalogenated (i.e., they contain several halogen atoms), are useful as flame retardants, insecticides, herbicides, and solvents.

When alkanes of more complex structures are used, it is found that tertiary hydrogens are replaced at a faster rate than secondary which, in turn, are replaced faster than primary hydrogens. It is frequently difficult to get clean replacement of one type to the complete exclusion of others and as a result, mixtures of products are commonly obtained. If these mixtures can be used directly, this poses no problem, but frequently very undesirable properties are associated with the impurities. An example of this can be found in the chlorination of an organic molecule called phenol. The desired product – 2,4,6-trichlorophenol is contaminated with another product called dioxin, which has the reputation, perhaps undeserved, of being one of the most toxic compounds known.

5.3 Electrophilic Addition to Alkenes: Our First Mechanism

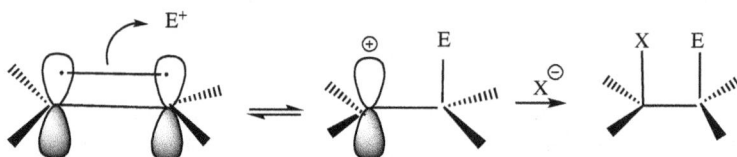

Fig. 5.1

Alkenes (olefins) are electron-rich molecules; that is, they contain more electrons than are required to hold the atoms together in the molecule. Therefore, they can be considered to be nucleophilic compounds. The two sp^2-hybridized carbon atoms of the double bond are held together with one σ and one π bond (Fig. 5.1). Reactions of these molecules usually involves the π bond since these electrons are held less strongly than the σ-bond electrons.

The major type of reaction that alkenes undergo is *electrophilic addition*. As mentioned above, alkenes are nucleophiles and one would expect them to react most readily with electrophiles. (At this point you should review the definitions of *nucleophile* and *electrophile* in Chapter 4).

Consider what might happen when an alkene is in the presence of some elec-trophile E^+ (Fig. 5.1). Since a positive ion (E^+) will attract negative charge (electrons) the pair of electrons in the π bond can be used to form a new bond as shown. Several things have happened here. Note that the right-hand carbon is now sp^3-hybridized. Also, since the new C–E bond uses both the electrons from the original π bond (a bond is always two electrons!), the left-hand carbon is now positively charged (you should convince yourself of this by counting the electrons). The p-orbital is now empty. The stability of E^+ must be quite delicately balanced. It must be stable enough to exist at least semi-independently of the counterion X^- which is also present (i.e., E^+ and X^- must not totally recombine into EX) and it must not be so stable that it has no interest in combining with the alkene. The "arch-typical" electrophile H^+ fulfils these quali-ties admirably. Conversely, ions such as Na^+, K^+, Ca^{2+}, etc., are too stable to react with alkenes.

A *positively charged, trivalent carbon atom is called a carbocation.* (The older name is "carbonium ion"). It is a discrete entity, but since it is of higher energy than the alkene (because of its charge), it fits the designation we learned for an *intermediate.* Note that it is also *very electrophilic.*

What can happen to this carbocation? Since it is less stable than the alkene, it must undergo further change. There are two possibilities. The newly formed C–E bond could break, reforming the alkene and E^+; i.e., the formation of the carbocation is reversible. Alternatively, if a negative species X^- (i.e., an anion and a nucleophile) is present in the reacting system, it could form a bond to the positively charged center (Fig. 5.1). In this step, the second carbon is rehybridized to the sp^3 form.

Practically speaking, it is not possible to add a positively charged species to a reac-tion mixture without a negatively charged one to balance the charges. For the common case of $E^+=H^+$, one normally adds H^+Cl^- or H^+Br^-, etc. The overall reaction shown in Fig. 5.1 is called an *electrophilic addition.* It is the most common reaction type of alkenes. Many different E^+X^- systems react in this way and we will consider specific examples shortly. Before proceeding, it is instructive to examine the energy diagram for this reaction type (Fig. 5.2). Since the carbocation is an intermediate, it must have a transition state on both sides of it: i.e., it is an energy minimum (cf. Chapter 4). The question then arises as to which barrier is higher – or – which step in the reaction is the slow step. Simple electrostatic principles would predict that the reaction between a positively charged and a negatively charged species (i.e., the second step) should be faster than that between a neutral molecule and an ion. The first step should be slower. This is found and is indicated in Fig. 5.2 where the first transition state is of higher energy.

What species undergo addition to alkenes? In general, some of the best are the hydrogen halides HBr and HCl (but *not* NaCl, NaBr, etc., for the reasons given on the previous page). Some examples are given in Fig. 5.3. Take another look at the last ex-ample in Fig. 5.3.

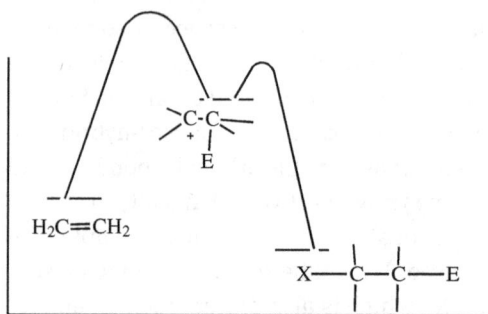

Fig. 5.2

Does anything appear puzzling? Why did the proton become attached to the end carbon atom to form 2-bromopropane instead of to the center one to form 1-bromopropane? The answer to this question is considered in the next section.

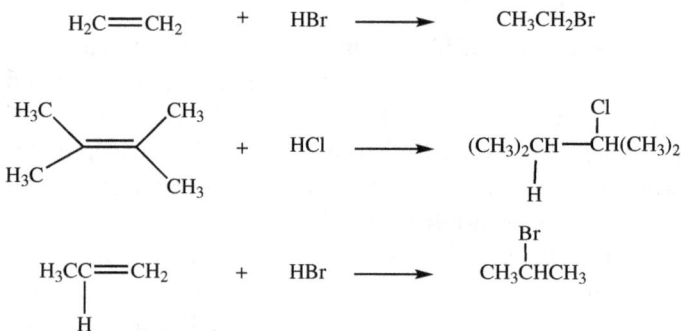

Fig. 5.3

5.4 Carbocation Stabilities (Markownikov's Rule)

In the early days of organic chemistry when mechanistic principles were poorly understood, many reactions were attempted and if a general trend could be detected, this trend was generalized as a "rule." As mechanistic principles became better understood, the underlying reason which is responsible for the observed results was recognized. The problem with simply memorizing the rule is that, like all "rules," exceptions occur. However, if the *mechanistic principles* are understood, the exceptions can be predicted.

The rationale for the formation of 2-bromopropane and not the isomeric 1-bromo isomer was first generalized as a rule called *Markownikov's Rule.*

In the addition to an unsymmetrically substituted alkene, the more positive part of the adding reagent (addend) becomes attached to the sp^2-carbon bonded to the most hydrogen atoms.

In the example in Fig. 5.3, the more positive part of the addend HBr is clearly the hydrogen atom. It becomes attached to the sp^2 carbon which is attached to two hydrogens (C1) and not to C2 which is attached to only one hydrogen.

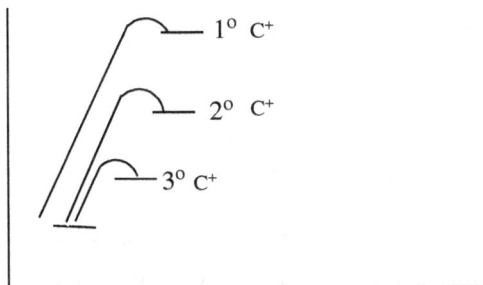

Fig. 5.4

What is behind the result generalized in Markownikov's Rule? It can best be explained by noting that, because of an effect called *hyperconjugation,* carbocations have different energies depending on how many carbon atoms are directly attached to the positive center. From a variety of different types of experiments, it is found that primary carbocations (i.e., $C–CH_2^+$) are less stable than secondary (i.e., $C–CH^+–C$), which in turn are less stable than tertiary ones (e.g., $(C)_3–C^+$). Consider what this means in terms of the activation energy for the electrophilic addition reaction (Fig. 5.4). It is clear that the smallest energy barrier that must be surmounted to reach an intermediate is that leading to the tertiary (3°) carbocation. Since nature, like organic chemists, prefers to follow the path of least resistance, the intermediate formed most readily is the more substituted one. In the specific case of addition to propene (Fig. 5.3), the two possible intermediates would be as shown below and the product is formed from the more stable intermediate. Shortly, we will see other factors that influence carbocation stabilities but for now, we will operate with simple ions where the effect just described controls the reaction pathway.

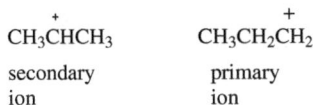

$CH_3\overset{+}{C}HCH_3$ $CH_3CH_2\overset{+}{C}H_2$

secondary   primary
ion   ion

5.5 Catalysis

When an alkene is dissolved in water, no reaction occurs. This is understandable when it is recognized that, although the elements required for addition to occur are present (i.e., H^+ and OH^-), in neutral water the concentration of H^+ is only 10^{-7} M. In this case the concentration of E^+ is so low that the first step in the mechanism has no chance of occurring. How could we overcome this? Obviously, if some strong acid is added, the concentration of protons will increase. However, you must remember that, at the same time, the concentration of the nucleophile (OH^-) will decrease from its already low level of 10^{-7} M in order to keep the product of $[H^+][OH^-] = 10^{-14}$.

Fortunately, the neutral molecule H_2O can act as a nucleophile. Oxygen has two pairs of unshared electrons and fits the definition of nucleophile. The series of reactions shown in Fig. 5.5 can then occur.

Several very important points need to be noted. The acid catalyst used in step 1 is regenerated in step 3: i.e., the acid is not consumed. Secondly, every step in the sequence is reversible. The importance of this will become apparent shortly. Finally, as noted above, the *nucleophile is* H_2O and *not* OH^-. If you have any difficulty seeing why the various atoms in Fig. 5.5 are positively charged, do an electron count on them.

Typical conditions for the *acid-catalyzed* addition of water to alkenes are the use of 10% H_2SO_4 in H_2O. Many beginning students intuitively feel that if a little of something does a good job of catalyzing a reaction, a lot of it will do a better job. However, if the acid concentration is increased to above 70%, it is found that almost no reaction occurs. The reason for this can be seen from the equilibrium shown in Fig. 5.5. In the presence of small amounts of acid, there is a large excess of water which will shift the equilibrium to the right and lead to the alcohol product.

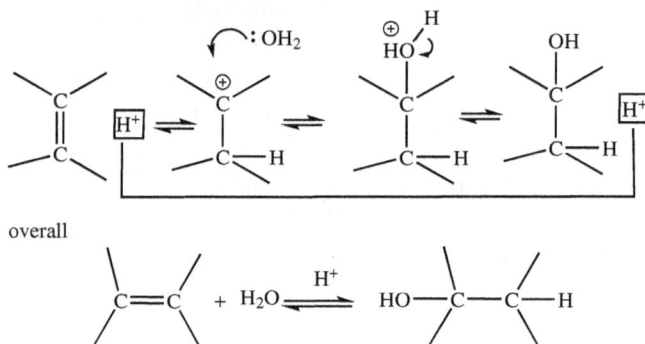

overall

Fig. 5.5

However, if excess acid is used, most of the water will be in the protonated form H_3O^+ which is *not* nucleophilic. Since the water concentration is now low, the equilibrium will shift to the left. This effect is illustrated in the following equations.

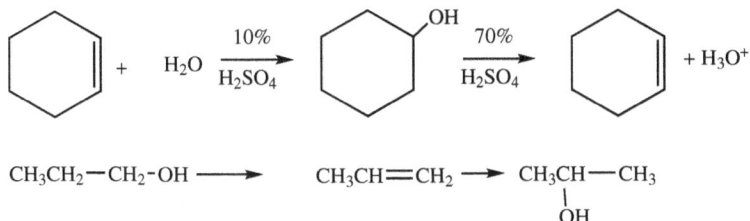

$$CH_3CH_2-CH_2-OH \longrightarrow CH_3CH=CH_2 \longrightarrow CH_3CH-CH_3$$
$$\underset{OH}{|}$$

The *Principle of Microscopic Reversibility* tells us that in any equilibrium reaction, the mechanistic steps going from A $\longrightarrow$ B are exactly the same as going from B $\longrightarrow$ A, but in the reverse order. Therefore, if the mechanism for the reaction of an alkene with water under acid catalysis is known, by definition we must know the mechanism for the reverse reaction.

5.6 Stereochemistry

Take another look at the sequence shown in Fig. 5.1. In the case where $E^+=H^+$ the intermediate ion is relatively bare: that is there are no extra electrons associated with nearby atoms that might interact with the positive center. The carbocation is sp^2-hybridized – i.e., planar with the now-empty p-orbital perpendicular to that plane. When the nucleophile (X^-) attaches to the carbon, it must do so using the only orbital available – the p-orbital. Therefore, X^- may attack either from the top or the bottom of the ion. Since the ion is planar, there is no reason why one of these modes should be preferred. This results in a random mixture of addition of both E^+ and X^- from *the same side* and from *opposite* sides of the molecule – i.e., *cis* and *trans* addition. Furthermore, the *initial* attack by the electrophile E^+ can also occur from either side of the alkene with equal probability for the same reasons. This will be true for any reaction at a planar carbon atom. This concept, which is of immense importance, will be developed much more fully in Chapter 6.

Now consider what would happen if E^+ was a large atom with unshared pairs of electrons. Such a situation is encountered when Br_2 is added to an alkene.

The molecule bromine (Br_2) can be visualized as existing as Br^+Br^-. This is a gross simplification of the actual situation but will suffice for our purposes here. If the Br^+ ion interacts with an alkene, we arrive at the situation shown in Fig. 5.6. The large bromine atom has three unshared electron pairs and its electronegativity is such that it is able to share one of these pairs with the positive center. The result is the cyclic form with a positive charge on the Br atom. Such an ion is called a *bromonium* ion. When this ion reacts with the nucleophile Br^-, the stereochemistry is such that the nucleophile

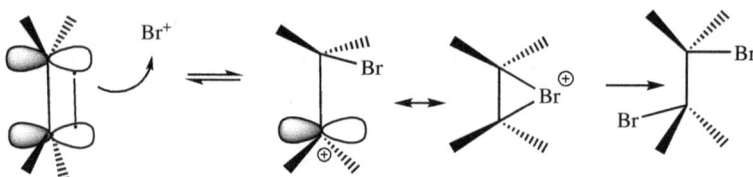

Fig. 5.6

attaches to the carbon atom from the *opposite side* to the bond it is replacing. This phenomenon will be discussed much more fully in the next chapter and for now, will have to be accepted. The overall result of this sequence is the addition of Br_2 to an alkene, one Br atom being on one side of the molecule, and the other Br atom being on the other side – i.e., a *trans addition*. Some examples are shown below. It must be emphasized that this effect will only be observed when the electrophile has lone pairs of electrons and a relatively low electronegativity which allows these to be shared.

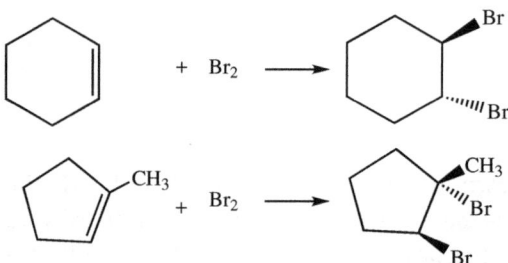

5.7 Functional Group Characterization

From time to time, we will come across a reaction that is useful, not only for preparing a particular type of molecule, but also in determining which functional groups are present in a molecule whose structure is unknown. These reactions have one thing in common. They produce a distinct *visually detectable change*. (You should keep track of such reactions as they form the basis of favored examination questions!) The reaction of bromine with alkenes is such a reaction.

Bromine is a brown liquid with a very high vapor pressure. It is soluble in the solvent carbon tetrachloride (CCl_4) and also to some extent in water. These solutions are brown in color. Alkenes are generally colorless and dissolve in CCl_4 to give colorless solutions. When a (brown) solution of Br_2 is added to a solution of an alkene, the addition product (a dibromide) is formed almost instantly. This material is also colorless. When Br_2 is added to a solution of an alkene, an instantaneous disappearance of the brown color is observed (but see problem 5.10). To see how this can be used consider the following problem:

Two bottles, one containing cyclohexane and the other 2-hexene are on the shelf but the labels have fallen off. How could you decide which bottle contained the cyclohexane? (Note that both have the molecular formula C_6H_{12}.)

The simplest answer to this is that addition of a (brown) solution of Br_2 in CCl_4 to a sample of 2-hexene will cause disappearance of the color, whereas similar treatment of cyclohexane will cause no disappearance of the color.

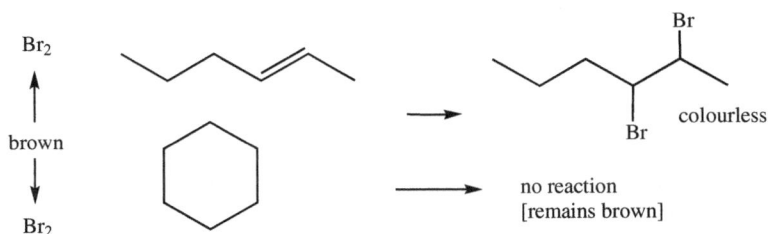

5.8 Solvent Effects (Regiochemistry)

Up to this point, we have been ignoring the effects of the reaction solvent: i.e., we have been assuming that it plays no part in any of the reactions. In fact, solvent effects are very common. These effects may be of two kinds.

1. Solvents may affect how fast a reaction proceeds.
2. Solvents may affect the type of product formed in a reaction.

We will pass over #1 temporarily, but one aspect of #2 can be conveniently discussed at this point.

Carbon tetrachloride is an ideal solvent in that it does not function as either an electrophile or nucleophile. When an alkene is reacted with bromine in CCl_4, the only electrophile present to react with the double bond is Br^+. Also, since CCl_4 is non-nucleophilic, the only nucleophile present in the reacting system is Br^- (bromide ion) (Fig. 5.6).

Consider what would happen if another nucleophile (X^-) was present in the system (Fig. 5.7). Since Br_2 is added, there will always be Br^- present and formation of the dibromide is always possible. However, the other nucleophile will compete in the second step and it is possible to form an addition product containing one bromine atom and the other nucleophile (X). Which of these two will be the major product (i.e., the product formed in the largest amount) is dependent on two factors which are:

1. Which of the two nucleophiles has a higher nucleophilicity. The better nucleophile will have a lower activation energy for the formation of the bond to carbon and will react faster. We will discuss what determines a species' nucleophilicity in the next chapter.

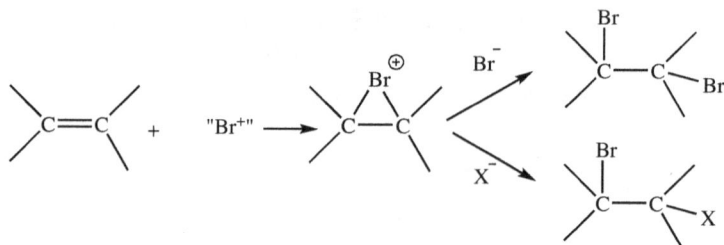

Fig. 5.7

2. Which of the two nucleophiles is present in the larger concentration. Since the two reactions of the positively charged ion with the two nucleophiles are bimolecular, their rates will be dependent on the nucleophile concentration. i.e.,
    Rate of formation of dibromide $\alpha[C^+][Br^-]$
    Rate of formation of X-C-C-Br $\alpha [C^+][X^-]$

Therefore, the relative rates are dependent on the relative concentrations of Br^- and X^-.

Let us look at a specific example. If cyclohexene is dissolved in carbon tetrachloride and Br_2 is added, the dibromide is formed. However, if the reaction is conducted in water, a large amount of a molecule with the formula $C_6H_{11}BrO$ is formed. This molecule is *trans*-2-bromocyclohexanol (Fig. 5.8). This product is formed because water is nucleophilic and can compete with bromide ion (Br^-). Since water is the solvent, its concentration is very high and its reaction is favored. (Another way of putting this is that, since there are many more molecules of water present than Br^-, the chances of

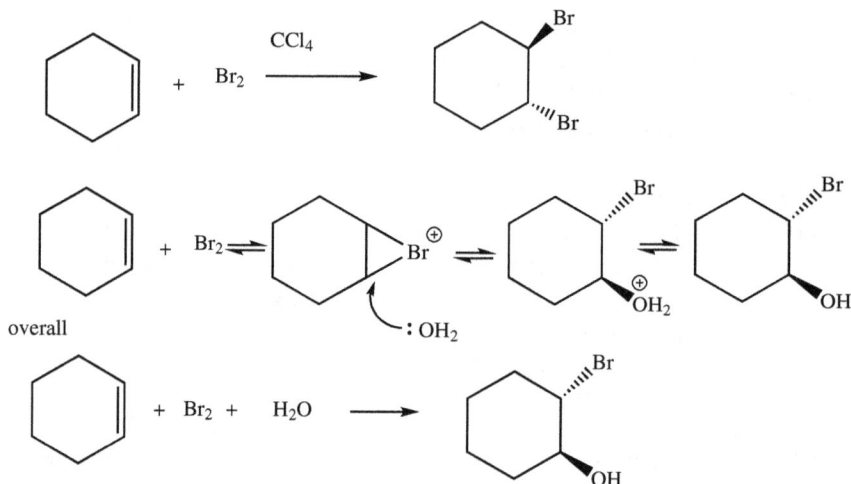

Fig. 5.8

the bromonium ion encountering and reacting with a water molecule are much larger than the same process occurring with a bromide ion.) Note that there is no change in the basic mechanism.

When the alkene undergoing reaction is not symmetrical (e.g., 2-methyl-2-butene) and the intermediate bromonium ion is opened by the nucleophile, the question arises – at which end of the cyclic ion does the final reaction occur? This can be answered by noting that the *resonance hybrid* shown in Fig. 5.6 will have the majority of its positive charge on the more highly substituted carbon atom (because positive charge is more stable on more highly substituted atoms). Another way of saying this is that the resonance form with the greater stability contributes more to the total resonance hybrid. Therefore, the electron-rich nucleophile will prefer to react at that center. In other words, Markownikov's Rule is still followed. Some problems of this type are included at the end of the chapter.

5.9 Resonance Effects on Electrophilic Addition Reactions

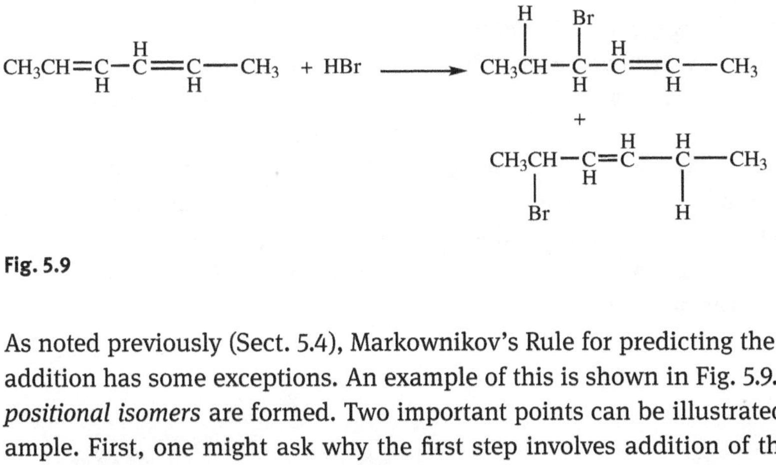

Fig. 5.9

As noted previously (Sect. 5.4), Markownikov's Rule for predicting the orientation of addition has some exceptions. An example of this is shown in Fig. 5.9. Note that *two positional isomers* are formed. Two important points can be illustrated from this example. First, one might ask why the first step involves addition of the electrophile H^+ to C2 and not C3, since both of these would generate secondary carbocations. A more detailed look at the intermediates formed by these two possible processes gives an immediate answer (Fig. 5.10). The intermediate generated by attack at C3 forms a secondary carbocation in which the positive center is separated from the remaining π bond by a sp^3-hybridized carbon. As pointed out in Chapter 4, this effectively prevents delocalization of the electrons into the vacant p-orbital of the carbocation. However, the intermediate generated by attack at C2 generates an intermediate in which the electrons can be delocalized, spreading these electrons and the positive charge over a three-atom system. This resonance effect will stabilize this intermediate relative to the other one and channel the reaction through the lower energy intermediate since it

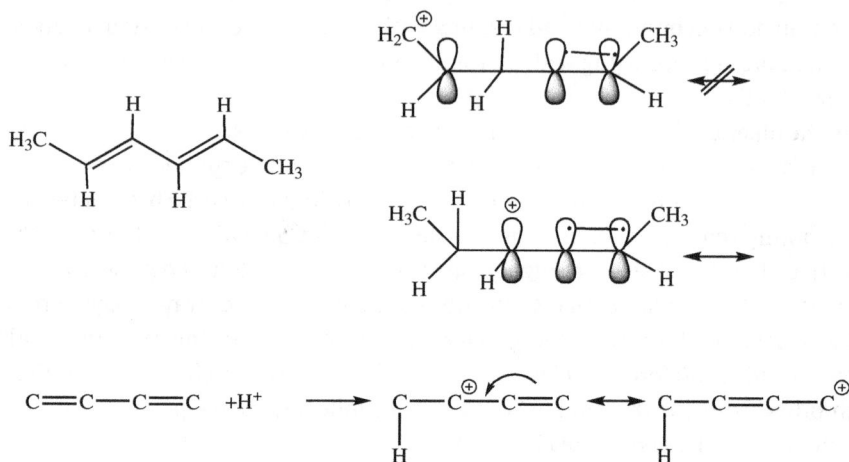

Fig. 5.10

has a lower activation energy leading to it. The energy for these processes is shown in Fig. 5.11. This effect is very general. *When a resonance-stabilized intermediate can be generated, it will control the reaction pathway.*

It is important to remember the meaning of the resonance forms shown in Fig. 5.10. It does *not* mean that the molecule is alternating its structure between A and B, but rather that in *the single entity represented by these two structures,* part of the positive charge resides on the two starred carbon atoms (below). In the second step of the reaction (i.e., the attack of the nucleophile) two sites of attack are possible, one leading to 4-bromo-2-hexene and the other leading to 2-bromo-3-hexene. The resonance effect explains not only the site of attack by the electrophile, but also the formation of two positional isomers as products.

$$\overset{*}{\text{CH}_3}\overset{}{\text{CH}_2}\overset{*}{\text{CHCHCHCH}_3}$$

CH₃CH₂CHCHCHCH₃

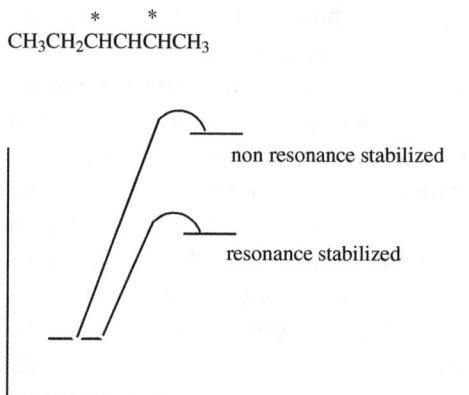

Fig. 5.11

Consideration of the preceding material will lead to the conclusion that the resonance effect of the type discussed will occur *only* if the starting diene has four sp^2-hybridized carbon atoms joined in a contiguous chain. Such dienes are called *conjugated* dienes. Nonconjugated dienes do not possess the structural properties that allow resonance effects to occur.

5.10 Other Reactions of Alkenes

Alkenes undergo many other reactions which, for one reason or another, do not exactly fit the mechanistic type we have been considering. We will look at some of these under special headings.

5.10.1 Hydroboration

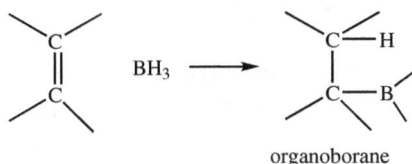

organoborane   **Fig. 5.12**

The molecule borane (BH$_3$) undergoes addition to the double bond of alkenes. Depending on the structure of the particular alkene used, one, two, or all three hydrogens may react with one, two, or three alkene molecules (i.e., the stoichiometry maybe 1:1, 1:2, or 1:3). The product of this reaction is called an organoborane (Fig. 5.12). This reaction does not fit the mechanism discussed in Sect. 5.3 since it is found that *cis* addition is always favored. Furthermore, in molecules with different degrees of substitution on the ends of the double bond, the boron atom becomes attached to the *less-substituted* carbon. (Contrast this with the addition of HBr.) For examples, see Fig. 5.13.

Organoboranes are frequently pyrophoric materials (that is they ignite spontaneously on contact with air). For this reason, they are seldom isolated. Rather the reaction mixture that contains them is treated directly with a mixture of hydrogen per-

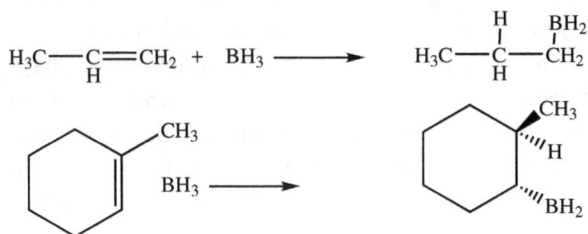

Fig. 5.13

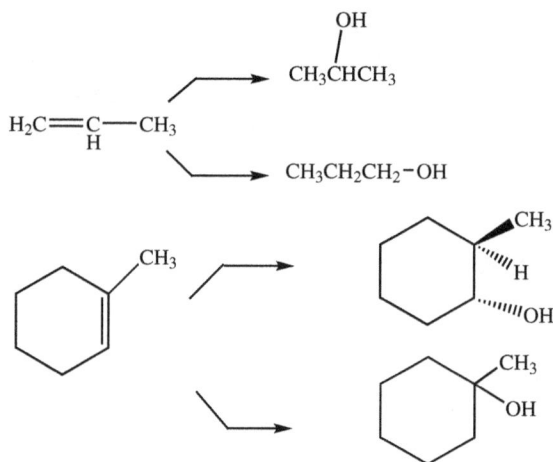

Fig. 5.14

oxide and sodium hydroxide. This combination of reagents replaces the boron atom with an OH group: i.e., an alcohol is formed. It has been found that the new OH group is always on the *same* side of the molecule as the boron atom was. The overall effect of addition of BH_3 followed by reaction with H_2O_2/OH^- is the addition of water in a *cis* manner. Furthermore, in unsymmetrically substituted alkenes, the orientation of addition is the *reverse* of that obtained when acid-catalyzed addition of water is carried out. For examples, see Fig. 5.14. This sequence allows the *anti-Markownikov(!) addition of the elements of water* to a double bond. (Note that water itself is not used in this reaction!)

5.10.2 Hydrogenation

Hydrogen (H_2) can be added to an alkene. The product of this reaction is an alkane. As pointed out previously (Chapter 4), this reaction does not occur because the activation energy is too high. However, in the presence of a catalyst (usually a finely divided metal like Pt, Pd, or Ni) the reaction proceeds smoothly (Fig. 5.15). This reaction is frequently referred to as catalytic reduction.

Alkenes are *unsaturated* compounds. Organic molecules that are composed of long chains of carbons with many double bonds are referred to as polyunsaturated. The triglycerides (fats) obtained from animals are solid materials (esters – see Chapter 9) which have long chains of carbons that have very few double bonds (i.e., they are saturated). Conversely, the triglycerides derived from vegetable oils are polyunsaturated. Modern dietary ideas suggest that, because of their metabolic fate, intake of polyunsaturated fats is preferable to intake of saturated ones. However, the physical characteristics of the polyunsaturated materials (they are liquids) makes them less appetizing (and therefore commercially less attractive) than the more saturated analogs. Commercially, many vegetable oils are hydrogenated to remove some of the double bonds and improve their characteristics. Instead of using peanut oil (a highly unsaturated oil) directly in the manufacture of peanut butter, the oil is partially hydrogenated to bring it to a consistency acceptable to the consumer. (Would you rather pour your peanut butter?) Check the label on a bottle and you will find one ingredient is frequently "partially hydrogenated peanut oil."

Another side effect of hydrogenating some of the double bonds is the formation of "*trans*-fatty acids," which may have negative health effects. By now you should be able to figure out the meaning of this phrase.

5.10.3 Oxidation

Oxidation of organic compounds can be defined as either the addition of oxygen atoms or the removal of hydrogen atoms from a molecule. Two oxidation reactions will be considered here.

A. Ozonolysis

Ozone (O_3) is a very powerful electrophile. It reacts with a carbon–carbon double bond to produce a compound called an ozonide. The ozonide is not isolated (because of its unpleasant habit of exploding!), but is treated directly with a reducing agent like zinc metal. The net result of these two operations is the cleavage of the carbon–carbon double bond into two pieces, each of which has a carbon–oxygen double bond. As we will see later, compounds containing this combination of atoms will form a large part of our interest in future chapters. Some examples of ozonolysis are shown in Fig. 5.15.

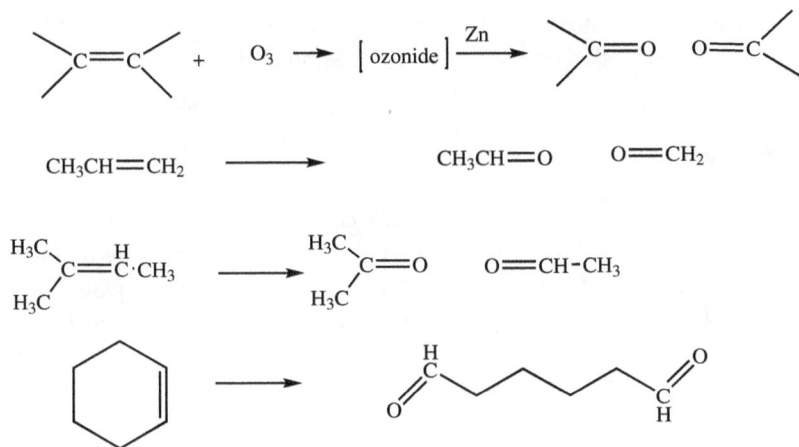

Fig. 5.15

B. Oxidation with Dichromate or Permanganate

The reaction of alkenes with potassium permanganate ($KMnO_4$) or sodium dichromate ($Na_2Cr_2O_7$) leads to different types of products depending on the pH of the solution. Examples are shown in Fig. 5.16 for $KMnO_4$. The reactions of dichromate are similar. From these examples, it can be seen that

1. Reaction at neutral or alkaline pH adds two OH groups to the double bond (i.e., the addition of H_2O_2).
2. Oxidation at acidic pH breaks the double bond in the same way as ozone does. However, if the sp^2-carbon of the alkene is attached to at least one hydrogen atom, it is changed into a COOH group while if no hydrogens are present, the carbon ends up with only one oxygen atom attached by a double bond.

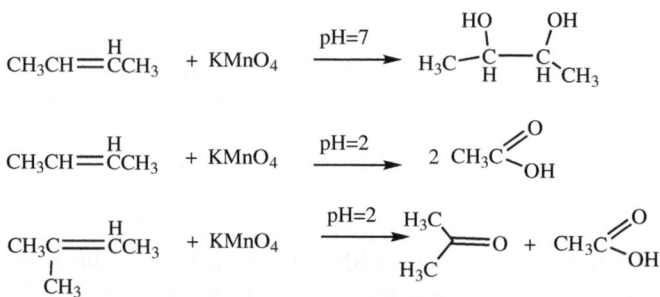

Fig. 5.16

5.11 Reactions of Alkynes

Alkynes are molecules that contain a carbon–carbon triple bond. The reactions of alkynes are somewhat different to those of alkenes, but some similarities exist. In this section, we will consider only two reactions: hydrogenation and hydration.

5.11.1 Hydrogenation

Just as an alkene reacts with H^+ in the presence of a catalyst, so too do alkynes. The product is an alkene that can be further hydrogenated to an alkane. The rates of these two reactions are sufficiently different to enable the process to be stopped when only one equivalent of hydrogen has reacted. Note that the alkene formed is in the *cis* or Z configuration.

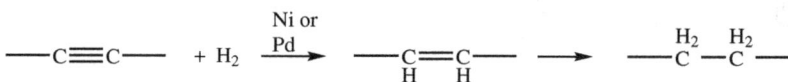

5.11.2 Hydration

The addition of water to alkenes under acid catalysis leads to alcohols. The same reaction with alkynes requires a more complex catalyst and leads to a compound that has a carbon–oxygen double bond. The intermediate in this reaction is the simple addition product which is unstable and immediately isomerizes to the observed product. For alkynes that are at the end of a chain (a terminal alkyne), the addition follows Markownikov's Rule.

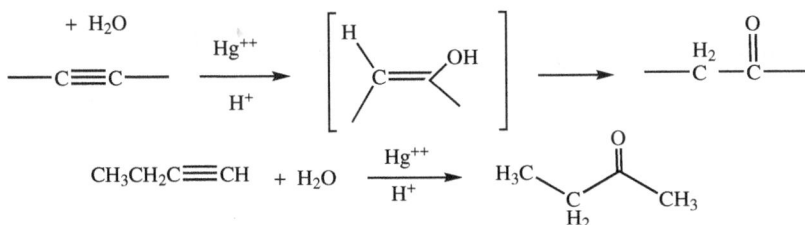

5.12 Summary

In this chapter, the first mechanistic reaction type has been introduced. We have seen how it can be used to prepare a number of different kinds of functional groups. In Sect. 5.1, it was pointed out how important it is to be able to recognize reaction types and to learn reactions forward and backward. Problems at the end of this and succeeding chapters are designed to help you sharpen your skills in this area. It is recommended that you do these and also start a list of reactions subdivided into "reactions of" and "preparations of" various functional group types. A suggested format is illustrated below.

		Reactions of		
Functional group	Reagent	Catalyst	Mech	Product
Alkenes(C=C)	H_2O	H^+	Addition	Alcohol

		Preparations of		
Functional group	Reactant	Reagent	Catalyst	Mech
Alcohol	Alkene	H_2O	H^+	Addn

5.13 Problems

5-1. Show the step-by-step mechanism for the addition of hydrogen bromide to 3-methyl-2-pentene. Be sure you show which steps are reversible.

5-2. Order the carbocation intermediates formed by ionization of the C–Br bond in terms of their stabilities. Consider both resonance and substituent effects.

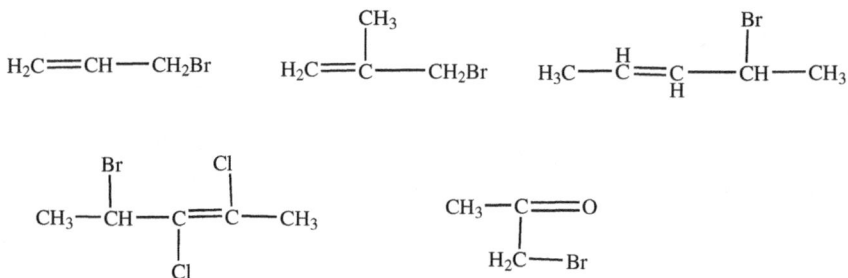

$$H_2C{=}CH{-}CH_2Br$$

$$H_2C{=}\overset{\overset{\textstyle CH_3}{|}}{C}{-}CH_2Br$$

$$H_3C{-}\overset{\overset{\textstyle H}{|}}{\underset{\underset{\textstyle H}{|}}{C}}{=}C{-}\overset{\overset{\textstyle Br}{|}}{CH}{-}CH_3$$

$$CH_3{-}\overset{\overset{\textstyle Br}{|}}{CH}{-}\overset{\overset{\textstyle Cl}{|}}{\underset{\underset{\textstyle Cl}{|}}{C}}{=}C{-}CH_3$$

$$CH_3{-}\overset{\textstyle }{C}{=}O \\ \overset{\textstyle |}{H_2C}{-}Br$$

5-3. What *stereochemical* result would you predict when Cl_2 is added to cyclohexene? Make a comparison with the stereochemical result found when Br_2 is added.

5-4. Which of the following are conjugated compounds?
(a) 1,4-hexadiene (b) 1,2,5-decatriene, (c) 1,3-cyclopentadiene
(d)

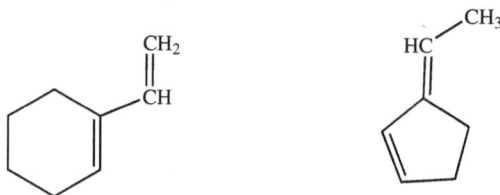

5-5. What is the hybridization state of C2 in the compound in question 5.4(b)?

5-6. Addition of HBr to propene gives 2-bromopropane as predicted by Markownikov's Rule, but addition of HBr to compound I (below) gives II.

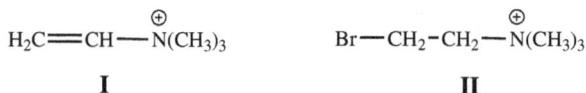

$$H_2C{=}CH{-}\overset{\oplus}{N}(CH_3)_3 \qquad\qquad Br{-}CH_2{-}CH_2{-}\overset{\oplus}{N}(CH_3)_3$$

$$\textbf{I} \qquad\qquad\qquad\qquad \textbf{II}$$

(a) Which reaction would you expect to be faster and why?
(b) Explain why the orientation of addition to II violates Markownikov's Rule.

5-7. The molecule iodine monochloride (ICl) adds to alkenes via an electrophilic mechanism. If 1-methylcyclohexene is used as the alkene, predict the structure of the product. Your answer should consider both stereochemistry (*cis*, *trans*, or mixture) and regiochemistry (i.e., which atom goes where).

5-8. The addition of I_2, IBr, and ICl to ethene occur with relative rates of $1 : 10^3 : 10^5$. Explain this observation.

5-9. The addition of water to propene requires an acid catalyst. Sulfuric acid is often used. Why is HBr a less acceptable choice?

5-10. A student was given a vial of a clear, colorless liquid and asked to determine if it was pentene or cyclopentane. He added a few drops of this liquid to a (brown) solution of bromine in CCl_4 and saw no color change. He concluded that the molecule was cyclopentane, but his report was marked wrong. Where did he go wrong?

5-11. Which of the following reagents would be expected to react with propene?
CN^-, NaCl, HBr, OH^-, NH_3, BH_3, H_2O_2, ICN

5-12. Addition of bromine to 1,2-dimethylcyclopentene gives the *trans* dibromide shown but addition of HBr to the same alkene gives two compounds. What are these and why are they formed?

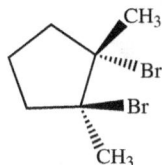

5-13. Assign the correct stereochemical descriptor (E or Z) to the following:

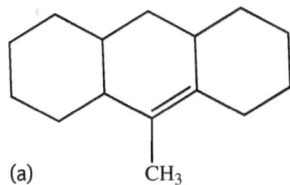

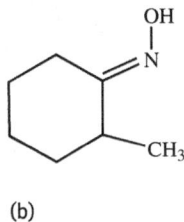

(a)   CH₃   (b)

5-14. There are two alkene structures with the molecular formula C_6H_{12} that would be named as methyl pentenes *and* for which there are E and Z isomers. Draw these and give the complete IUPAC names.

5-15.

(a) Write the structure of the products formed when each of the following is treated, first with ozone and then with zinc metal.

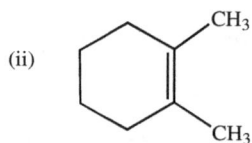

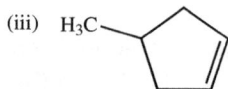

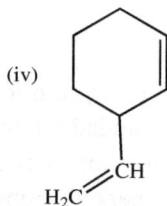

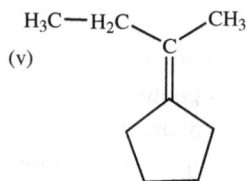

(i) $(CH_3)_2CH=CHCH_2CH_3$

(ii)

(iii) H_3C—

(iv)

(v) $H_3C—H_2C$ C CH_3

(b) What alkene, when ozonized and the ozonide reduced with zinc, will give the following products?

(i) $H_2C=O$ + $CH_3CH=O$

(ii) 2 moles of

(iii) $H_2C=O$ + $CH_3C—C—CH_3$ + $CH_3CH=O$

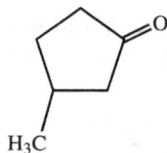

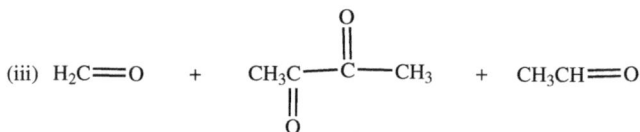

5-16. Complete the following equations by replacing the ? with the proper reagent, starting material, or product. If a catalyst is necessary, indicate this and also show any aspect of stereochemistry that is important.

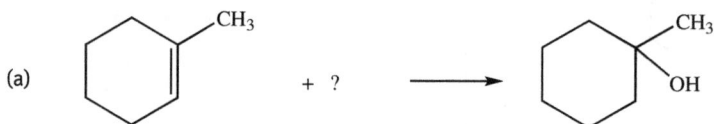

(a)

+ ? ⟶

(b) $(CH_3)_2CH=CH—CH_3$ + ? ⟶ $(CH_3)_2CH—CH—CH_3$ with OH

(c)   + Br$_2$ ⟶ ?

(d) H$_2$C=CH(CH$_2$)$_4$CH=C(CH$_3$)$_2$ + Br$_2$ ⟶ ?

(e)   + ? ⟶

assign stereochemistry here

(f) ? + BH$_3$ ⟶

(g) CH$_3$CH$_2$O—C(H)=CH$_2$ + HCl$_{aq}$ ⟶ ?

(h) ? + ICl ⟶

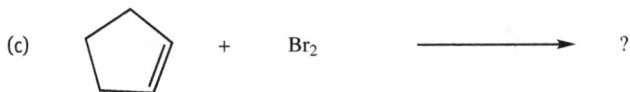

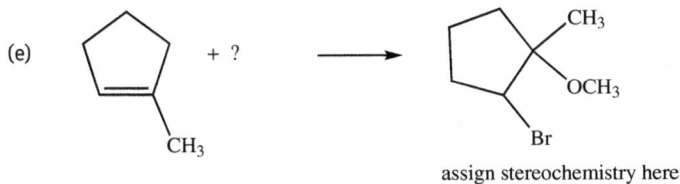

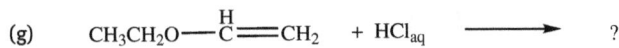

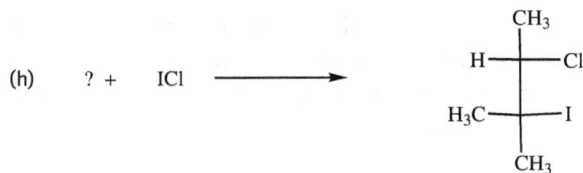

5-17. Suggest how you could convert propene into each of the following:

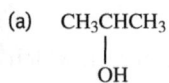

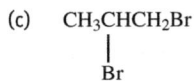

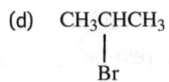

(a) CH$_3$CHCH$_3$
     |
    OH

(b) CH$_3$CH$_2$CH$_2$OH

(c) CH$_3$CHCH$_2$Br
     |
    Br

(d) CH$_3$CHCH$_3$
     |
    Br

(e) CH$_3$CH=O

(f) CH$_3$CH$_2$CH$_3$

6 More Stereochemistry and Another Reaction Type – Nucleophilic Substitution

6.1 Introduction

In Chapter 5, we saw that alkenes that have an excess of electrons and are therefore nucleophilic, react with electrophiles. In this chapter, we will see another reaction of nucleophiles and electrophiles which involves a different class of organic molecule and leads to a totally different reaction type. However, before considering this, we must take a look at another aspect of a topic which we considered in Chapter 3 – stereochemistry. This is necessary to understand the important points, both of the new mechanism we will learn, and also many other effects in organic chemistry.

6.2 Optical Isomerism (or Chemistry with Mirrors)

We will now introduce another, very common type of stereoisomerism. This type is perhaps the most subtle and difficult to learn and because of its significance some time will be spent on it. In order to understand the material, we must first spend some time considering what properties mirrors have.

> An excellent book that is both instructive and enjoyable and which explains the properties of mirrors has been written by Martin Gardner. It is entitled "The Ambidextrous Universe." Reading the initial chapters of Gardner's book will give you a good understanding of the intriguing properties of mirrors.

If you stand in front of a mirror and extend your right hand toward the mirror, which hand does your "mirror image" extend? The left one of course, but the direction in which it points is opposite to you. Similarly, if you point away from the mirror, your image points in the opposite direction. If, however, you point sideways, your reflection points in the same direction. The difference lies in the direction you point relative to the plane of the mirror. Actions *perpendicular* to the mirror plane are reversed, but those in the plane are not.

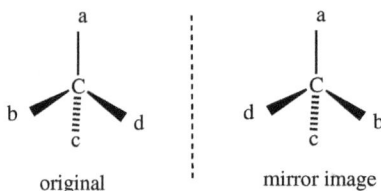

original   mirror image

Fig. 6.1

https://doi.org/10.1515/9783110778311-006

What has all this to do with organic chemistry? Consider a molecule with one sp^3-hybridized carbon attached to four different atoms (Fig. 6.1). This molecule if it was placed in front of a mirror would appear as shown in Fig. 6.1. Now try to superimpose the original on the mirror image. You will find that it cannot be done. When any two of the pairs of substituents a, b, c, d are superimposed, the other two pairs are opposed. There are two forms of the molecule. They are clearly stereoisomers of each other, but of a new type. To handle these types, we must define and remember some new terms. Two of these are:

1. *Enantiomers* – stereoisomers related as nonsuperimposable mirror images.
2. *Chiral* or *asymmetric carbon* – a carbon atom bonded to four different atoms or groups of atoms.

(Note that #1 says nothing about requiring four different groups. The reason for this will become apparent shortly.) The two forms of the molecule in Fig. 6.1 are *different configurations* of the molecule. (Review the terms configuration and conformation from Chapter 3 if you are not absolutely sure of them.) Note that a, b, c and d can be groups as well as atoms. All the molecules in Fig. 6.2 can exist in two configurations. For practice, try drawing the mirror images of these compounds and showing they are not superimposable on the originals.

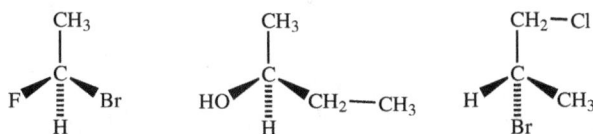

Fig. 6.2

All the types of isomers we have considered *before* this section have one characteristic in common. Positional and geometric isomers are distinguishable on the basis of their physical properties: that is, they have different melting points, boiling points, solubilities, etc. Conversely, enantiomers have the same physical properties with one exception. They differ *only* with respect to how they interact with a beam of polarized light. If a beam of polarized light is passed through a solution of an enantiomer, the plane of polarization will be twisted either to the right or left. How much it will be twisted depends on

–   the specific molecule being used;
–   the number of molecules of the enantiomer in the solution – i.e., the concentration;
–   the wavelength of light used and the temperature.

The important point to note is that the mirror image molecule will rotate the same beam of light an equal amount in the *opposite direction*. The reason for this difference is beyond the scope of this book. Because of this optical property of enantiomers, such compounds are frequently called optical isomers and the molecules are said to be *optically active*.

Compounds that rotate light in a clockwise direction (i.e., to the right) are called dextrorotatory (the symbols d or + are used) and those that rotate light in an anticlockwise direction (to the left) are called levorotatory (the symbols l or – are used).

Only enantiomers (i.e., molecules with nonsuperimposable mirror images) are optically active. The *necessary and sufficient* condition for optical activity to occur is that the mirror image not be superimposable on the original.

Another term we will need is *racemic mixture*, which describes a 1:1 mixture of a chiral compound and its mirror image. A racemic mixture will always be optically inactive since the rotation in one direction due to one enantiomer is exactly cancelled by the rotation in the other direction due to its mirror image.

Note – Do not confuse d and l with D and L which are terms you may have seen previously. We will consider these shortly.

6.3 Polarimetry

How is the direction and amount of rotation of polarized light measured? The instrument is called a polarimeter and is shown schematically in Fig. 6.3. A monochromatic light source (usually the sodium D line) is passed through a fixed polarizer. With the solution cell empty, the moveable polarizer is rotated until all the light coming to the eye is cut off and the reading in degrees of arc is taken on the scale. The solution cell is then filled with a solution of the compound. If it is optically active, the plane of polarization of the light will be twisted while passing through the solution. In order to cut off the light coming to the eye the moveable polarizer must be rotated. The number of degrees that you must turn it is called α, the observed rotation (positive if clockwise; negative if anticlockwise). Since this value depends on the concentration of the compound, a concentration-independent number $[\alpha]$, the specific rotation is defined as

$$[\alpha]_\lambda^t = 100\alpha/l \cdot c$$

where t = temperature, λ = wavelength of light used, l = cell length in *decimeters*, and c is the concentration in grams per 100 ml of solution. $[\alpha]$ is a constant for any given compound.

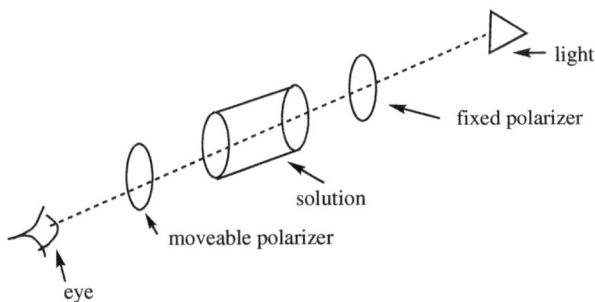

Fig. 6.3

A natural question arises. What is the relationship between the configuration of a molecule and the direction it rotates the plane of polarized light? It would be very convenient if the configuration (R or S) could be determined by simply measuring the rotation. Unfortunately, no relationship between either the direction of rotation or its size and the absolute configuration (R or S) exists. The actual chore of determining which of the two possible configurations one has is frequently a long and difficult task.

In the early days of organic chemistry, one molecule called (+)-glyceraldehyde was arbitrarily assigned the configuration shown below. The stereochemistry of many molecules was related to it by chemical means and if the arbitrary assignment was actually correct, the configuration of these related molecules could also be said to be known: i.e., the *relative configuration* (relative to (+)-glyceraldehyde) was known. Later, the actual structure of (+)-glyceraldehyde was determined by X-ray crystallography and found to be as assigned. Therefore, the *absolute configuration* of glyceraldehyde and all the molecules that had been related to it were known. (It is interesting to note that there was a 50% chance of the original assignment of configuration being correct and for once, Murphy's Law failed and the correct guess was made. Organic chemists are wont to point out that when physicists had the same odds in assigning the charge on an electron, the results were not so fortunate!)

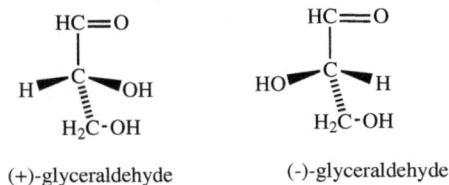

(+)-glyceraldehyde   (-)-glyceraldehyde

6.4 Nomenclature – The R and S System

Since enantiomers are isomers, we must be able to refer to them unambiguously. A system which was originally devised for referring to carbohydrates (sugars) uses the terms D and L, but as more and more types of chiral compounds became known, the system began to break down and a new system was devised. This modern system is based on the Cahn–Ingold–Prelog method we have already seen in connection with the stereochemistry of double bonds (Sect. 3.6). (At this time, it is very useful to review the Sequence Rules and their applications.) However, a few extensions of the rules are required.

To assign the configuration to a chiral center using this system, the four groups *attached to the chiral carbon atoms* are ordered in priority according to the sequence rules already set out. The molecule is then oriented in such a way that the lowest priority group is on the side of the molecule *remote from the viewer*. The order of priority of groups 1, 2, and 3 is then determined to be either a clockwise (R) (from the Latin Rectus = right) or anti-clockwise (S) (from the Latin Sinister = left) sequence and the appropriate symbol (R or S) is added to the name. For example, in Fig. 6.4(a) a three-dimensional drawing of one configuration of 2-bromobutane is shown. To determine the configuration, assign the priorities (b) and rotate the molecule to get priority group 4 at the back. The sequence 1 > 2 > 3 describes a right-handed (clockwise) arc and the molecule is therefore (R) 2-bromobutane.

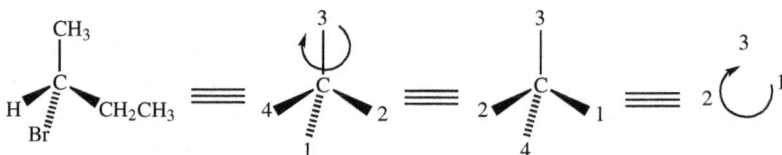

Fig. 6.4

One further addition to the sequence rules is useful.

Rule #4. When groups under consideration have multiply bonded atoms, those atoms are *replaced* by an equal number of singly bonded atoms of the same type.

For example:

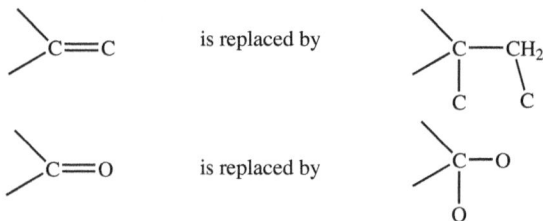

As a final example, Fig. 6.5 shows how to determine the configuration of a molecule with two (or more) chiral centers. The two centers are considered separately. It is necessary that you be able to take a drawing and name it correctly as well as be able to take a name and draw the correct three-dimensional structure. The stereochemical de-

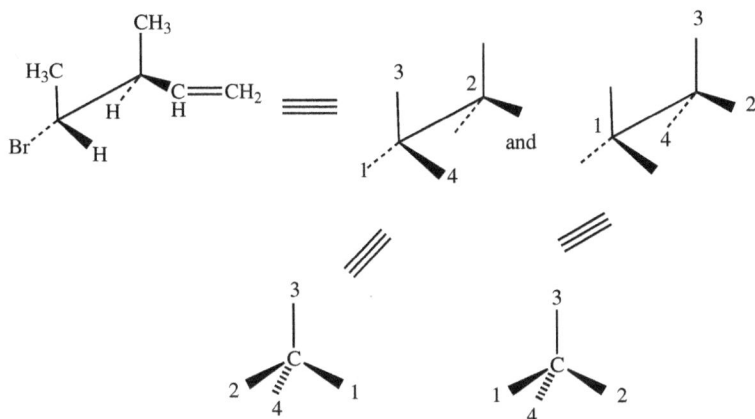

Fig. 6.5

scriptors R and S are incorporated into the name in brackets preceding the name. The complete name for the molecule in Fig. 6.5 would then be (3S,4R) 4-bromo-3-methyl-1-pentene.

6.5 Fischer Projections

The drawings in Fig. 6.5 suggest that, as molecules become larger and more chiral centers are encountered, the molecules become more difficult to draw. To avoid this problem, another type of projection drawing called the Fischer Projection has been developed. If you learn the rules of manipulation, these drawings will make the handling of complex molecules relatively simple and straightforward. The system is as follows:

1. Molecules are shown in the eclipsed conformations with bonds in the horizontal plane in front of the plane of the carbon chain and bonds in the vertical plane in or behind the plane of the carbon chain.

For example (Fig. 6.6a), drawing a is ready for conversion to the Fischer projection but drawing b is not and must be reoriented to c.

2. Once the molecule is correctly oriented, the chiral carbon(s) is represented by an intersection of two lines (the shadows of these bonds when a light is placed in front of the molecule) and the substituents are placed in their correct positions.

Using this, the drawings in Fig. 6.6a are converted to those shown in Fig. 6.6b. Some more complex examples are shown in Fig. 6.7. What is the advantage of using Fischer projections? Provided that the rules are followed, one can draw mirror images directly (e.g., Fig. 6.8). Frequently it is useful to be able to change Fischer projections into another orientation. *This is equivalent to reorienting the three-dimensional drawings* (i.e.,

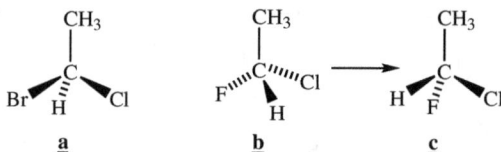

Fig. 6.6a

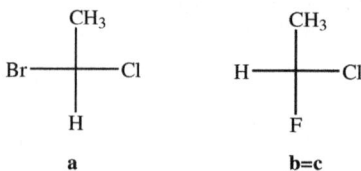

Fig. 6.6b

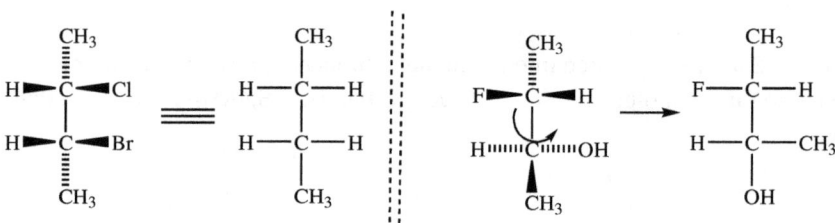

Fig. 6.7

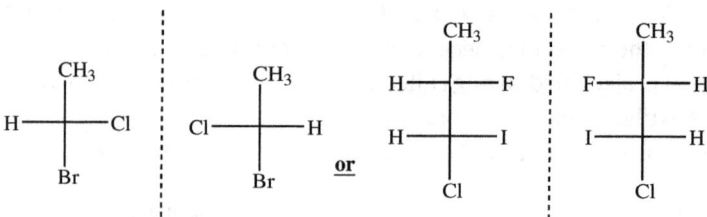

Fig. 6.8

changing the conformation *but not the configuration* of the molecule) and can be done if the following three rules are *strictly* observed.

1. Exchange of any two groups around the same chiral center converts one Fischer projection into its enantiomer.

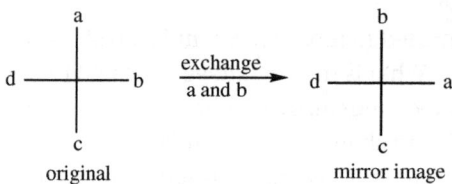

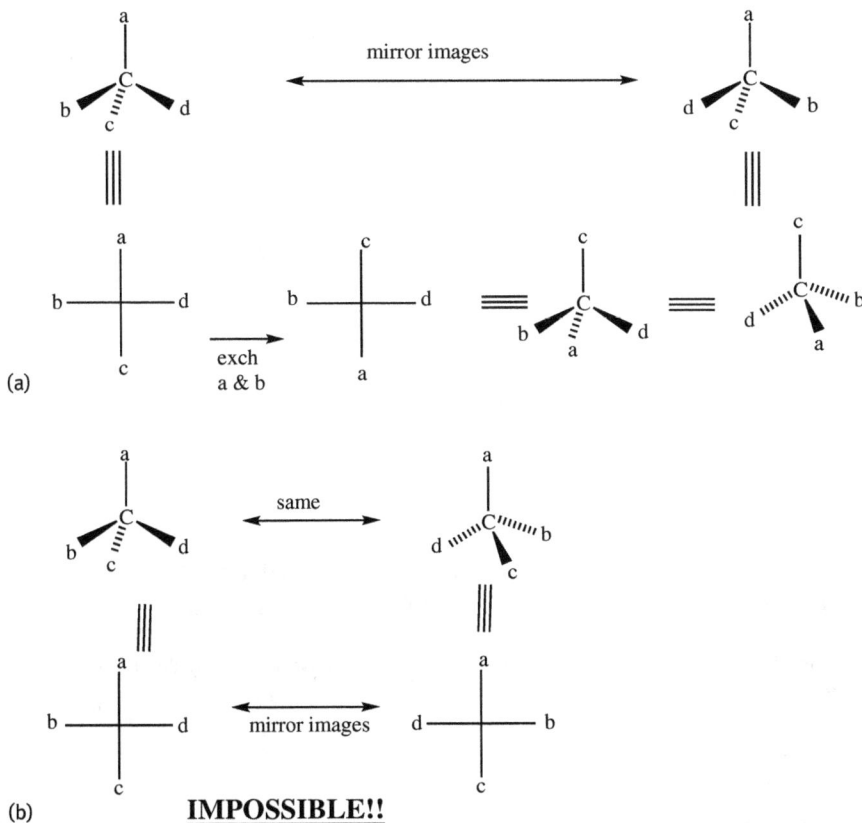

Fig. 6.9

Figure 6.9(b) shows why the molecule must be oriented properly before the Fisher projection can be drawn. Because the three-dimensional drawing on the right is oriented incorrectly, the same molecule leads to two different answers – clearly an impossible situation.

2. Since the mirror image of a mirror image must be the same as the original, exchange of any *two pairs* of groups around the same chiral carbon must generate the Fischer projection with the *same* configuration (Fig. 6.10).

3. Exchange of any three groups in sequence around the same chiral carbon will generate the same molecule. (This is a special case of Rule #2, where one group has been moved twice.) This is illustrated in Fig. 6.11 and simply represents a conformational change in the molecule.

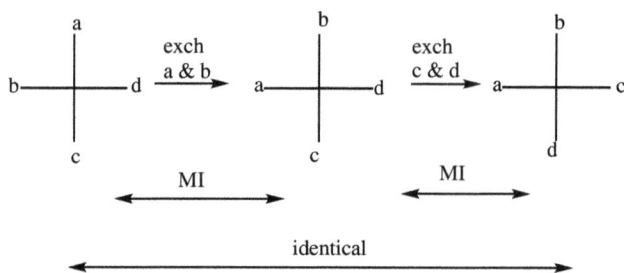

Fig. 6.10

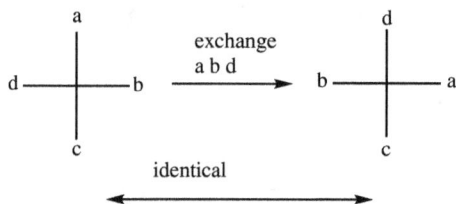

Fig. 6.11

There are three other ways of rearranging Fisher projections that might be attempted. The consequences of these are outlined in the following paragraphs.

1. Rotation of a Fisher projection 180° about a vertical axis converts a molecule into its mirror image.

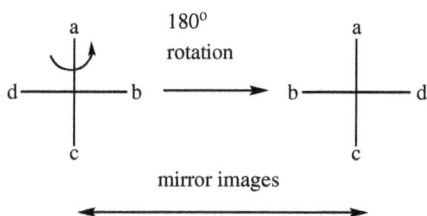

2. Rotation of a Fischer projection 90° about an axis perpendicular to the plane or the paper converts a molecule into its mirror image.

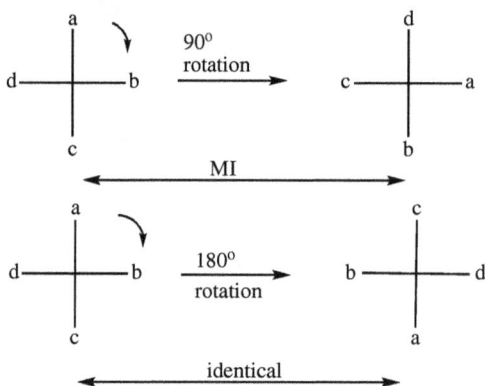

3. Since the mirror image of a mirror image is the original molecule, a further rotation of 90° about the same axis (i.e., a total of 180° rotation) gives the same molecule.

We will now start to use these concepts to solve a variety of more complex problems.

6.6 Assigning Configurations from Fisher Projections

If a chiral center is correctly converted into a Fischer projection, determining its R or S configuration is simple. First the molecule must be converted into a form with the lowest priority group at either the top or the bottom *without changing its configuration*. Once this is done the sequence of the remaining groups can be read directly. For example, consider the Fischer projection shown at the left in Fig. 6.12. Since the lowest priority group is not at either the top or the bottom, the Fischer projection must be redrawn in a form that fulfils this requirement. This can be accomplished by exchanging three groups in sequence (Rule #3) as shown. Since the priorities of the remaining three groups is Br > Cl > CH$_3$, the configuration is R.

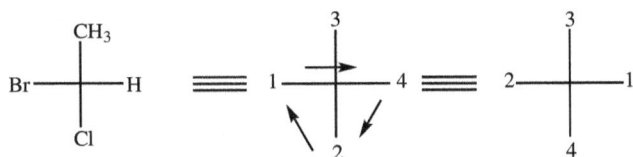

Fig. 6.12

You should be able to assign configurations from Fischer projections and also to draw the Fischer projection of a given configuration. Some problems at the end of the chapter will give you practice in doing this.

6.7 Compounds With More Than One Chiral Center

Consider the molecule 2-bromo-3-chlorobutane. We can draw four possible arrangements of this molecule (Fig. 6.13). Figure 6.13 shows the three-dimensional drawings and the corresponding Fischer projections. What is the relationship between these four compounds, which are all stereoisomers of each other? Examination will show that A and C are enantiomers as are B and D. However, A and B, A and D, and B and C are not. To identify these we need a new term.

Diastereomers – stereoisomers not related as mirror images.

Note that this means that all stereoisomers are either enantiomers or diastereomers. Also, you will recognize that the definition we had for geometrical isomers defines these as a particular type of diastereomer. It is also always true that *compounds with the designations cis and trans are diastereomers of each other.*

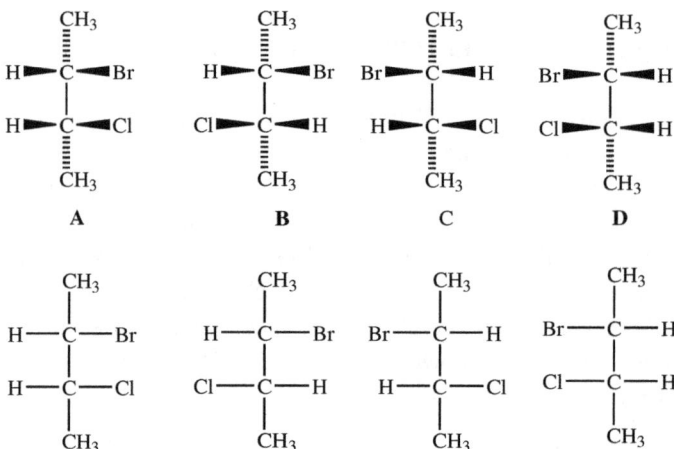

Fig. 6.13

For a compound with n different chiral centers, there will be 2^n stereoisomers related as $2^n/2$ pairs of enantiomers.

We will see the reason why the word "different" is important in the above definition shortly. Note that each of A, B, C, and D in Fig. 6.13 are optically active since each of these has a mirror image that is not superimposable on the original molecule.

Diastereomers are not like enantiomers in that they do differ in physical properties: e.g., melting point, solubility, etc. As pointed out previously, *the only class of isomers that do not differ in these properties is enantiomers.*

To assign the configuration to compounds with more than one chiral center, each center is considered separately. For compound A in Fig. 6.13, the chiral carbons have the priorities as indicated in Fig. 6.14. Rearranging each of these so that the priority 4 group is at the bottom gives the drawings shown (Fig. 6.14) and the whole name of this stereoisomer is (2S,3R) 2-bromo-3-chlorobutane.

It should be evident that the mirror image of an S configuration is R, and vice versa. The enantiomer of A in Fig. 6.13 would necessarily be (2R,3S) 2-bromo-3-chlorobutane.

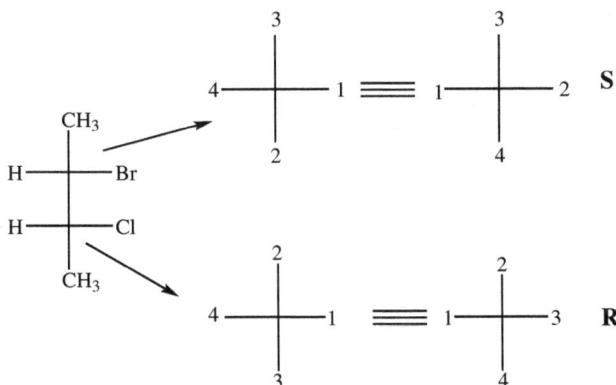

Fig. 6.14

6.8 Meso Forms

Consider the molecule 2,3-dibromobutane. Again, if we draw the Fischer projections it appears that there are four possible isomers (Fig. 6.15).

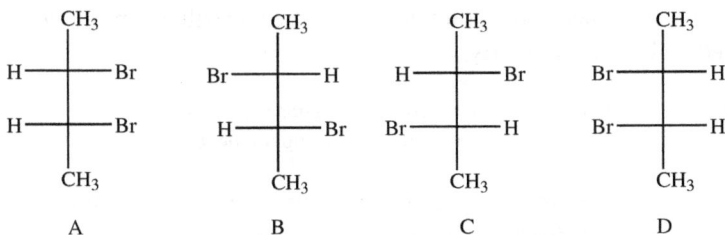

Fig. 6.15

Again, it appears that A and D are enantiomers as well as B and C. There is no question that A and D are mirror images. *However*, they are *not nonsuperimposable* (or better said, they ARE superimposable). If D is rotated 180° as shown in Fig. 6.16 (recall that this is an "allowed" rotation of a Fischer projection), it will be seen that A and D are superimposable:- i.e., they are the same molecule! Note that this cannot be done with compounds B and C in Fig. 6.16.

> *Meso forms* are compounds that *are* superimposable on their mirror images even though they contain chiral carbons. Since the mirror images are superimposable, they are *not* optically active.

Therefore, there are only three stereoisomers of 2,3-dibromobutane: the two enantiomers B and C and the meso form A.

It now becomes clear why the word "different" was required in the mathematical expression for calculating the number of possible stereoisomers (Sect. 6.7). It is be-

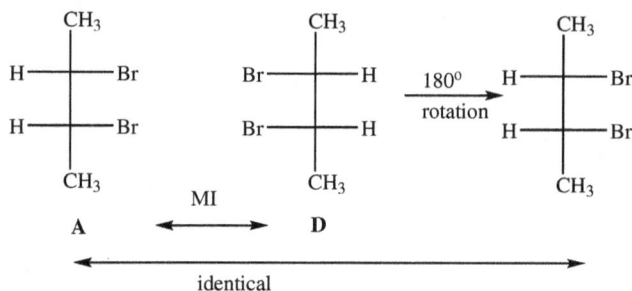

Fig. 6.16

cause the two chiral carbons in 2,3-dibromobutane have the same substituents that a meso form is possible. It should also be clear now why the definition of enantiomer (Sect. 6.2) did not refer to the presence of chiral carbons. Compound A (Fig. 6.15) does have two chiral carbons, but the whole molecule is not optically active because it *is* superimposable on its mirror image.

The definition of a meso form has been given. However, in some cases, a short-cut can be used to identify these forms. If a plane can bisect a molecule such that one half of the molecule is the mirror image of the other half, the molecule will be a meso form. Such a plane is called a plane of symmetry.

Caution – The Fischer projection must be drawn in the right orientation to be able to pick this plane out. When in doubt, draw the mirror image and see if they can be superimposed.

Some examples of meso forms with the plane of symmetry indicated are shown in Fig. 6.17. Note that the plane may bisect atoms so that one half of the atom is above the plane and the other half is part of the mirror image below the plane.

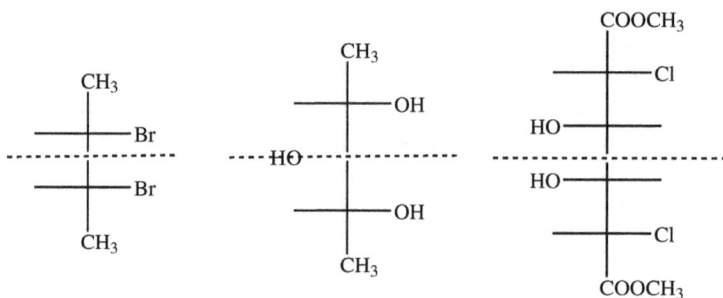

Fig. 6.17

Another example is shown in Fig. 6.18. No obvious plane of symmetry exists and one would be tempted to say that it is not a meso form. However, note that it can be rearranged as shown to give the same molecule in another arrangement which is clearly a meso form. This illustrates why it is risky to rely on spotting planes of symmetry in order to identify meso forms.

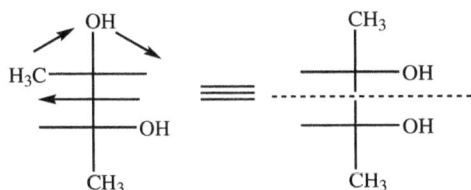

Fig. 6.18

6.9 Resolution and Racemization

The question arises – if enantiomers have identical physical properties, how can you separate them? Obviously, the traditional methods such as recrystallization and distillation, etc., which rely on differences in physical properties, are useless.

The separation of mixtures of enantiomers (racemic mixtures) is a process called *resolution*. It requires the use of an optically active compound which must be already available in the pure state. Fortunately, Mother Nature supplies us with an abundance of these. The process is illustrated in Fig. 6.19. The reaction of a racemic mixture with a pure enantiomer gives two diastereomers that do have different physical properties and can therefore be separated by the usual techniques. Subsequent decomposition of these separated diastereomers leads to the recovery of pure enantiomers.

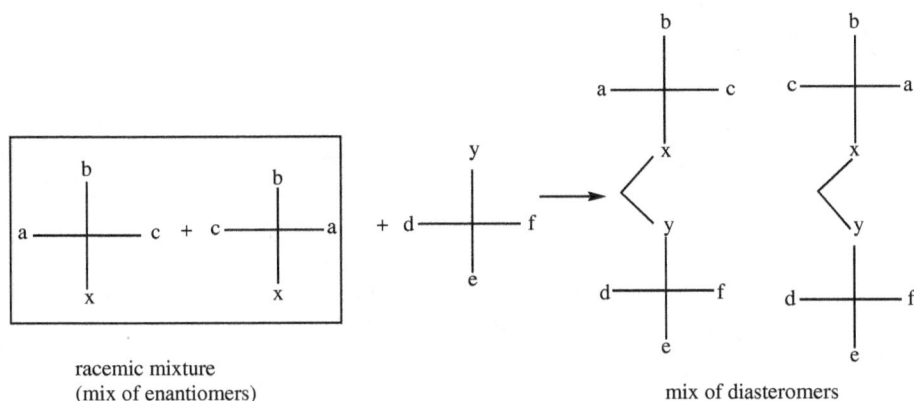

racemic mixture
(mix of enantiomers)

mix of diasteromers

Fig. 6.19

Racemization is the reverse of resolution. It is the conversion of one enantiomer into a racemic mixture.

6.10 Why Is All This Stereochemistry Important? (Subtitle: What Am I Doing Here?)

Most molecules in nature are chiral. Some obvious examples of very common biologically important molecules are the amino acids from which proteins are made. All the amino acids except glycine have one chiral carbon. Proteins, which may have hundreds of amino acids therefore have hundreds of chiral carbons and an astronomical number of possible stereoisomers. However, all naturally occurring amino acids have the S configuration. If on a trip into outer space you visited a planet where all the amino acids had the R configuration, you might eat very well – but you would starve to death since your body could not use these. A more practical example can be found in the area of carbohydrates (sugars). The only difference between starch and cellulose is a stereochemical one. Your body will accept starch as a nutrient, but paper is not very useful as a foodstuff! Two other examples of biologically active molecules with chiral carbons are shown in Fig. 6.20.

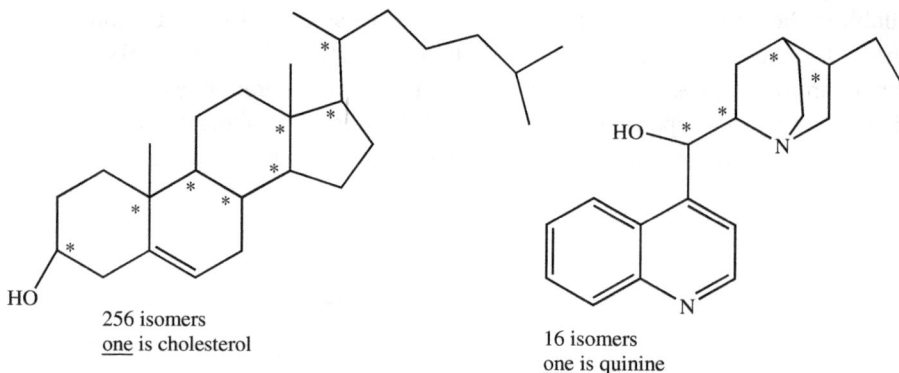

256 isomers
<u>one</u> is cholesterol

16 isomers
<u>one</u> is quinine

Fig. 6.20

We are now ready and equipped to deal with the promised second type of mechanism. The concepts developed in Chapters 4 and 5 and the first part of Chapter 6 will all come together in the discussion of the mechanism of Nucleophilic Substitution.

6.11 Nucleophilic Substitution

Consider the reaction

$$CH_3Br + I^- \longrightarrow CH_3-I + Br^-$$

The iodide ion has replaced or *substituted* for the bromine atom. It should be clear that iodide ion is a nucleophile: that is, it has a pair of electrons available to form a bond. That the carbon atom in bromomethane is electrophilic may not be so obvious. Recall that bromine is more electronegative than carbon. The bond joining these two atoms is a *polar covalent* one (Sect. 1.4) and this bond can be written as $^{\delta+}C-Br^{\delta-}$. If iodide ion, which is negatively charged, is to attack this molecule, it obviously will have to do so at the positive center. Since this atom, unlike the situation in alkenes, is already bonded to four atoms, if something is to become bonded to it, something else must leave. In this case it is the bromine atom (as bromide ion). The reaction is therefore called a *nucleophilic substitution* (Sn) reaction. The question arises, how does this reaction proceed? Consideration of this question leads to the conclusion that there are in fact two limiting possibilities that do not violate any rules of valency. These are considered in the next sections.

6.12 The Carbocation Route – S_N1 Reactions

One way such a reaction could proceed is shown in Fig. 6.21. Ionization of the C–Br bond would lead to bromide ion and a carbocation – exactly the same type of intermediate seen in Chapter 5. This carbocation could then combine with any nucleophile present. Recombination with bromide ion regenerates the starting material, but reaction with iodide ion produces the organic iodide compound. Note that this *route has an intermediate.*

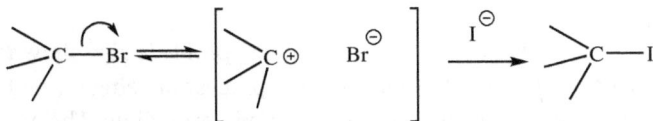

Fig. 6.21

Which of the two steps in Fig. 6.21 would proceed faster? Separation of two fragments with unlike charges should be more difficult than bringing together two oppositely charged fragments. Therefore, the first step should be the slower of the two. The energy profile should then look as shown in Fig. 6.22. Recalling that the rate equation for a reaction involves only those species that are involved in the mechanism in or before the slow step (Sect. 4.7), the rate equation for this mechanism would look like

$$\text{rate } \alpha[\text{C–Br}]$$

Therefore, this reaction is first order. Its rate is independent of the concentration of iodide ion. Since the reaction is a first-order, nucleophilic substitution, it is termed an S_N1 *reaction*.

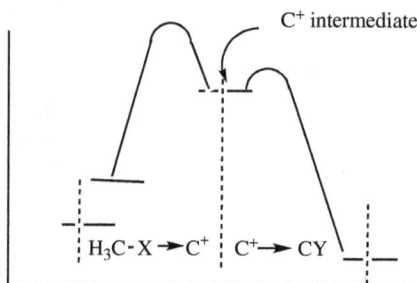

Fig. 6.22

6.13 The Concerted Route – S_N2 Reactions

In the S_N1 reaction, the reaction proceeds in two steps. An alternate pathway would involve a different timing of the steps: i.e., a formation of the new C–I bond at the same time as the C–Br bond is breaking. Such reactions are said to be *concerted reactions*.

> The rate of a reaction that follows the S_N1 mechanism is independent of the concentration of the nucleophile.

Such a mechanism would look as shown in Fig. 6.23. The energy profile is also shown there. The dotted line represents partial bonds – bonds in the process of being formed or broken. Note that this reaction mechanism does not have an intermediate. The entity enclosed in brackets is a transition state. Since both reacting species are involved in the single step, the rate equation looks like

$$\text{rate } \alpha[\text{CH}_3\text{Br}][\text{I}^-]$$

i.e., the rate is dependent on two concentrations and the reaction is second order. This mechanism is called an S_N2 *mechanism*.

It is important to realize that these two mechanisms, the S_N1 and the S_N2, are limiting cases. A bond might start to break, but before it is totally broken, the nucleophile might start to bond. This case is intermediate between the S_N1 and the S_N2 case. There

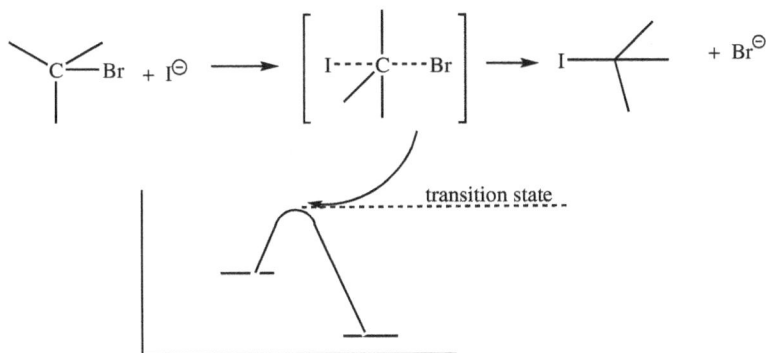

Fig. 6.23

are an infinite number of intermediate possibilities, each differing from the others by a slight difference in the timing of the events. However, it is most convenient to discuss reactions in the black and white terms of S_N1 and S_N2 mechanisms.

6.14 Stereochemistry Again

A major difference between the S_N1 and the S_N2 pathways lies in the stereochemical relationship of the products and reactants in each case. In the S_N1 case, ionization of the bond between the carbon and the group that is leaving produces an intermediate that has a positively charged sp^2-hybridized carbon (i.e., a carbocation). As was pointed out previously (Sect. 5.3), this means that the reacting center is planar with a vacant p-orbital perpendicular to the plane of the remaining three substituents. Suppose we had started with one enantiomer of a chiral compound. The intermediate now has lost its chirality (Fig. 6.24). The same intermediate would be generated from the mirror image of the starting material.

When the new bond between the intermediate and the nucleophile (Nu⁻) begins to form, it must use the p-orbital on the intermediate. The nucleophile must approach

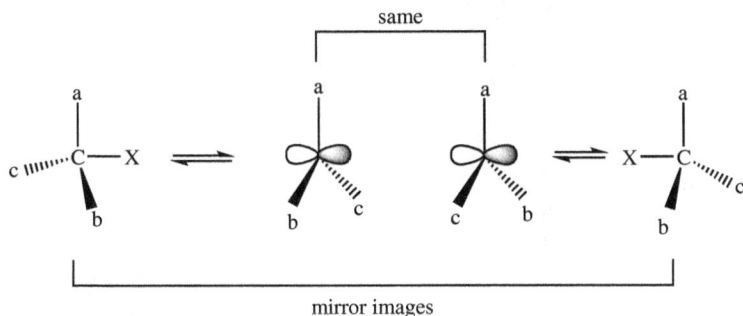

Fig. 6.24

the carbocation along the axis of this orbital. Since the intermediate is planar, there is no reason why attack should occur from the top rather than from the bottom side. The result of this is that two products are formed (Fig. 6.25) in equal amounts. The two products are mirror images of each other and, if the compound is chiral, a racemic mixture results. It is important to note in this context that the word that describes what is obtained from an S_N1 reaction – or from an addition or any other reaction – depends entirely on what you started with. For example, *addition* of HBr to 3-methyl-3-hexene will lead to a racemic mixture. However, if another chiral center exists in the molecule, the stereochemical results may be either two diastereomers or two pairs of enantiomers depending upon whether the starting material was one enantiomer or a mixture. You must always keep this in mind. However, it *is* safe to say that the product of reaction of a carbocation and a nucleophile will lead to a 1:1 mixture of mirror images and will therefore be optically inactive *unless* other chirality exists in the molecule.

The situation in the S_N2 mechanism is somewhat different. Here no intermediate exists. However, the attacking nucleophile must again utilize the orbital that was originally used to form the bond to the group which is leaving as no other is available. This requires that the nucleophile attack the carbon either from the same side or the opposite side of the carbon from which the other group is departing. This is shown in Fig. 6.26 along with the stereochemical implications of each mode of attack. Three situations can be visualized.

1. Attack of the nucleophile is random: i.e., equal amounts occur from the same and opposite sides as the leaving group.
2. The nucleophile attacks from the same (front) side.
3. The nucleophile attacks from the opposite (back) side.

As can be seen from Fig. 6.26, if a single enantiomer is used as a starting point, attack from the front side results in a single chiral molecule with the *same* configuration as the starting material. Conversely, attack from the rear requires that the molecule being formed have the opposite configuration to the starting material. A random attack would result in a racemic mixture.

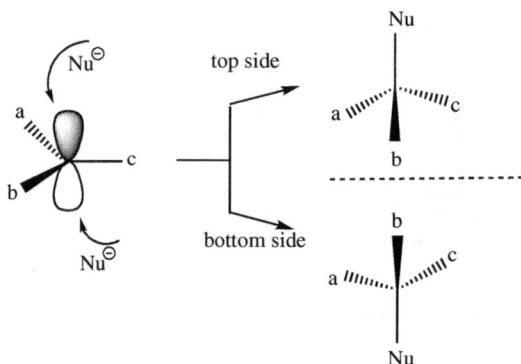

Fig. 6.25

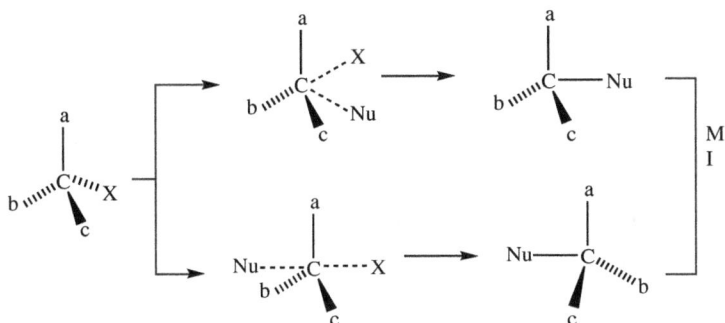

Fig. 6.26

Many investigations of reactions that proceed by the S_N2 mechanism have shown that the operative mechanism is *inversion of configuration*: i.e., backside attack. This process is frequently called *Walden Inversion*. A closer look at the transition state corresponding to this inversion process is useful. In the process leading from the starting material to product, the carbon and its three attached substituents must turn itself "inside out." An analogy with an umbrella is illustrative (Fig. 6.27). Obviously at the transition state the carbon and its three substituents must approach planarity. The results of this inversion process and the effect of the planarity of the transition state are worthy of elaboration. Consider the molecule (S) 2-bromobutane (Fig. 6.28). Nucleophilic substitution via an S_N2 process will always lead to a molecule with the R-configuration (provided that the priorities do not alter). The reverse is also true so that the R isomer will always afford the S product.

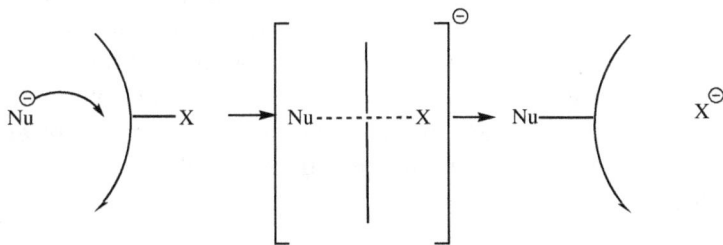

Fig. 6.27

This process will also always change a *cis*-disubstituted cyclic compound (e.g., 4-methyl-1-bromocyclohexane) into a *trans* configuration (Fig. 6.28) or the reverse. Furthermore, this means that an axial substituent will be converted to an equatorial one and vice versa in a cyclohexane system. (The molecule may, of course, invert its conformation subsequently to change this result, but the relative orientations of the substituents, *cis* or *trans*, is not altered by this process.) It is well known that substi-

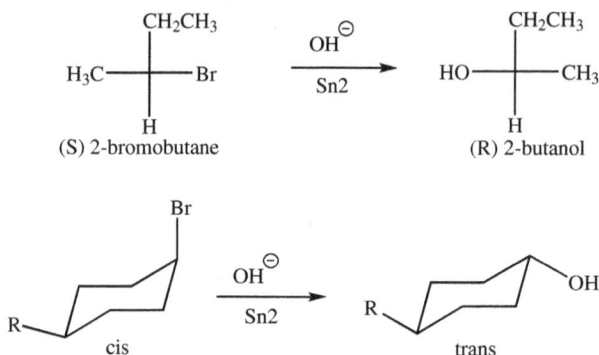

Fig. 6.28

tution processes on cyclohexane rings, by either S_N1 or S_N2 mechanisms, are more difficult than in acyclic compounds. This fact can be attributed, at least in part, to the requirement for a planar transition state (S_N2) or intermediate (S_N1). The angles for the planar carbon are 120°. The presence of a ring constrains the ability of the molecule to adapt to this angle and makes achieving the planar state more difficult than in acyclic cases: i.e., it raises its energy. Therefore, the activation energy is higher and the reactions are slower.

6.15 Which Mechanism Operates? – The Basic Factors

Nucleophilic substitution reactions are perhaps the most common reactions in organic chemistry. They are applicable to the formation of a tremendous range of different classes of compounds. We have seen the consequences of the possibility of two mechanistic types of reaction and a natural question that arises is – "How can I tell which mechanism will operate in a specific case?" To answer this question we must look at the effect of four reaction variables – the carbon center, the nucleophile, the leaving group, and solvent effects.

6.15.1 The Carbon Center

In the S_N1 reaction, an intermediate carbocation is formed. As the stability of this intermediate increases, the activation energy leading to it decreases and so its formation is more likely. *We* have already seen what factors influence the stability of carbocations in connection with electrophilic addition reactions and these principles apply here as well.

In general, a reactant that can give a tertiary or resonance-stabilized carbocation intermediate will tend to undergo substitution by an S_N1 mechanism.

In the S_N2 mechanism, the nucleophile must start to form a bond to the sp^3-hybridized carbon. To do so, it must approach this carbon close enough to all the orbitals to interact. If the carbon has large groups attached to it, they will interfere with this approach and the larger the groups are, the more they will interfere. This is an excellent example of a *steric effect*. Since the smallest groups possible are hydrogen atoms, it follows that S_N2 reactions will proceed best at reacting carbons that have the largest number of hydrogen atoms attached: i.e., the reactivities should be primary > secondary > tertiary. This is the *exact opposite of the carbocation stabilities* and the ease of reaction via the S_N1 mechanism. In general, S_N2 reactions proceed best at primary *centers*. It will be noted that the secondary centers are intermediate in both cases. Nevertheless, a secondary reactant is *more likely* to undergo S_N2 reactions than a tertiary and more likely to undergo S_N1 reactions than a primary reactant.

It is important to note that the reacting carbon in both of the mechanistic cases we have been studying has *always* been sp^3-hybridized. Reactions of these types do not normally occur on sp^2 or sp-hybridized carbons.

6.15.2 The Leaving Group

The atom or group of atoms that is being replaced is called the leaving group. The strength of the bond between this group and the carbon will determine how easily it is broken. It would be expected that, as this bond strength decreases, both S_N1 and S_N2 reaction rates should increase. This is found to be so. However, since the leaving group must leave unassisted in the S_N1 case, but is assisted in its departure by the incoming nucleophile in the S_N2 case, the effect is larger in the S_N1 case. The question then arises as to how to predict "leaving group abilities." This can be done rather simply by looking at the strengths (pKas) of the corresponding protonated form (HX) or the leaving group X^-. For example,

$$NH_3 < H_2O < HCl < HBr < HI$$

in terms of their acidities and so

$$NH_2^- < HO^- < Cl^- < Br^- < I^-$$

in terms of leaving group abilities. Note that the acidities of HCl, HBr, and HI refer to the *pure* compounds and *not* to the aqueous solutions (see Sect. 4.2).

6.15.3 The Nucleophile

In the S_N1 reaction, the rate of reaction is independent of the nature and concentration of the nucleophile (Sect. 6.12) This is clearly not so for the S_N2 mechanism. It would be expected that a strong nucleophile would be better able to "push off" the leaving group than a weak one and increase the reaction rate. This raises the question as to what makes a strong nucleophile. Two generalizations can be used as a basis for predicting nucleophilic strength (nucleophilicity).

In the same *horizontal row* of the periodic table, the nucleophilicity of an atom parallels it basicity.

NH_3 is a stronger base than water (the nucleophilic atoms O and N are in the same row) and NH_3 is also a stronger nucleophile. Similarly OH^- is a stronger base and a better nucleophile than water. Another comparison is that between an alcohol (ROH) and water. The K_{eq} of these are

$$H_2O \rightleftharpoons H^+ + OH^- \quad K_{eq} = 10^{-14}$$
$$ROH \rightleftharpoons H^+ + RO^- \quad K_{eq} = 10^{-17}$$

So hydroxide is a weaker base than RO^- (an alkoxide) and also a poorer nucleophile.

In the same vertical column (period) of the periodic table, the nucleophiles at the bottom of the column are stronger than those at the top.

In the situations usually encountered in organic chemistry, this comparison applies to the halogens. Iodide is a stronger nucleophile than bromide which, in turn, is stronger than chloride. *notice that this is the reverse of the situation found in the comparison of horizontally related elements.* Iodide ion is a weaker base than bromide ion (HI is a stronger acid than HBr) but is a stronger nucleophile. Notice also that $I^- > Br^- > Cl^-$ in *both* nucleophilicity and leaving group ability (Sect. 6.15.2) This will have some interesting consequences. As pointed out previously, in order for a nucleophile to start a bonding process with a carbon, it must come within bonding distance. We have seen (Sect. 6.15.1) that as the degree of substitution on the reacting center increases, the ease with which the nucleophile can approach decreases. This effect is most pronounced in the S_N2 mechanism. This same steric effect will occur also if the nucleophile increases in size in the vicinity of the nucleophilic atom. For example, the alkoxide ion derived from methanol (methoxide) is a potent nucleophile but that derived from 2-methyl-2-propanol (commonly called tertiary butanol) is essentially non-nucleophilic. The presence of the three methyl groups prevents the nucleophilic oxygen atom from approaching the reacting carbon to within bonding distance. This is another excellent example of a steric effect.

6.15.4 Solvent Effects

The most important property of a solvent in terms of its effect on substitution reactions that we will consider is its dielectric constant. Other properties such as solvation and hydrogen bonding are also very important but beyond the scope of this text. This property is a measure of its ability to isolate electrical charges from each other. Organic chemists usually refer to the "polarity" of a solvent; the more polar solvents being those with the higher dielectric constants. Since the S_N1 mechanism requires species with unlike charges to be separated in the rate-controlling step, any effect that reduces the natural attraction of the charges for each other should make this step easier and increase the rate of reaction. Table 6.1 gives a list of some common solvents and their dielectric constants.

> Reactions proceeding by an S_N1 mechanism are favored by an increase in the polarity of the reaction solvent.

For reactions proceeding via the S_N2 mechanism, the effect of solvent polarity varies with the electrical nature of the nucleophile. To see why this is so, consider the reactions shown in Fig. 6.29. In the first case a negatively charged nucleophile reacts to form, at the transition state, a species in which partial negative charge is located on both the nucleophile and the leaving group. Since these charges repel each other, a solvent of low dielectric constant will speed this process by allowing the nucleophile to "push" the leaving group away. In the second case shown in Fig. 6.29, the neutral nucleophile generates a transition state where opposite developing charges exist on the nucleophile and the leaving group. Use of a solvent of high dielectric constant will speed this process by reducing the attractive forces between these oppositely charged species.

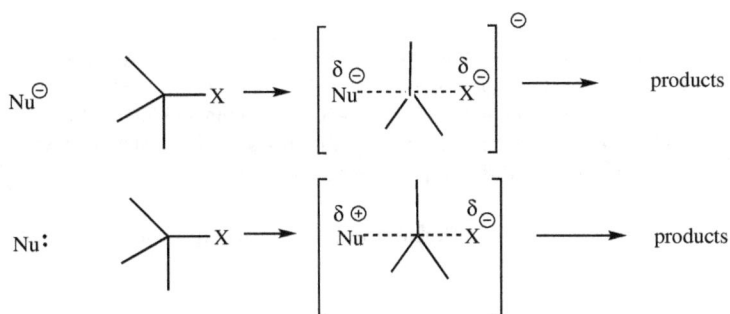

Fig. 6.29

> Solvents of high dielectric constant (polarity) favor S_N2 reactions of neutral nucleophiles. Solvents of low dielectric constant favor S_N2 reactions of negatively charged nucleophiles.

6.16 Putting It All Together

We can now apply the foregoing principles to the problem of predicting the relative rates of reactions. With limited experience, it is difficult to make absolute decisions as to which mechanism (S_N1 or S_N2) or about how fast a particular reaction will proceed. However, comparisons between two reactions are much more feasible. Several "rules" can be formulated.

Substitutions will tend to go via the S_N1 route only if the intermediate that would be involved is relatively stable – i.e., is tertiary or resonance-stabilized.

Comparing the following reactions, the one listed first will be more likely to proceed via the S_N1 mechanism.

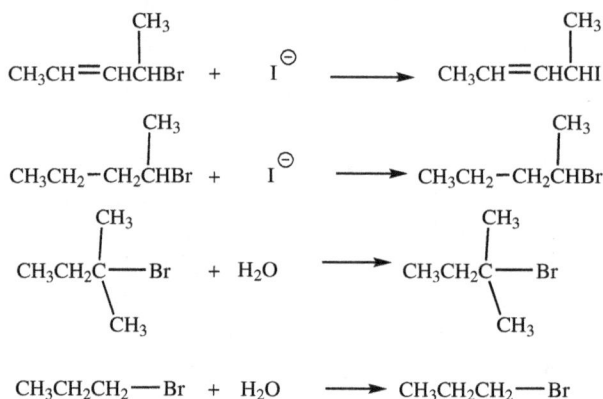

Fig. 6.30

If a reaction is going to proceed via the S_N1 mechanism, changing the nucleophile will have no effect on the rate, but a better leaving group will speed it up. Since increasing the strength of the nucleophile has no effect on S_N1 reactions but speeds up S_N2 reactions and increasing the strength of the leaving group favors S_N1 reactions, reactions with the stronger nucleophile will tend to be S_N2 and those with the better leaving groups will tend toward S_N1.

Therefore, in the reactions shown in Fig. 6.30, the one listed first will proceed fastest.

6.17 The Scope of Nucleophilic Substitution

As previously noted, nucleophilic substitution reactions are perhaps the most commonly used reactions in organic chemistry. The type of product formed depends on the specific nucleophile used. Shown in Table 6.2 are some typical nucleophiles and the product types they form. Examine this table carefully and note each reaction as a preparation of a functional group type. (Are you keeping your list up-to-date?). We will meet many of these as well as many others in future chapters.

6.18 Reversibility and Catalysis

All reactions are, in principle, reversible. This was thoroughly discussed in Chapter 4. However, if the free energy difference between the reactants and products is very large, the activation energy for the reverse reaction is much larger than for the forward reaction and the forward reaction becomes irreversible. Consider the reaction

$$CH_3Br + Na^+OH^- \rightleftharpoons CH_3OH + Na^+Br^-$$

The hydroxide ion is a good nucleophile and bromide ion is a good leaving group. The reaction proceeds to give methanol and sodium bromide. For the REVERSE reaction, bromide is a decent nucleophile but hydroxide is a very poor leaving group (water is a weak acid). This reverse reaction does not proceed and the reaction becomes irreversible. It is therefore fruitless to attempt to prepare bromomethane (or any alkyl bromide) from methanol (or any alcohol) using sodium bromide. However, consider the following reaction which does work:

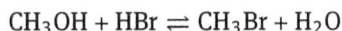

$$CH_3OH + HBr \rightleftharpoons CH_3Br + H_2O$$

Why should this reaction produce bromomethane while the sodium bromide one did not? Obviously the difference must lie in the presence of the proton. Consider the following equilibrium:

$$ROH + H \rightleftharpoons ROH_2^+$$

This is an acid-base reaction where the oxygen atom of the alcohol uses one of its lone pairs of electrons to form a bond to the proton. In the process it becomes positively charged. Note what has happened to the leaving group ability in the process! The protonic acid of OH^- is H_2O, but that of OH_2 is H_3O^+, the actual species present in aqueous solutions of mineral acids. Since leaving group abilities are related to the acid strengths, *the protonation of the alcohol has increased its leaving group ability* to the point where bromide ion can replace it. This shows the effect of an *acid catalyst*. Yet another effect pushes the equilibrium to the right. The equilibrium constant has the form $[CH_3Br][H_2O]/[CH_3OH][HBr]$. If the concentration of water is kept low by removing it from the reaction, Le Chatlier's principle tells us that the equilibrium will be shifted toward the right. In this case, this happens automatically since the water will react with the strong acid HBr as:

$$H_2O + H^+ \rightleftharpoons H_3O^+$$

The equation for the process is

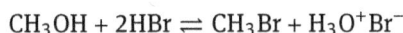

$$CH_3OH + 2HBr \rightleftharpoons CH_3Br + H_3O^+Br^-$$

We will see many more examples of acid-catalyzed reactions in upcoming chapters. The material we have covered in this chapter has set the stage for a look at some functional groups which are most frequently prepared by substitution reactions. In Chapter 7, we will look at the classes called alcohols, ethers, and alkyl halides, all of which can be prepared by *nucleophilic substitution* reactions (S_N1 or S_N2).

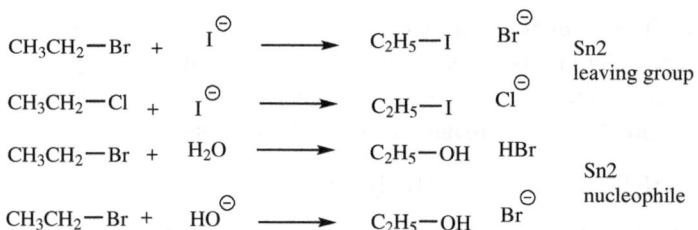

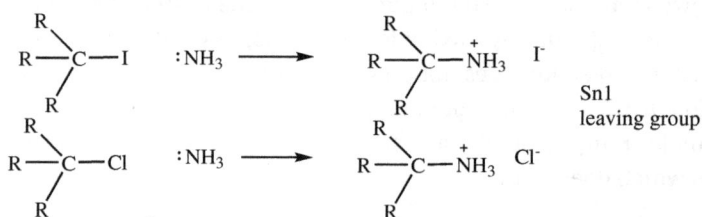

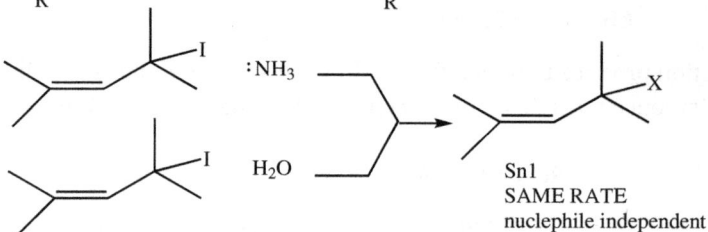

$$CH_3CH_2-Br \ + \ \overset{\ominus}{I} \longrightarrow C_2H_5-I \quad \overset{\ominus}{Br}$$

Sn2
leaving group

$$CH_3CH_2-Cl \ + \ \overset{\ominus}{I} \longrightarrow C_2H_5-I \quad \overset{\ominus}{Cl}$$

$$CH_3CH_2-Br \ + \ H_2O \longrightarrow C_2H_5-OH \quad HBr$$

Sn2
nucleophile

$$CH_3CH_2-Br \ + \ \overset{\ominus}{HO} \longrightarrow C_2H_5-OH \quad \overset{\ominus}{Br}$$

Sn1
leaving group

Sn1
SAME RATE
nuclephile independent

Tab. 6.1: Common solvents and their dielectric constants.

Solvent	Dielectric (D)
H_2O	81
CH_3CN	38
CH_3OH	33
CH_3CH_2OH	24
$(CH_3)_2C=O$	23
CH_3COOH	5
$(C_2H_5)_2O$	4
benzene	2
CCl_4	2
$CHCl_3$	2

Tab. 6.2: Common nucleophiles and the products they give.

R'–L + Nu⁻ ⟶	R'–Nu + L⁻
Nucleophile	Product
H_2O, OH^-	R'OH (an alcohol)
ROH, RO^-	R'–O–R (ether)
R–C(=O)–OH, R–C(=O)–O⁻	R–C(=O)–OR' (ester)
NH_3	$R'NH_2$ (amine)
RNH_2	RR'NH (amine)
R_2NH	R_2NR'
N_3^-	$R'N_3$ (azide)
H_2S	R'SH (thiol)
CN^-	R'CN (nitrile)
H^-	R'H (alkane)
Cl^-, Br^-, I^-	R'–Cl, R'–Br, R'–I

6.19 Problems

6-1. Draw the Fischer projection of (R) 2-bromobutane. Then draw its mirror image. What is the absolute configuration of this molecule? The rotation of the R isomer is +15°. What is the rotation of the mirror image? Which one of these would be referred to as (+)-2-bromobutane? Which one is the *d* isomer? On the basis of the above information, can you say which enantiomer of 2-chlorobutane has a (+) rotation?

6-2. Convert the following 3-D drawings to valid Fischer projections. Then redraw them (if necessary) in the form required to assign the absolute configurations and determine their configuration.

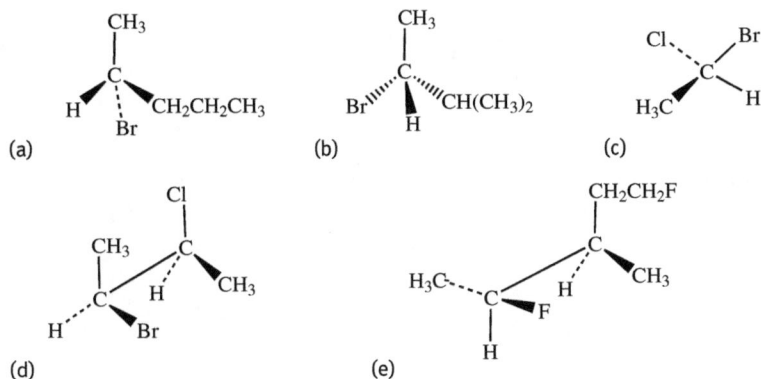

(a)   (b)   (c)

(d)   (e)

6-3. Convert the Fischer projections of the following molecules into valid 3-D drawings.

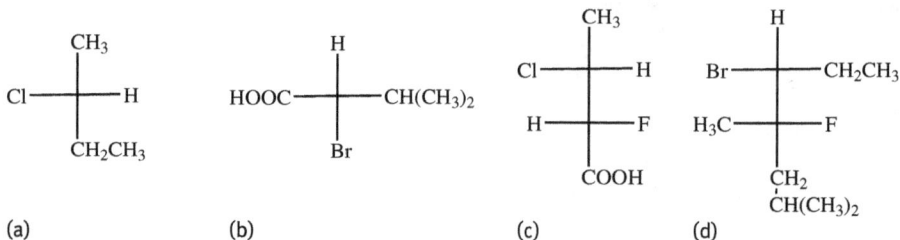

(a) (b) (c) (d)

6-4. Draw the Fischer projections of the following compounds:
(a) (2R, 3R) 2,3-dibromopentane
(b) (R) 2-chlorobutane
(c) (2R, 3R, 4S) 2,3,4-trichlorohexane

6-5. Are *cis* and *trans* 2-butene enantiomers? Will they be optically active?

6-6. Using 3-D drawings, explain why CH_2ClBr does not affect polarized light.

6-7. Choose the term (enantiomer, diastereomer, positional isomer, identical) that correctly describes the relationship between the following pairs of compounds.

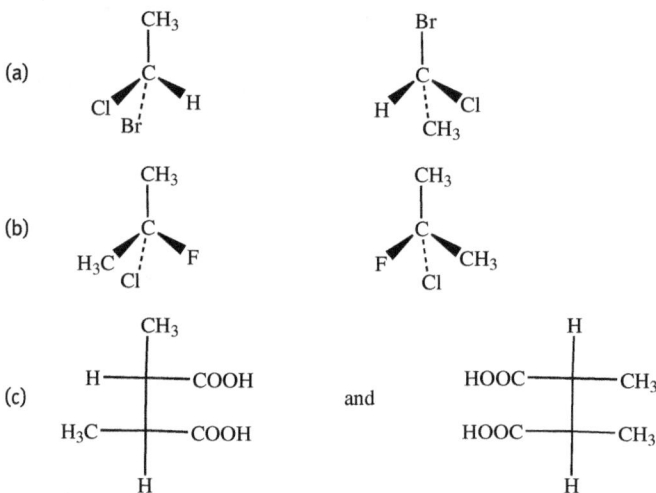

(a)

(b)

(c)

and

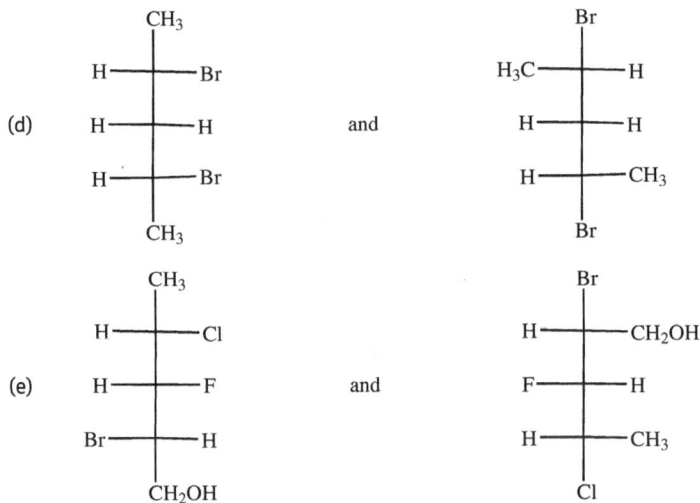

(d)

```
        CH3                         Br
   H ───── Br              H3C ───── H
   H ───── H      and        H ───── H
   H ───── Br                 H ───── CH3
        CH3                         Br
```

(e)

```
        CH3                         Br
   H ───── Cl               H ───── CH2OH
   H ───── F      and        F ───── H
  Br ───── H                 H ───── CH3
       CH2OH                        Cl
```

6-8.
(a) Which of the 14 structures in 6.7 are meso forms?
(b) Assign the R or S configurations to each of the chiral centers of the left drawing in 6.7 (a), (c), and (d).

6-9. Order the following species according to their abilities as leaving groups. (Refer to Table 4.1 if necessary).

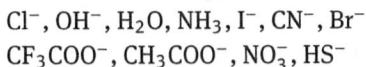

Cl^-, OH^-, H_2O, NH_3, I^-, CN^-, Br^-
CF_3COO^-, CH_3COO^-, NO_3^-, HS^-

6-10. Within each series, order the species according to their ability as nucleophiles.
(a) H_2O, NH_2^-, OH^-, NH_3, CN^-
(b) Br^-, I^-, F^-, Cl^-, HCl

6-11. Choose the best of the indicated solvents for each of the following reactions.

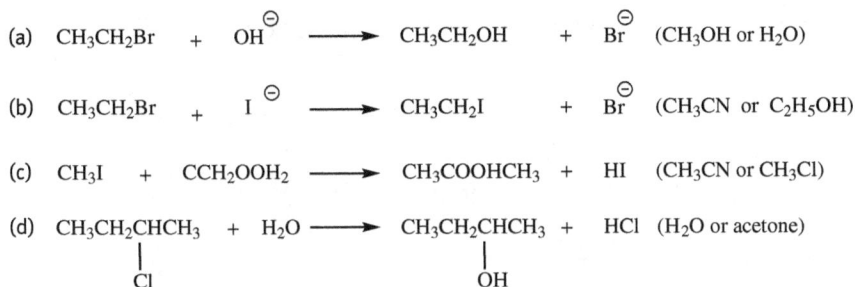

(a) CH_3CH_2Br + $OH^{\ominus}$ ⟶ CH_3CH_2OH + $Br^{\ominus}$ (CH$_3$OH or H$_2$O)

(b) CH_3CH_2Br + $I^{\ominus}$ ⟶ CH_3CH_2I + $Br^{\ominus}$ (CH$_3$CN or C$_2$H$_5$OH)

(c) CH_3I + CCH_2OOH_2 ⟶ $CH_3COOHCH_3$ + HI (CH$_3$CN or CH$_3$Cl)

(d) $CH_3CH_2CHCH_3$ + H_2O ⟶ $CH_3CH_2CHCH_3$ + HCl (H$_2$O or acetone)
    | |
    Cl   OH

6-12. For each of the following equations, decide whether the K_{eq} will be greater or less than 1. Consult the tables of acidities, etc., where necessary

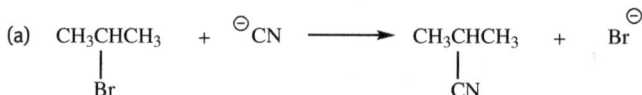

(a) CH_3CHCH_3 + $^{\ominus}CN$ ⟶ CH_3CHCH_3 + $Br^{\ominus}$
       $|$
      Br CN

(b) $CH_3CH_2CH_2OH$ + $^{\ominus}OH$ ⟶ $CH_3CH_2CH_2O^{\ominus}$ + H_2O

(c) $CH_3C{\equiv}CH$ + $CH_3CH_2O^{\ominus}$ ⟶ $CH_3C{\equiv}C^{\ominus}$ + CH_3CH_3OH

6-13. Predict which one of the following pairs of reactions is *more likely* to proceed via an Sn1 mechanism.

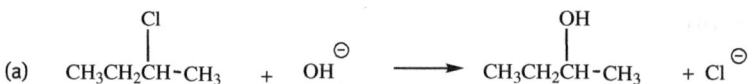

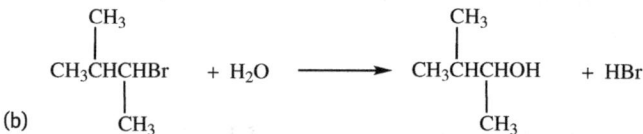

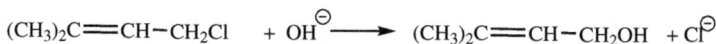

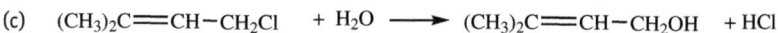

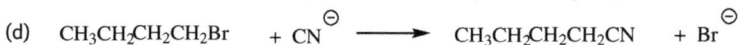

(a)
         I OH
         $|$ $|$
$CH_3CH_2CH{-}CH_3$ + $^{\ominus}OH$ ⟶ $CH_3CH_2CH{-}CH_3$ + $I^{\ominus}$

          Cl OH
          $|$ $|$
(a) $CH_3CH_2CH{-}CH_3$ + $^{\ominus}OH$ ⟶ $CH_3CH_2CH{-}CH_3$ + $Cl^{\ominus}$

          CH_3 CH_3
          $|$ $|$
CH_3CH_2CBr + H_2O ⟶ CH_3CH_2COH + HBr
          $|$ $|$
          CH_3 CH_3
      CH_3 CH_3
      $|$ $|$
$CH_3CHCHBr$ + H_2O ⟶ $CH_3CHCHOH$ + HBr
          $|$ $|$
(b) CH_3 CH_3

(c) $(CH_3)_2C{=}CH{-}CH_2Cl$ + $OH^{\ominus}$ ⟶ $(CH_3)_2C{=}CH{-}CH_2OH$ + $Cl^{\ominus}$

(c) $(CH_3)_2C{=}CH{-}CH_2Cl$ + H_2O ⟶ $(CH_3)_2C{=}CH{-}CH_2OH$ + HCl

    $CH_3CH{=}CHCH_2Br$ + $CN^{\ominus}$ ⟶ $CH_3CH{=}CHCH_2CN$ + $Br^{\ominus}$

(d) $CH_3CH_2CH_2CH_2Br$ + $CN^{\ominus}$ ⟶ $CH_3CH_2CH_2CH_2CN$ + $Br^{\ominus}$

6-14. In each of the following reactions, assume that the mechanism is either pure Sn1 or Sn2 for the substitution reactions or the usual electrophilic addition for the alkene reactions. Draw the structures of the products including stereochemistry where appropriate and indicate whether the product will be optically active or not.

(a) (R)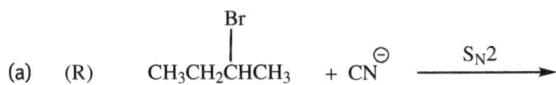

(b) (R,R)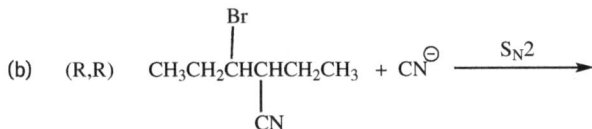

(c) (S)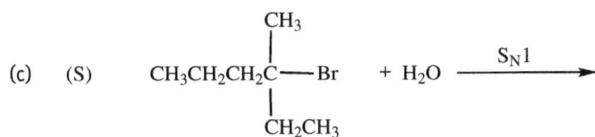

(d)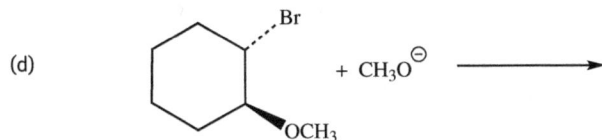

(e)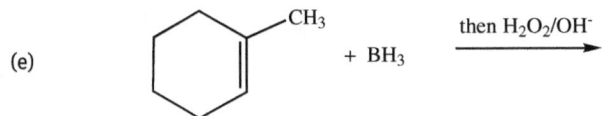

6-15. The R isomer of compound I rotated light in a clockwise direction when placed in a polarimeter (i.e., it is the (+)-isomer). In a series of reactions, *none of which affected the bonds to the chiral center,* I was converted into II. II rotated light in an anticlockwise direction (i.e., it was the (-)-isomer). The (+)-isomer of II was also obtained by oxidation of the (-)-isomer of III, a reaction which again does not affect the chiral center of III.

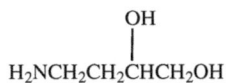

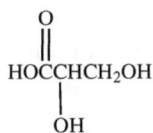

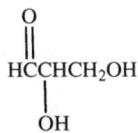

I   II   III

(a) Draw the Fischer projection of the R isomer of I.

(b) Deduce and draw the structure of the (-)-isomer of II and assign it the R or S configuration.

(c) Deduce the configuration of III and assign it the correct stereochemical descriptor.

6-16. Replace the "?" in the following equations with the correct reagent or product. If a catalyst is required for the reaction, indicate what it is. Show stereochemistry if it is important.

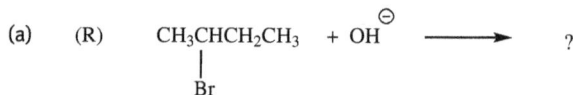

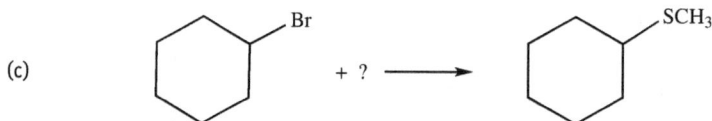

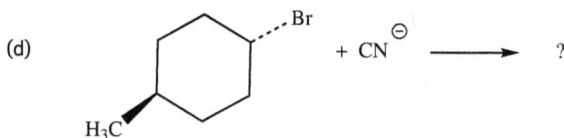

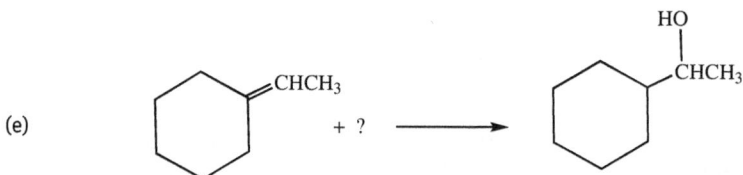

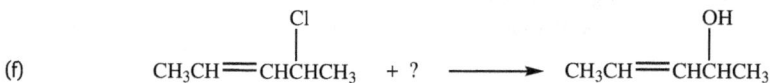

(a)   (R) $CH_3CHCH_2CH_3$ + $\overset{\ominus}{OH}$ ⟶   ?
            |
            Br

(b)   (Z) $CH_3CH=CHCH_3$ + Br_2 $\xrightarrow{H_2O}$?

(c)

+ ? ⟶

(d)

+ $\overset{\ominus}{CN}$ ⟶ ?

(e)

+ ? ⟶

(f) $CH_3CH=CHCHCH_3$ + ?   ⟶ $CH_3CH=CHCHCH_3$
               | |
               Cl   OH

(g) NH_3 + ?   ⟶ $CH_3CH_2\overset{\oplus}{NH_3}$ $\overset{\ominus}{Br}$

(h) H_2O + ? $\xrightarrow{\hspace{2cm}}$ $\begin{array}{c} OH \\ | \\ CH_3CCH_2CH_3 \\ | \\ CH_3 \end{array}$

(i)   + $I^\ominus$ $\xrightarrow{\hspace{2cm}}$?

(j)   + $CN^\ominus$ $\xrightarrow{\hspace{2cm}}$?

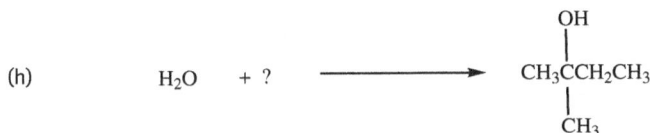

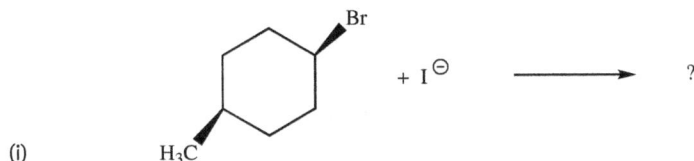

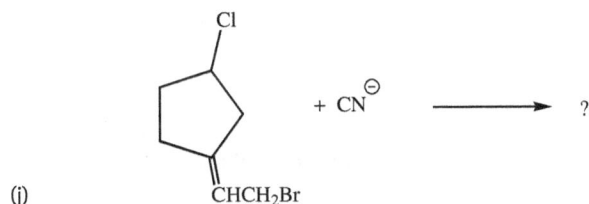

6-17. When *cis* 2-butene is reacted with Br_2 in CCl_4, a dibromide is isolated which boils at 161 °C. When *trans* 2-butene is used in the same reaction, the dibromide isolated boils at 157 °C at the same pressure.
(a) Draw the mechanistic pathway for both reactions using *Newman* projections.
(b) Convert the final product in each case into a Fischer projection (remember the Fischer projections represent molecules in *eclipsed* conformations).
(c) Assign each product an IUPAC name including the correct R and S configuration for each chiral center.
(d) Identify the relationship between the two products.
(e) Which (if either) of the products obtained from the two reactions will be optically active? For any which are not optically active, indicate *why*.

6-18. When a solution of Br_2 in methanol is added to 2,3-dimethyl-2-hexene, how many products (including stereoisomers) might be expected?

6-19. Go back and examine question 3.11 again. Provide a new answer that addresses the question of stereochemistry – i.e., how many compounds in your original list can have stereoisomers?

7 Alcohols and Ethers, Amines, and Alkyl Halides. Introduction to Carbonyl Compounds (Aldehydes, Ketones)

7.1 Introduction

Alcohols and ethers are the molecules in which one or both of the hydrogen atoms of water have been replaced by carbon atoms. It is useful to consider these classes together. In the same way, amines are derived from ammonia by replacing one, two, or all three N–H bonds with N–C bonds. Finally, alkyl halides are just what the name implies – alkyl groups attached to a halogen atom (F, Cl, Br, or I). The chemistry of these classes is closely related and will be grouped together. At the end of this chapter, the chemistry of another type of functional group that can be derived from alcohols will be considered. As usual, we will bypass the nomenclature of these materials and use the nomenclature text to teach this.

7.2 Alcohols and Ethers – Chemical and Physical Properties

Alcohols, being capable of forming hydrogen bonds, both as donors and acceptors are much higher boiling compounds than alkanes of comparable molecular weight (Table 7.1). They are also moderately soluble in water (C1 to C3 are miscible with water in all proportions) until the size of the hydrocarbon chain reaches a sufficient size to overcome the effect of the hydroxyl (OH) group. Ethers cannot be hydrogen-bond donors and are therefore much lower boiling than the corresponding alcohols. Their boiling points are very similar to those of the alkanes where oxygen has been replaced by a methylene (CH_2) group (Table 7.1). They are insoluble in water.

We have already seen two reactions of water several times: i.e.,

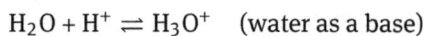

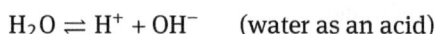

$$H_2O + H^+ \rightleftharpoons H_3O^+ \quad \text{(water as a base)}$$
$$H_2O \rightleftharpoons H^+ + OH^- \quad \text{(water as an acid)}$$

If the structure of an alcohol is examined it will be seen that these two reaction types are still possible:

$$\text{e.g., } ROH + H^+ \rightleftharpoons ROH_2^+ \quad \text{(alcohol as a base)}$$
$$ROH \rightleftharpoons H^+ + RO^- \quad \text{(alcohol as an acid)}$$

It should be recalled that water is a stronger acid than an alcohol. There is very little difference in the base strengths of water and alcohols. Conversely, ethers cannot function as acids, but can function as bases.

$$ROR + H^+ \rightleftharpoons R_2OH^+$$

https://doi.org/10.1515/9783110778311-007

Alcohols have two kinds of bonds as part of the functional group: a C–O and an O–H bond. We will see examples of the reactions of both types. Many reactions of alcohols are those of the O–H bond. Since the reactions of water are reactions of this same O–H bond, it should not be surprising that much of the chemistry of alcohols is very reminiscent of the reactions of water. Conversely, ethers have only C–O bonds and their chemistry is quite different from that of water.

7.3 Preparation of Alcohols and Ethers

Alkenes can undergo the *addition* of the elements of water by hydroboration or acid-catalyzed hydration, alkyl halides can undergo *nucleophilic substitution* by water or OH⁻, and carbonyl compounds can undergo *nucleophilic addition*. All of these processes lead to the formation of alcohols. (You should recall the importance of Markownikov's Rule in some of the above reactions!) Showing the analogy of water and alcohols, acid-catalyzed addition of alcohols to alkenes or nucleophilic substitution of alkyl halides by alcohols proceeds by exactly the same mechanism as the reactions of water, but the products are *ethers*.

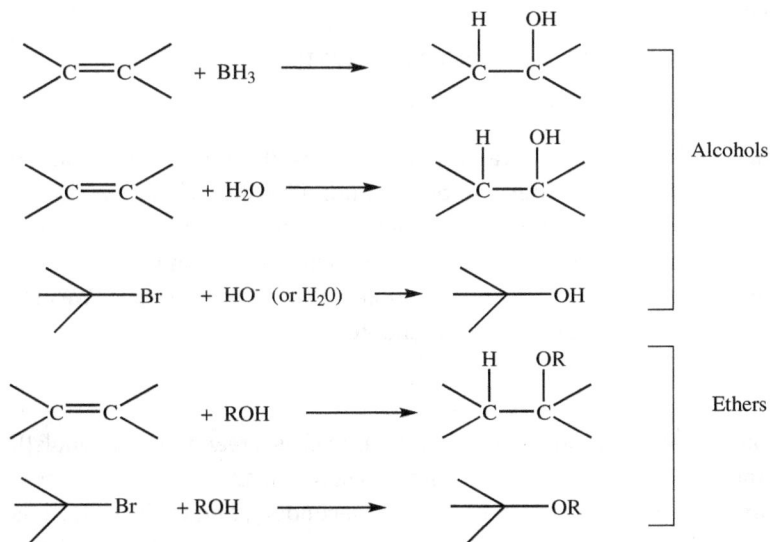

In exactly the same way, but using ammonia (NH_3) as the nucleophile, the formation of *amines* can take place. Recall that nitrogen atoms are better nucleophiles than the corresponding oxygen atoms, so the formation of amines is a faster reaction than the formation of alcohols *using water as the nucleophile*. There are some other things we must consider about the formation of amines, but these will be postponed until later in this chapter.

Another reaction which we saw in Chapter 5 and which is applicable to the preparation of alcohols is hydrogenation. Compare the following two reactions.

As you will observe, the reaction is the same, but the product is an alcohol. Hydrogenation of a ketone gives a secondary alcohol and hydrogenation of aldehydes gives primary alcohols.

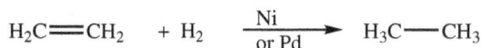

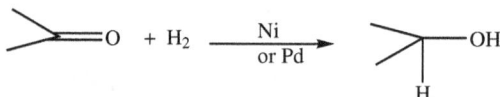

$$H_2C{=}CH_2 \ + \ H_2 \quad \xrightarrow[\text{or Pd}]{\text{Ni}} \quad H_3C{-}CH_3$$

7.4 Reactions of Alcohols

As noted, alcohols have two different kinds of bonds: O–H and C–O bonds. As might be expected, these behave differently. As noted above, the similarity in the structure of water and alcohols suggests a similar type of reactivity. One very useful alcohol reaction is that with sodium.

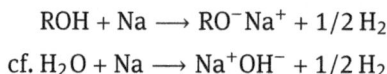

$$ROH + Na \longrightarrow RO^-Na^+ + 1/2\,H_2$$
$$cf.\ H_2O + Na \longrightarrow Na^+OH^- + 1/2\,H_2$$

Just as the product of the reaction of water and sodium (sodium hydroxide) can be thought of as the sodium salt of water, the product (an alkoxide) of the reaction of sodium with an alcohol can be thought of as the salt of that alcohol. The alkoxide is a useful nucleophile and base. We have seen it used in a synthesis of ethers (see figure, Sect. 7.3). We will see other uses for this type of molecule in future sections. We also saw a reaction of the C–O bond in the previous chapter: i.e.,

$$ROH + HBr \longrightarrow RBr + H_2O$$

(Note this reaction as a preparation of alkyl halides as well as a reaction of alcohols!). This reaction is the basis for a useful classification test for alcohols called the Lucas Test. It is based upon the rate of reaction of primary, secondary, and tertiary alcohols with hydrochloric acid. If an alcohol is added to conc. hydrochloric acid, it will dissolve. (Note that many alcohols are insoluble in water. However in conc. hydrochloric acid, protonation of the alcohol will occur giving ROH_2^+, which is ionic and therefore water soluble.) As the reaction shown above proceeds, the alkyl halide formed is insoluble and precipitates out of solution (usually as a second liquid phase). This process happens most rapidly for tertiary or allylic alcohols where the carbocation formed is tertiary or resonance-stabilized and most slowly for primary alcohols. (In practice, some $ZnCl_2$ is usually added to the solution. This gives more reliable results.)

Tab. 7.1: Boiling points of some related organic compounds (grouped by molecular weight).

Mol Wt	Alkane	Alcohol	Ether	Aldehyde	Ketone
30–30	ethane (−88°)	methanol (65°)			
44–46	propane (−44°)	ethanol (78°)	dimethyl (−25°)	ethanal (−21°)	
58–60	butane (0°)	1-propanol (97°)	methyl ethyl (7°)	propanal (49°)	propanone (56°)
72–74	pentane (36°)	1-butanol (117°)	diethyl (35°)	butanal (76°)	2-butanone (80°)

The rate at which the second phase is formed is dependent on the carbocation stabilities. Therefore, alcohols that form resonance-stabilized or tertiary intermediate carbocations will react fastest, those that proceed through a secondary carbocation will react at an intermediate rate, and those that proceed through a primary carbocation will be the slowest.

Other methods for the conversion of alcohols to alkyl halides exist. The two most common are shown below.

$$ROH \quad + \quad SOCl_2 \quad \longrightarrow \quad RCl + \quad SO_2 \quad + \quad HCl$$

$$ROH \quad + \quad PBr_3 \quad \longrightarrow \quad RBr \quad + \quad H_3PO_4$$

These two reactions constitute the best *preparative* methods for converting alcohols to alkyl halides.

7.5 Reactions of Ethers

Ethers are relatively unreactive. For this reason they are frequently used as solvents in reactions. However, they will undergo the same carbon–oxygen bond reactions as alcohols. For example:

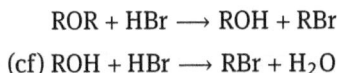

$$ROR + HBr \longrightarrow ROH + RBr$$
$$(cf) ROH + HBr \longrightarrow RBr + H_2O$$

An interesting question arises when ethers with different alkyl groups attached to oxygen are cleaved by acid. Which alkyl group gives the alkyl halide? Let us see if what we have learned can help us predict this. For example, which reaction occurs (Fig. 7.1): To answer this type of question, *always begin with the mechanism*, which in this case is shown in Fig. 7.2.

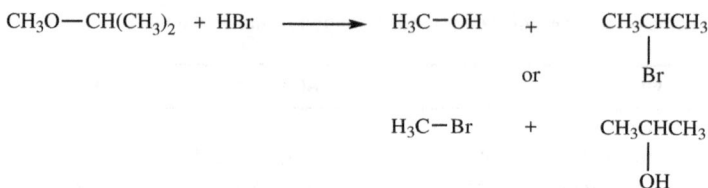

Fig. 7.1

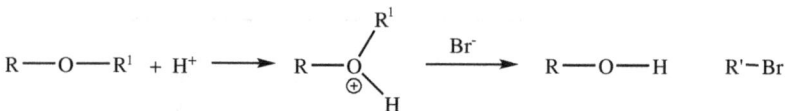

Fig. 7.2

In the intermediate, the positively charged oxygen atom will be pulling the electrons in the two C–O bonds towards it. The two carbon atoms will be positively polarized (Fig. 7.3). Since increased substitution of a carbon atom makes it more stable in a positive form (cf. carbocations) the secondary carbon in our example will be more willing to release its share of electrons than will the methyl group and so the isopropyl group will have a larger partial positive charge than the methyl group. Since bromide ion is negatively charged, it will be attracted to the site of highest positive charge The products will be methanol and 2-bromopropane. *Simple isn't it!*

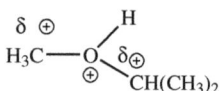

Fig. 7.3

7.6 Preparations of Alkyl Halides

We have already seen the following reactions which lead to the formation of a C-hal bond. Alkyl halides are the substrates for nucleophilic substitution reactions. As such, they are the raw materials for a huge number of functional group preparations. At this point you should review Table 6.2 to refresh your memory as to the types of compounds that can be prepared by nucleophilic substitution reactions.

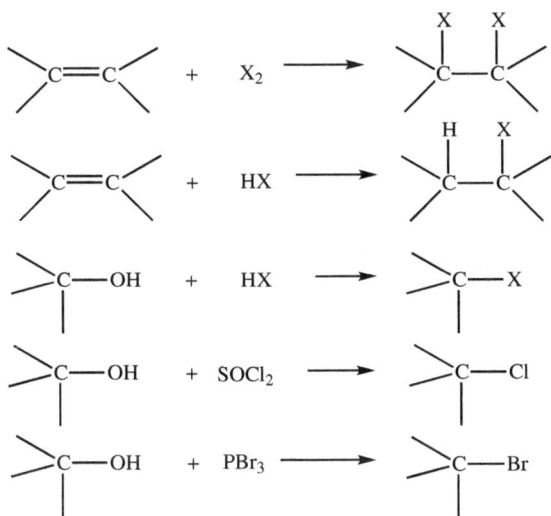

7.7 Amines

As noted previously, amines are derived from ammonia by replacement of one or more of the hydrogen atoms. The material in Chapter 10 of this text treats the subject of organic compounds containing nitrogen in much more detail, but an introduction to this subject is best considered with the material on alcohols.

The nitrogen atom in ammonia has a lone pair of electrons that can be used to form a bond in a nucleophilic substitution reaction. The product of this process is an ammonium salt (a positively charged, tetravalent nitrogen atom; cf. $NH_4^+Cl^-$). Because ammonia is a base, a proton transfer can occur leading to an amine (Fig. 7.4). Just like ammonia, amines are nucleophilic and can take part in subsequent substitutions to replace the second and third hydrogen atoms (Fig 7.4a) The products are also amines.

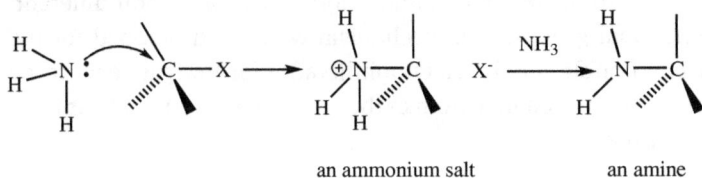

an ammonium salt   an amine

Fig. 7.4

There is a very unfortunate contrast between the naming of alcohols and amines. The terms primary, secondary, and tertiary when applied to **alcohols** refers to the number of hydrogen atoms *on the carbon attached to the oxygen atom.* The same words are

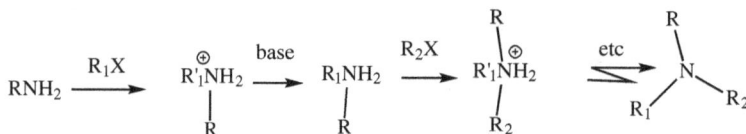

$$RNH_2 \xrightarrow{R_1X} \overset{\oplus}{R'_1}\underset{R}{NH_2} \xrightarrow{base} \underset{R}{R_1NH_2} \xrightarrow{R_2X} \overset{\oplus}{R'_1}\underset{R_2}{NH_2} \xrightarrow{etc} $$

Fig. 7.4a

used for amines, but in this case they refer to the number of hydrogen atoms *on the nitrogen atom*. These names are summarized in Fig. 7.5.

alcohols	RCH_2OH	R_2CHOH	R_3COH	
	primary	secondary	tertiary	
amines	RNH_2	R_2NH	R_3N	$R_4\overset{\oplus}{N}$
	primary	secondary	tertiary	quaternary ammonium salt

Fig. 7.5

Refer back to the figure in Sect. 7.3. The acid-catalyzed addition of water or alcohols to alkenes can be used for the preparation of alcohols or ethers in addition to the routes using nucleophilic substitution. This route is *not* applicable to the formation of amines. The nitrogen atom of ammonia and amines is *basic* and will react with the acid catalyst to form an ammonium salt. This means the acid catalyst is gone and the nitrogen atom, being positively charged, is no longer nucleophilic.

7.8 Introduction to Carbonyl Groups

The molecules to be studied now all contain a carbon–oxygen double bond. This group is called a *carbonyl* (carbon-eel') *group* and it forms part of several different, but closely related, functional groups. The mechanism we have developed for the reaction of carbon-*carbon* double bonds (electrophilic addition) might seem to be applicable here also. However, the change from carbon to oxygen is a significant one and the process must be revised.

7.9 The Carbonyl Group – Nucleophilic Addition

The carbonyl group consists of one sp^2-hybridized carbon atom attached by a double bond to one oxygen atom (Fig. 7.6). Let us look at this structure more closely to see if we can predict what type of reactions it might undergo. Since sp^2-hybridized carbon is planar, the four atom system A, B, C, and O must all lie in the same plane. Oxygen is more electronegative than carbon and therefore the C=O bond will be polarized toward oxygen leaving a partial positive (δ^+) charge on the carbonyl carbon atom. We would then predict that nucleophiles would tend to react at carbon and electrophiles at oxygen. This basic fact controls almost all of the chemistry of groups that *contain* the carbonyl group.

Fig. 7.6

If a nucleophile attacks the positive carbon and forms a bond to it with its pair of electrons, one of the existing four bonds must be broken. However, when the carbonyl π bond is broken, the oxygen atom remains attached (Fig. 7.7). Notice that the carbon changes its hybridization from sp^2 to sp^3. This is an *addition* reaction but it is different from the addition reactions studied in Chapter 5. This one is initiated by a *nucleophile* and is called a nucleophilic addition.

Alkenes are attacked by electrophiles. Carbonyl carbon atoms are attacked by nucleophiles.

What happens to the intermediate in Fig. 7.7 depends on what the nucleophile is and the nature of A and B. In this chapter we will look at the reactions of one type of compound and in Chapter 8, another related group of compounds.

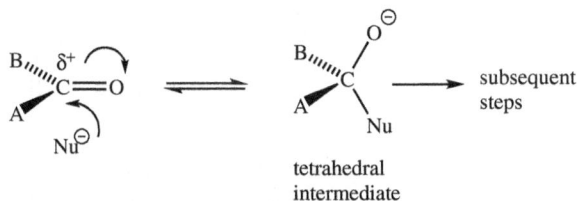

Fig. 7.7

If an electrophile attacks the carbonyl group, it must do so at the *oxygen atom* (Fig. 7.8). One of the nonbonding pairs of electrons on oxygen can be used to form a bond to the electrophile. In the process the oxygen becomes positively charged. This positive charge can be delocalized to the carbon atom via a *resonance* process. This is a stabilizing factor since carbon, being less electronegative than oxygen, can better accommodate the charge. We will see the process depicted in Fig. 7.8 many times as a mechanism for catalyzing reactions of carbonyl groups.

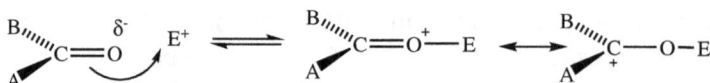

Fig. 7.8

The process illustrated in Fig. 7.7 is very similar to that illustrated in Fig. 5.5 in that a nucleophile is attacking a planar, sp²-hybridized carbon. In that case, we saw that attack from either side of the planar center was possible and the result was a *racemic mixture*. The same situation exists when a nucleophile attacks a carbonyl group. The product obtained is a 1:1 mixture of enantiomers (provided, of course, that four different groups are attached and no other chiral centers are present. If another chiral center *is* present, the product will be a mixture of diastereomers).

7.10 Aldehydes and Ketones – Structure and Physical Properties

Aldehydes are molecules that contain a carbonyl group which is attached to at least one hydrogen atom. They contain the structural unit –CH=O. Ketones (kee'-tones) are compounds in which a carbonyl group is attached to two carbon atoms. (Notice that there is no "y" in the word ketone!!) Shown in Fig. 7.9 are some simple aldehydes and ketones with their trivial names where they are commonly used.

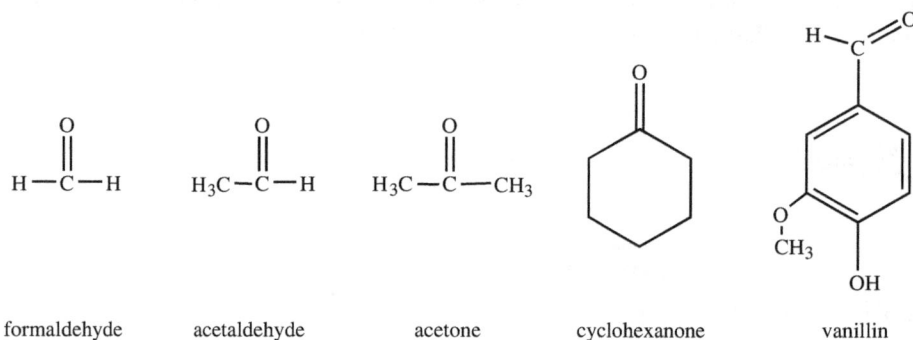

formaldehyde   acetaldehyde   acetone   cyclohexanone   vanillin

Fig. 7.9

Aldehydes and ketones are molecules with very characteristic odors. They are lower boiling than their related alcohols since they cannot self-hydrogen bond. They are also water soluble up to about C5 but essentially insoluble above this point.

7.11 Preparations of Aldehydes and Ketones

To date we have seen two methods of preparing aldehydes and ketones: i.e., ozonolysis of alkenes (Sect. 5.10.3) and addition of water to alkynes (Sect. 5.11.2). The oxidation of alcohols with reagents such as chromium trioxide, sodium dichromate, or potassium permanganate is a standard method for preparing aldehydes and ketones. (Note these as reactions of alcohols in your list). Oxidation of secondary alcohols occurs smoothly to give ketones. Oxidation of primary alcohols initially gives aldehydes but, unless special care is taken, these are further oxidized to carboxylic acids (see Chapter 8). To stop at the aldehyde stage, the reaction must be done in the presence of a molecule called pyridine (C_5H_5N). There are other methods for preparation of aldehydes and ketones which will be considered in subsequent sections and chapters.

7.12 Reactions of Aldehydes and Ketones

7.12.1 General Considerations

As pointed out in Sect. 7.9, the characteristic reaction of aldehydes and ketones is *nucleophilic addition*. The beginnings of this mechanism were shown in Fig. 7.7. Notice that the intermediate formed in this reaction is an alkoxide and is therefore quite basic. Therefore, if this process took place in a solvent like water or an alcohol, the *complete* mechanism would look as shown in Fig. 7.10. The slowest step is the first one, and generally the reactions are reversible. The importance of this will be seen shortly. Overall, this reaction results in the addition of H–Nu across the carbonyl double bond.

Since hydrogen atoms are the smallest possible substituents, aldehydes are clearly less sterically hindered than ketones.

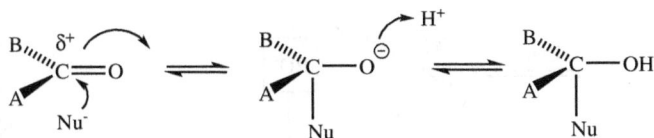

Fig. 7.10

Rates of reaction and K_{eq} of reactions involving aldehydes are larger than those involving ketones due to the steric differences around the carbonyl carbon.

7.12.2 Catalysis

One might ask why should the nucleophile attack the carbon? In fact, many times they don't! For example, when an alcohol and a ketone are mixed, essentially nothing happens. It is convenient to describe this situation in one of two ways. *Either* the nucleophile (ROH) is not nucleophilic enough to attack the electrophile (the carbonyl carbon) *or* the electrophile is not electrophilic enough to attract the nucleophile. In order to get a reaction between these we must *either* increase the nucleophilicity of the alcohol *or* the electrophilicity of the carbonyl carbon. How can this be done? The answers are shown in Fig. 7.11.

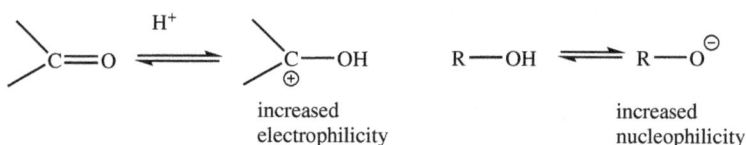

Fig. 7.11

Remember that a negatively charged species is a stronger nucleophile than the corresponding neutral one. Therefore, conversion of the alcohol into the alkoxide will increase its nucleophilicity. However, if *all* the alcohol was converted into alkoxide, there would be no proton available to complete the reaction (Fig. 7.10) and the initial equilibrium would be established but no neutral product could be formed. Further examination of the mechanism shows that only a very small (i.e., catalytic) amount of alkoxide is required to cause the reaction because it is regenerated in the second step: i.e., *it is a catalyst* (Fig. 7.12). Since alkoxides are basic, this mechanism is referred to as a *base-catalyzed mechanism*.

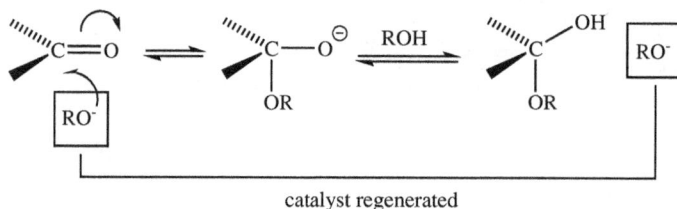

catalyst regenerated

Fig. 7.12

Alternatively, we have seen that the carbonyl oxygen atom can react with an electrophile (e.g., H$^+$) to form a species in which the original positively polarized carbon is made much more positive (Figs. 7.8 and 7.11). This increases its electrophilicity to the point where it can react with an alcohol that is in the neutral state (Fig. 7.13). The product of this step has a positively charged oxygen atom that can lose its charge *by loss of a proton*. This mechanism is referred to as an *acid-catalyzed mechanism*. Note the similarity of some of the intermediates in this scheme to those in the acid-catalyzed addition of water to alkenes (Chapter 5).

It is important to note that in the acid-catalyzed mechanism, the nucleophile is the alcohol and *not* the alkoxide. The pKa's of most alcohols are over 15, so very little alkoxide is present under neutral conditions and even less would be present when the acid (catalyst) is added.

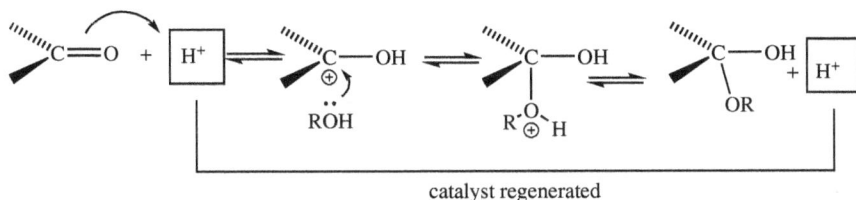

catalyst regenerated

Fig. 7.13

The product of both the acid- and base-catalyzed reactions of aldehydes and ketones with alcohols have a carbon bonded to one hydroxyl (-OH) and one ether (-OC) type oxygen. Such molecules are called hemiacetals (if they are derived from aldehydes) or hemiketals (if they are derived from ketones). It is important to note that *both* mechanistic pathways are completely reversible. If the reaction of 2-propanone (acetone) with methanol is catalyzed by the addition of a few drops of sulfuric acid or by the addition of a small amount of sodium metal (why not NaOH??), an equilibrium amount of the hemiketal will be formed according to the equation shown below. However, if you attempt to isolate the hemiketal by, for example, distilling away the excess alcohol and ketone, the hemiketal will simply slowly dissociate back into its component pieces in order to keep the K_{eq} satisfied (Fig. 7.14).

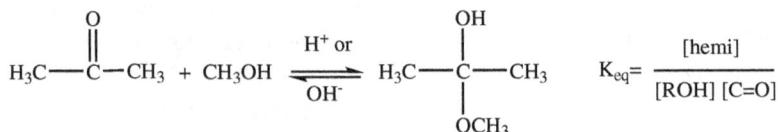

$$K_{eq} = \frac{[\text{hemi}]}{[\text{ROH}]\,[\text{C=O}]}$$

Fig. 7.14

In certain types of compounds that contain an aldehyde or ketone group four or five carbons away from a hydroxyl group, hemiacetal or hemiketal formation can take place in an *intramolecular* (within the same molecule) fashion. The best example of this is found in molecules called carbohydrates (see Sect. 7.12.4)

7.12.3 Ketals and Acetals

While it is true that hemiacetals and hemiketals are formed under acid and base conditions, a further reaction of these occurs *under acid-catalyzed conditions only* (Fig. 7.15). How does the second step of this process go? First, it should be noted that the acid catalyst is necessary. The reaction does *not* occur under basic conditions. This implies that an acid-base reaction of some type, where the substrate provides the basic site, initiates the process. There are two basic sites in the hemiacetal (ketal): the two oxygen atoms. The steps shown in Fig. 7.15 are possible.

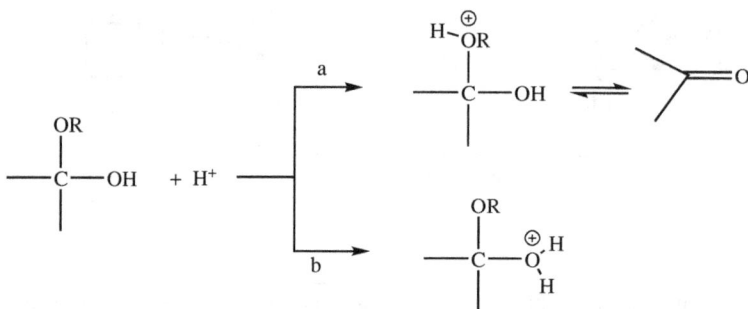

Fig. 7.15

Route *a* should be recognized as the reverse of the last step of the formation of the hemiacetal (ketal) (Fig. 7.13). Step *b* is a new one that converts a poor leaving group (OH) into a much better one (OH_2^+). Compare this intermediate with that discussed in the addition of water to alkenes or the reverse reaction which is the dehydration of alcohols (Chapter 5). If the water molecule departs, it leaves a carbocation which is stabilized in the same way as the first intermediate in the acid-catalyzed formation of a hemiacetal (ketal) (Fig. 7.8): e.g., Fig. 7.16. This stabilization makes the breaking of the $C–OH_2^+$ bond more facile than the corresponding reaction with simple alcohols. We have already seen that carbocations of this type will react with nucleophilic alcohols (Fig. 7.13). Repetition of this process leads to the *acetal* or *ketal* (acetals and ketals are molecules that have *two* ether-type linkages attached to the same carbon atom). The whole mechanistic process is assembled in Fig. 7.17. Again each step of the mechanism, and therefore the whole process, is reversible. However, the situation is somewhat different from the hemiacetal formation in that the equilibrium can be *pushed* in the desired direction by control of the concentration of water and/or alcohol. Why does the reaction not occur under basic conditions? In this case, the leaving group would have to be OH^- and we have already seen several cases that illustrate the reluctance of this group to act as a leaving group. This instance is no exception. The acetal (ketal), unlike the hemiacetal, is stable under basic conditions since OR^- is an even poorer leaving group than OH^-.

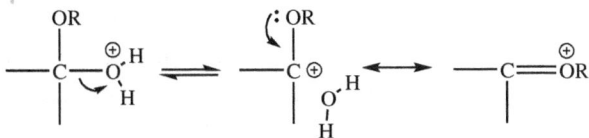

Fig. 7.16

The reverse of acetal formation – hydrolysis to the aldehyde – occurs under *much* milder conditions than the cleavage of ethers (Sect. 7.5) to which it is apparently related. If one examines the stabilities of the intermediates in both of these reactions, the reason for this difference should be apparent.

7.12.4 Protecting Groups

Suppose you have a molecule that contains both an aldehyde and an alkyl halide and you wish to react this molecule with a nucleophile *only* at the carbon–halogen bond. Aldehydes react with cyanide (CN^-) ion (a good nucleophile) by a nucleophilic addition mechanism (next section) and halides by a substitution mechanism. How could you prevent reaction at the carbonyl?

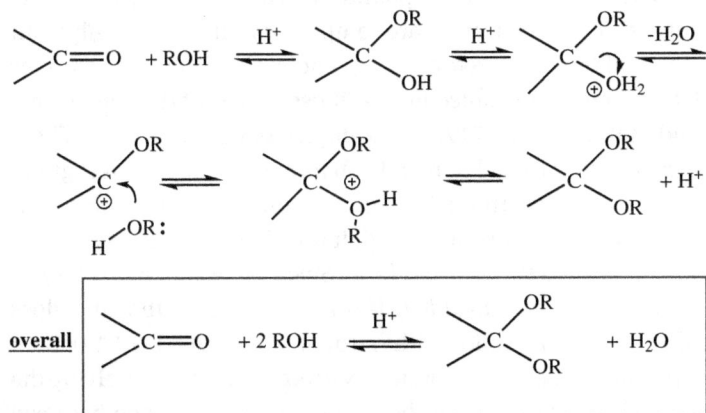

Fig. 7.17

Hemiacetals and hemiketals are formed reversibly under acidic or basic catalysis, but acetals and ketals are formed reversibly only under acidic catalysis and are stable under basic conditions. Therefore, carrying out the formation of an acetal (ketal) under acidic conditions with a large excess of alcohol to drive the equilibrium to the right, followed by raising the pH to above 7, allows isolation of the product.

To achieve this, the aldehyde can be converted into its acetal. This changes the nature of the functional group to one that is not attacked by nucleophiles. The alkyl halide will not be affected by the acid-catalyzed condition of acetal formation. The substitution can then be safely carried out on the alkyl halide part of the molecule. Since the formation of the acetal is reversible under acidic conditions, treatment of the substituted acetal with water and acid re-establishes the equilibrium shown in Fig. 7.14 and, since the concentration of water is now high pushing the equilibrium to the left, the aldehyde is reformed. The acetal is used to *protect* the aldehyde during the substitution process. This process is depicted in Fig. 7.18. The acetal (ketal) is a protecting group for aldehydes (ketones). This concept is very important and will become of great interest in future courses in organic chemistry.

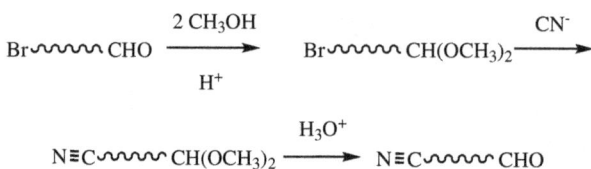

$$Br\sim\sim\sim CHO \xrightarrow[H^+]{2\ CH_3OH} Br\sim\sim\sim CH(OCH_3)_2 \xrightarrow{CN^-}$$

$$N\equiv C\sim\sim\sim CH(OCH_3)_2 \xrightarrow{H_3O^+} N\equiv C\sim\sim\sim CHO$$

Fig. 7.18

A prominent class of biologically important compounds are called *carbohydrates* (frequently referred to as sugars). These molecules are compounds with (commonly) four to six carbon atoms with an aldehyde or ketone group and each of the other carbons bearing one hydroxyl (OH) group. Such molecules as ribose (a part of the genetic material RNA), glucose, and mannose (Fig. 7.19) are examples of carbohydrates. These molecules commonly exist in a hemiacetal or hemiketal form with the hydroxyl group three or four carbons away from the carbonyl group providing the nucleophilic OH group. If the oxygen-containing ring is six-membered, it is called the *pyranose* form, if it is a five-membered ring, it is called the *furanose* form. *Since the hemiacetal is formed reversibly, these molecules can react as if an aldehyde was present*, even though it does not appear to be present on simple inspection of the structural formula (Fig. 7.19). Both *starch* and *cellulose* are polymers of glucose, formed by *acetal* formation involving the carbonyl group and one hydroxyl group from the same molecule and one hydroxyl group from another glucose molecule. The *only* chemical difference between starch and cellulose is a stereochemical one! The biochemical differences are significantly larger! Figure 7.19 shows the Fischer projections of the three simple sugars mentioned above as well as the hemiacetal form of glucose and the repeating structure of the glucose polymer cellulose. A fuller discussion of these molecules can be found in the texts mentioned in the preface.

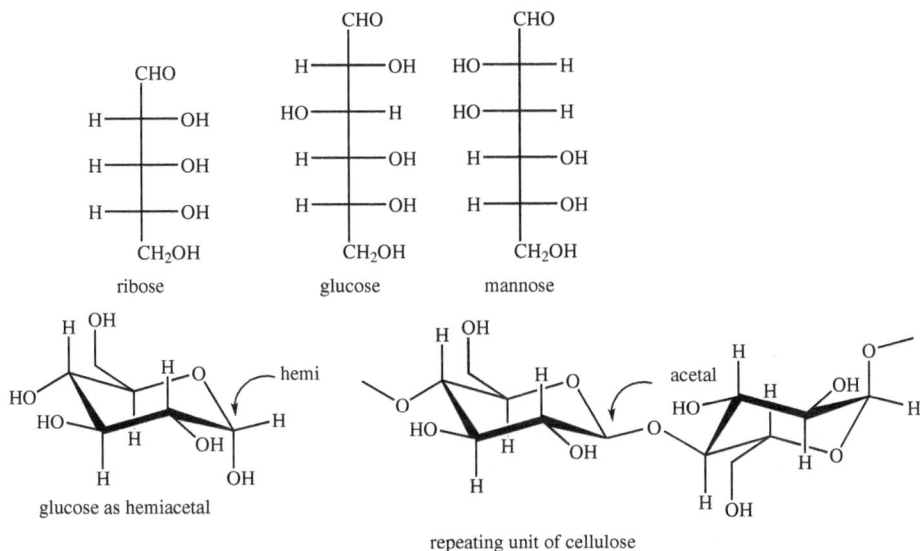

Fig. 7.19

7.12.5 Other Nucleophilic Addition Reactions

The principles developed in the foregoing sections can be generally applied to a number of reactions involving other nucleophilic species. However, some differences appear that are due to the different properties of the particular nucleophile used. We will consider some examples of these now.

7.12.5.1 Cyanohydrins

Hydrogen cyanide is a weak acid ($pK_a = 9$), but cyanide ion ($C\equiv N^-$) is a powerful nucleophile. Because it is a weak acid, there is very little cyanide ion present when HCN is dissolved in water. In the un-ionized HCN, there is no nucleophilic species present. (Contrast this with the nucleophilic properties of unionized alcohols.) Therefore, if an aldehyde is treated with HCN, almost no reaction occurs (because of the low concentration of nucleophile). Adding an acid catalyst increases the reactivity of the carbonyl, but reduces even further the concentration of CN^- due to the equilibrium

$$HCN \rightleftharpoons H^+ + CN^-$$

which is shifted to the left by the addition of acid and so acid is ineffective as a catalyst in this case.

Addition of base (OH^-) causes a reaction with the acid HCN to produce more cyanide ion which can then attack the carbonyl in the usual manner. Protonation of the (basic) alkoxide by the acid HCN regenerates the cyanide ion (Fig. 7.20). Since

HCN + OH⁻ ⇌ H₂O + CN⁻

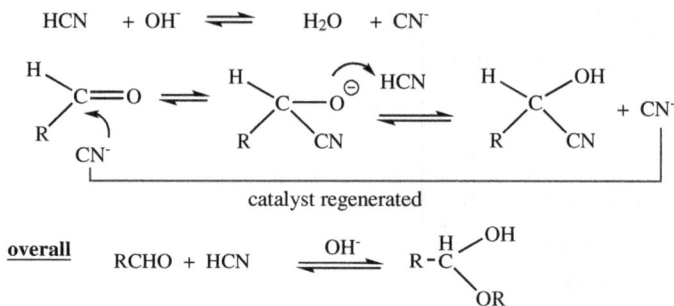

catalyst regenerated

overall   RCHO + HCN ⇌$\xrightarrow{\text{OH}^-}$ R−C⟨H, OH, OR⟩

Fig. 7.20

the pKas of alcohols and HCN differ by about 6, the second step, and therefore the overall reaction, goes essentially to completion. Note that this reaction could also be catalyzed by the addition of a catalytic amount of cyanide ion (CN⁻) since the function of the base is to generate this ion.

Molecules with a hydroxyl and a cyano group attached to the same carbon are called cyanohydrins.

7.12.5.2 Hydrogen Halides and Water

Hydrogen chloride is a strong acid. In a mixture of an aldehyde or ketone with HCl, there is sufficient acid to catalyze the usual addition reaction. Chloride ion (e.g., from NaCl) is not a strong enough nucleophile to attack the unprotonated carbonyl and therefore does not undergo addition.

The same reaction as observed for addition of alcohols can be written using water as the nucleophile. (After all, water is just a special case of an alcohol). The product is called a hydrate. These two types of reactions are grouped together here because, in *both cases*, the K_{eq} for the reactions are very small: i.e., the equilibrium lies far to the left and no appreciable amount of product is formed.

In general, carbon atoms with one hydroxyl and one halogen group *or* two hydroxyl groups are unstable and revert to carbonyl groups.

Certain structural groups can shift these equilibria. For example, trichloroethanal forms a stable hydrate with water. The product is known chemically as chloral hydrate but may be more familiar to aficionados of detective literature as knock-out drops or "Mickey Finn."

7.13 The Grignard Reaction – Introduction

Perhaps the most important nucleophilic addition reaction in a practical sense is known as the Grignard (Grin-yard) reaction. It is a method of forming new carbon–carbon bonds in a very specific way and allows the preparation of larger molecules by the joining of two smaller ones. Up to this point, none of the reactions we have seen, except the nucleophilic reactions of cyanide ion, have resulted in the formation of new C–C bonds. This difficulty will now be remedied.

7.13.1 The Grignard Reagent

Reaction between an alkyl halide and magnesium metal occurs to give a molecule in which the magnesium atom has been inserted into the carbon–halogen bond. The product is an organomagnesium halide and is commonly called a *Grignard reagent*. Since metals are more *electropositive* than carbon, the C-Mg bond is polarized towards carbon as shown in the figure. Notice that the carbon has reversed its polarity in forming the carbon-metal bond. How this reaction occurs will not concern us here; rather the reactions of the Grignard reagent with aldehydes and ketones will be the focus of our interest.

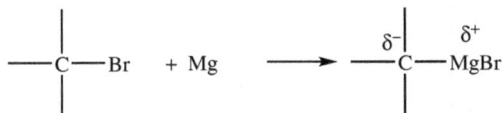

$$\underset{|}{\overset{|}{-}}\text{C}-\text{Br} \quad + \text{ Mg} \quad \longrightarrow \quad \overset{\delta^-}{\underset{|}{\overset{|}{-}}}\text{C}-\overset{\delta^+}{\text{MgBr}}$$

It is worth noting however that the C-Mg bond is *not* ionic, but rather polar-covalent. That this is so can be judged from the fact that Grignard reagents are soluble in ether solution. They are very strong bases and must be prepared with the exclusion of water, alcohols, or any even faintly acidic materials and also in the absence of air as all these materials cause their decomposition. They are normally prepared in ether solution under a nitrogen atmosphere and the solutions are used directly without isolation of the Grignard reagent.

7.13.2 The Grignard Reaction

As pointed out in the foregoing section, Grignard reagents (RMgX) are not ionic. However, it is convenient to write them as R^-MgX^+ to be able to illustrate their reactions more clearly. The negative carbon (a carbanion), which is a powerful nucleophile, will attack the positively polarized carbon of an aldehyde or ketone in the usual manner. (Note that, since the pKa of an alkane is about 40, the removal of a proton from an alkane by a base is not a feasible process. The generation of carbanions by the ploy of using magnesium and an alkyl halide circumvents this difficulty.) The product of the attack on a carbonyl group is the halomagnesium salt of an *alcohol* (Fig. 7.21). This reaction is strongly exothermic and therefore is irreversible (cf. Sect. 4.4). The alkoxide precipitates from the ether solvent as a gray sludge. Addition of an acid (e.g., HCl) causes protonation of the basic alkoxide forming an *alcohol* and a magnesium salt which is water soluble and easily separated.

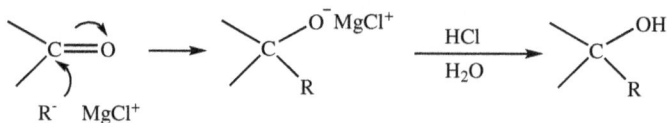

Fig. 7.21

The overall result of the Grignard reaction is the formation of a new C–C bond to the carbonyl carbon of an aldehyde or ketone with the formation of an alcohol. Aldehydes (except formaldehyde) give secondary alcohols whereas ketones give tertiary alcohols.

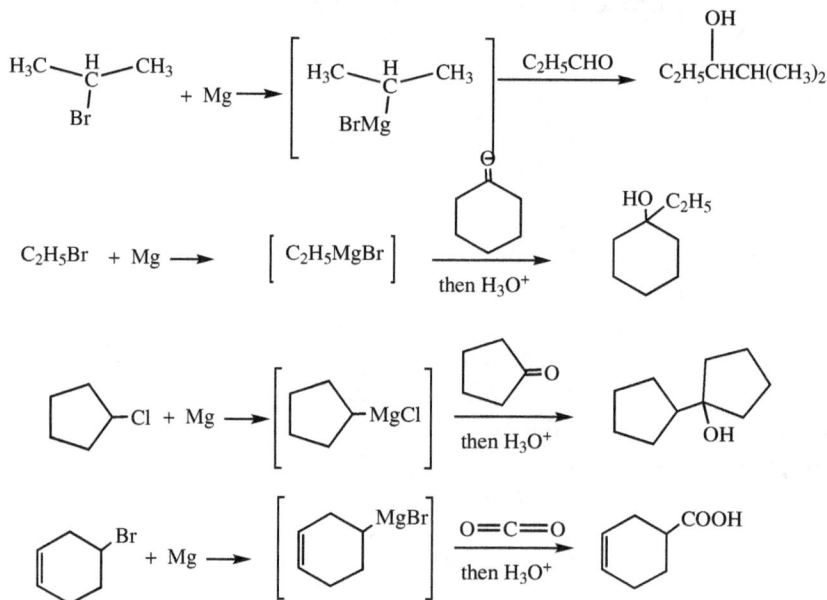

Fig. 7.22

Some specific examples of the Grignard reaction are shown in Fig. 7.22. The last equation in Fig 7.22 illustrates one reason why air must be excluded from the reaction. Carbon dioxide will react rapidly with Grignard reagents to form a class of compounds containing the functional group CO_2H. These are called carboxylic acids and are the subject of Chapter 8.

The formal consideration of the carbon part of a Grignard reagent as a carbanion is a gross oversimplification. That this is so can easily be seen from the fact that Grignards do not react well with alkyl halides which, as was pointed out in Chapter 6, are quickly displaced by strong nucleophiles. The elevated reactivity of Grignard reagents with carbonyls can best be accommodated by a simultaneous coordination of the Mg atom (electrophile) with the carbonyl oxygen atom. This essentially holds the Grignard reagent *in place* to allow the negative end of the reagent to react with the carbon of the carbonyl. No such orienting effect is possible with alkyl halides.

Examination of the reactants and products in Fig. 7.22 shows that the new C–C bond is always one of the bonds attached to the C–OH group. Therefore, if a question such as "How could you prepare 2-butanol using the Grignard reaction?" was asked, you should approach this problem in the following way (Fig. 7.23). 2-Butanol has *two* C–C bonds connected to the carbinol carbon atom which is marked by a #. Either one of these can be formed by the Grignard reaction. Disconnecting bond *a* leads to ethanal and ethylmagnesium bromide while disconnecting bond *b* leads to propanal and methylmagnesium bromide. Either of these methods works equally well.

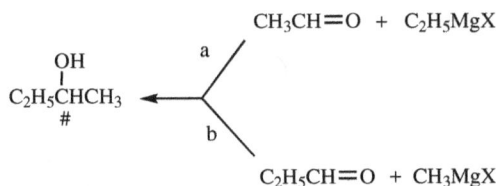

$$CH_3CH{=}O \ + \ C_2H_5MgX$$

$$\begin{array}{c} OH \\ | \\ C_2H_5CHCH_3 \\ \# \end{array} \quad \overset{a}{\underset{b}{\diagdown}}$$

$$C_2H_5CH{=}O \ + \ CH_3MgX$$

Fig. 7.23

Using the Grignard reaction, tertiary alcohols can be formed in three ways, secondary alcohols in two ways, and primary alcohols in only one way (using $H_2C{=}O$).

Therefore, to start a problem such as the one above, the steps should be:

1. Locate the carbinol carbon atom.
2. Identify the alcohol as being primary, secondary, or tertiary.
3. Sequentially disconnect all the bonds between this carbinol carbon atom and any other carbon atoms to determine the fragments required.

Some problems on which you can practice are included at the end of the chapter.

7.13.3 Acetylide Ions as Grignard Reagents

Terminal alkynes can ionize to form a proton and a negatively charged carbon ion called an acetylide ion. (Recall that the old name for alkynes is acetylenes). The equilibrium constant for this ionization is very small (ca. 10^{-35}). Therefore, in order to get an appreciable concentration of the acetylide ion, a very strong base is required. Sodamide (NaNH$_2$ – the sodium salt of ammonia) is frequently used. The acetylide ion is a powerful nucleophile and will react with alkyl halides in substitution reactions and with aldehydes and ketones to give products with the C(OH)C≡CR group. Both of these can be useful in organic synthesis.

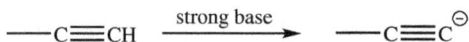

$$-\!\!-C\!\!\equiv\!\!CH \quad \xrightarrow{\text{strong base}} \quad -\!\!-C\!\!\equiv\!\!C^{\ominus}$$

7.14 Reductions with Metal Hydrides

Compounds with hydrogen directly bonded to a metal are known as metal hydrides. The hydride ion (H$^-$) is not stable unless it is bonded to an electropositive atom like a metal. Two common reagents in organic chemistry that have metal-hydrogen bonds are shown below.

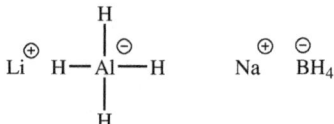

$$\overset{\oplus}{Li} \quad H\!-\!\!\overset{\overset{\displaystyle H}{|}}{\underset{\underset{\displaystyle H}{|}}{Al}}\!\overset{\ominus}{-}\!H \qquad \overset{\oplus}{Na} \quad \overset{\ominus}{BH_4}$$

The analogy between these compounds and Grignard reagents is strong. In Grignard reagents, the negative carbon attacks the carbonyl of an aldehyde or ketone to give alcohols. Sodium borohydride or LAH do the same thing except that hydrogen becomes attached (Fig. 7.24). The difference between them is that, because hydrogen is attached and not carbon, aldehydes give *primary* alcohols and ketones give *secondary* ones. Notice that the result of hydride addition to aldehydes and ketones is the *same* as using hydrogenation (Sect. 5.11.1). Both are reductions of the carbonyl group (see the definition of oxidation and reduction, Sect. 4.12) and should be noted as preparations of alcohols. However, it must be noted that the metal hydrides do NOT attack carbon-carbon double bonds and so, when reduction of a carbonyl group in the presence of an alkene is required, LAH borborohydride are the reagents of choice.

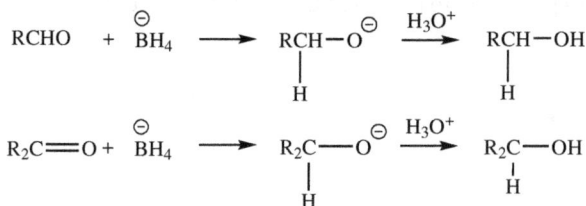

$$RCHO \;+\; \overset{\ominus}{BH_4} \quad\longrightarrow\quad \underset{\underset{\displaystyle H}{|}}{RCH\!-\!O^{\ominus}} \quad\xrightarrow{H_3O^+}\quad \underset{\underset{\displaystyle H}{|}}{RCH\!-\!OH}$$

$$R_2C\!\!=\!\!O \;+\; \overset{\ominus}{BH_4} \quad\longrightarrow\quad \underset{\underset{\displaystyle H}{|}}{R_2C\!-\!O^{\ominus}} \quad\xrightarrow{H_3O^+}\quad \underset{\underset{\displaystyle H}{|}}{R_2C\!-\!OH}$$

Fig. 7.24

7.15 Other Reactions of Aldehydes and Ketones

7.15.1 Oxidation

Aldehydes are easily oxidized, a process that gives carboxylic acids. Conversely, ketones are quite stable to these kinds of conditions. Any number of oxidants will accomplish the oxidation of aldehydes (e.g., CrO_3 or $KMnO_4$). However, these reagents will also oxidize primary and secondary alcohols. Any oxidant which changes color when it oxidizes a functional group in an organic molecule will indicate the presence of an oxidizable group. Such reactions as Cr^{+6}(orange) $\longrightarrow$ Cr^{+3}(green) or MnO_4^-(purple) $\longrightarrow$ Mn^{+2}(colorless) all indicate the presence of a group like -CHOH, C=C, C=O because all of these are easily oxidized. Since ketones are *not* easily oxidized, such color changes can be used to distinguish between an aldehyde and a ketone. When molecules which have *more than one* oxidizable group present (e.g., CH=O AND CH–OH or ketone + CH–OH) must be distinguished, simple treatment with, for example, CrO_3 is not enough since both will give the color change due to oxidation of the OH groups. There are some selective reagents – i.e., those that will oxidize aldehydes but not alcohols or ketones and two of these can be used for a *classification test for aldehydes.*

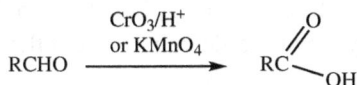

$$RCHO \xrightarrow[\text{or } KMnO_4]{CrO_3/H^+} RC\overset{O}{\underset{OH}{\diagup}}$$

1. The reaction of an alkaline solution of Ag^+ ion with aldehydes leads to the formation of acids and silver metal. The metal is formed as a silver mirror on the walls of the reaction flask or, as a gray precipitate (if the flask is not clean). The appearance of the mirror or precipitate is indicative of the presence of an aldehyde. This test is called the *Tollens' Test.*

2. The same reaction can be performed using an alkaline solution of Cu^+ ion. The presence of the aldehyde is confirmed if a reddish precipitate of Cu^0 is seen. This test is called the *Fehling's Test.*

These tests are frequently used in carbohydrate chemistry to distinguish between what are known as reducing and nonreducing sugars. Reducing sugars are aldehydes and nonreducing sugars are usually ketones (nonreducing since they do not reduce Ag^+ to Ag^0). Note that carbohydrates that have an aldehyde group masked as a hemiacetal will give a positive aldehyde test. The formation of the intramolecular hemiacetal is reversible. Therefore, in water, there is always an equilibrium amount of aldehyde present to react. As it is oxidized, more of the hemiacetal will open to the aldehyde to satisfy the equilibrium constant until, finally, all the aldehyde is oxidized. (What would happen with acetals?)

7.15.2 The Iodoform Test

Ketones, but not aldehydes which contain a methyl group attached to the carbonyl group, undergo a diagnostic reaction with alkaline iodine solutions. The appearance of a dense yellow solid called iodoform (triiodomethane) confirms the presence of a methyl ketone. Secondary alcohols that also have one methyl group substituent also give the reaction because iodine oxidizes them to ketones first (Fig. 7.25).

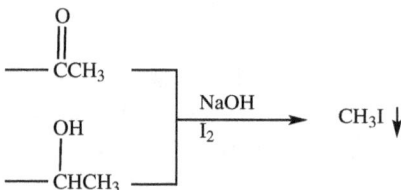

Fig. 7.25

7.16 Introduction to Multistage Syntheses

At this point we have accumulated enough reactions to be able to carry out chemical transformations that involve several steps. For example, how could you carry out the transformation shown in Fig. 7.26?

In such problems, it is always best to *start with the desired product* and identify what *kind* of molecule it is and *whether you have added any carbons* during the synthesis. In this case, the product is an alcohol and one carbon has been added.

Fig. 7.26

A mental (or written) list of the methods you have learned for the preparation of alcohols includes several items (are you keeping your lists up to date?) but only one of these allows the addition of an alkyl group – the Grignard reaction. The given starting material is an aldehyde and therefore will be useful in this type of synthesis. However, note that the product alcohol from a Grignard synthesis from this aldehyde is *not* what is required, but rather an isomer.

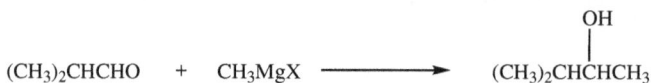

However, we do have the correct number and arrangement of carbon atoms. The problem now changes to how the following can be done.

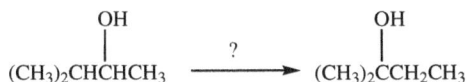

$$\underset{(CH_3)_2CHCHCH_3}{\overset{\overset{\displaystyle OH}{|}}{}} \quad \xrightarrow{\;?\;} \quad \underset{(CH_3)_2CCH_2CH_3}{\overset{\overset{\displaystyle OH}{|}}{}}$$

What we must now look for is a reaction of alcohols that will give a product which will allow the re-introduction of the hydroxyl group in the correct place. One solution to this is to recognize that dehydration will lead to an alkene, which when hydrated (acid-catalyzed addition of water) will lead to an alcohol under the control of Markownikov's Rule. An example is shown in Fig. 7.27. This type of problem is introduced here because it draws together many of the topics we have considered so far and emphasizes the inter-relationships between reactions we have been considering separately. Each chapter from now on will include some problems of this type. It should be noted that there is usually no single *right* answer. Very often, several routes from one material to another are possible and equally *correct*.

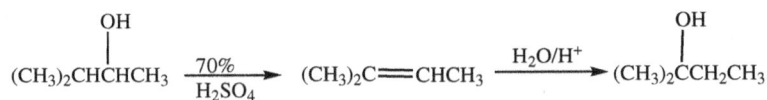

$$\underset{(CH_3)_2CHCHCH_3}{\overset{\overset{\displaystyle OH}{|}}{}} \quad \xrightarrow[H_2SO_4]{70\%} \quad (CH_3)_2C{=}CHCH_3 \quad \xrightarrow{H_2O/H^+} \quad \underset{(CH_3)_2CCH_2CH_3}{\overset{\overset{\displaystyle OH}{|}}{}}$$

Fig. 7.27

7.17 Problems

7-1. How could you distinguish between samples of the following pairs of compounds? Describe what you would do and what you would observe in each case.
(a) cyclohexanol and cyclohexene
(b) 2-methyl-2-butanol and 3-methyl-2-butanol
(c) 2,3-dimethyl-1-hexanol and 3-methyl-2-heptanol
(d) 3-hexen-2-ol and 5-hexen-2-ol

7-2. The use of sodium hydroxide as a catalyst in the following reaction does not lead to a successful reaction. Why?

$$CH_3CH_2CH_2OH \;+\; \underset{\overset{\displaystyle |}{Br}}{CH_3CHCH_3} \quad \xrightarrow{\hspace{2cm}} \quad CH_3CH_2CH_2OCH(CH_3)_2$$

7-3. Which of the indicated pair of reagents would you choose to best effect the desired reaction? Why?

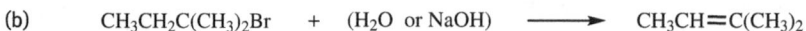

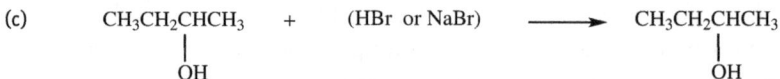

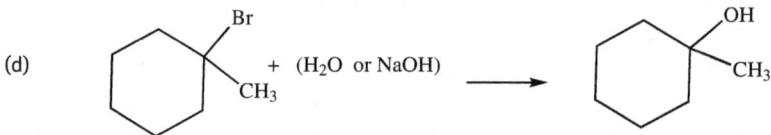

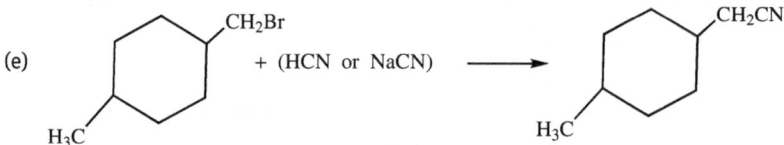

(a) $CH_3CH_2CH_2Cl$ + $(NH_3$ or $NaNH_2)$ ⟶ $CH_3CH_2CH_2Cl$

(b) $CH_3CH_2C(CH_3)_2Br$ + $(H_2O$ or $NaOH)$ ⟶ $CH_3CH=C(CH_3)_2$

(c) $CH_3CH_2CHCH_3$ + $(HBr$ or $NaBr)$ ⟶ $CH_3CH_2CHCH_3$
$\quad\quad\quad$ |
$\quad\quad\quad$ OH $\quad\quad\quad\quad\quad\quad\quad\quad\quad\quad\quad\quad\quad\quad\quad\quad\quad$ OH

(d) [cyclohexane with Br and CH₃] + $(H_2O$ or $NaOH)$ ⟶ [cyclohexane with OH and CH₃]

(e) [cyclohexane with CH₂Br and H₃C] + $(HCN$ or $NaCN)$ ⟶ [cyclohexane with CH₂CN and H₃C]

(f) $CH_3OCH(CH_3)_2$ + $(HCl$ or $NaCl)$ ⟶ CH_3OH + $(CH_3)_2CHCl$

7-4. When compound **I** is treated with an acid catalyst and water in which the oxygen atom is the isotope ^{18}O, the products are two alcohols, *one* of which has an ^{18}O atom incorporated. Write the product structures and indicate which product has the ^{18}O atom.

$$I = (CH_3)_3C{-}O{-}CH_3$$

7-5. When optically active I (below) is treated with aqueous acid, II is obtained. Would you expect alcohol II to be optically active? Explain your reasoning.

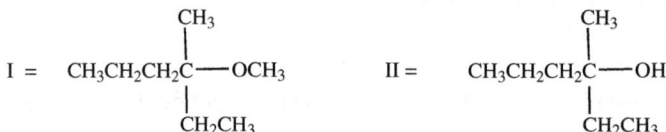

$\quad\quad\quad\quad\quad CH_3 \quad\quad\quad\quad\quad\quad\quad\quad\quad\quad\quad CH_3$
$\quad\quad\quad\quad\quad |\quad\quad\quad\quad\quad\quad\quad\quad\quad\quad\quad\quad\quad |$
I = $\quad CH_3CH_2CH_2C{-}OCH_3 \quad\quad$ II = $\quad CH_3CH_2CH_2C{-}OH$
$\quad\quad\quad\quad\quad |\quad\quad\quad\quad\quad\quad\quad\quad\quad\quad\quad\quad\quad |$
$\quad\quad\quad\quad\quad CH_2CH_3 \quad\quad\quad\quad\quad\quad\quad\quad\quad\quad CH_2CH_3$

7-6. Replace the "?" with the correct reagent or product. If a catalyst is necessary, indicate its nature. Show stereochemistry if this is important.

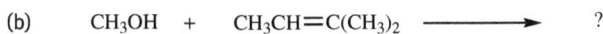

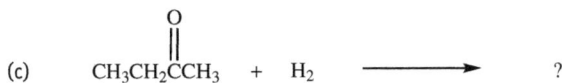

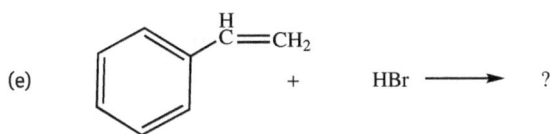

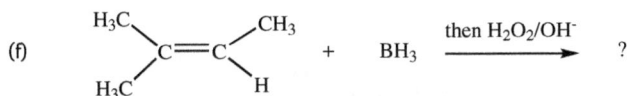

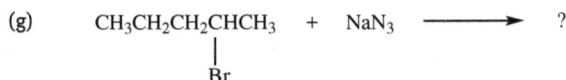

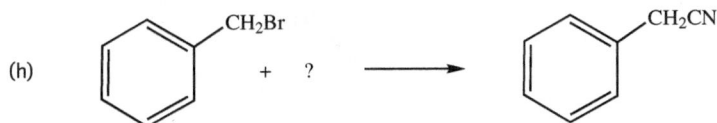

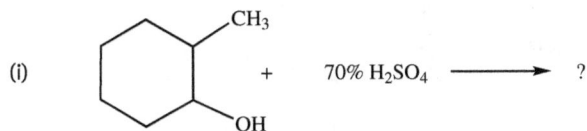

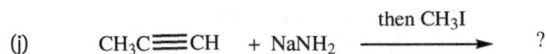

(a)
$$\underset{\text{Br}}{CH_3CH_2\overset{|}{C}(CH_3)_2} \quad + \quad ? \quad \longrightarrow \quad \underset{\text{OH}}{CH_3CH_2\overset{|}{C}(CH_3)_2}$$

(b) $CH_3OH \quad + \quad CH_3CH{=}C(CH_3)_2 \quad \longrightarrow \quad ?$

(c) $CH_3CH_2\overset{\overset{\displaystyle O}{\|}}{C}CH_3 \quad + \quad H_2 \quad \longrightarrow \quad ?$

(d) $CH_3CH_2CH_2CH_2OH \quad + \quad ? \quad \longrightarrow \quad CH_3CH_2CH_2CH_2Cl$

(e)

$+ \quad HBr \quad \longrightarrow \quad ?$

(f)

$+ \quad BH_3 \quad \xrightarrow{\text{then } H_2O_2/OH^-} \quad ?$

(g) $\underset{\overset{|}{Br}}{CH_3CH_2CH_2\overset{|}{C}HCH_3} \quad + \quad NaN_3 \quad \longrightarrow \quad ?$

(h)

$+ \quad ? \quad \longrightarrow$

(i)

$+ \quad 70\% \ H_2SO_4 \quad \longrightarrow \quad ?$

(j) $CH_3C{\equiv}CH \quad + \quad NaNH_2 \quad \xrightarrow{\text{then } CH_3I} \quad ?$

7-7.

(a) A student who did not have the benefit of taking Chemistry 230 was asked to plan a route from I to II. The answer given is shown below. Each step contains an error. Identify these and show what the expected product would be from each step.

Projected Synthesis

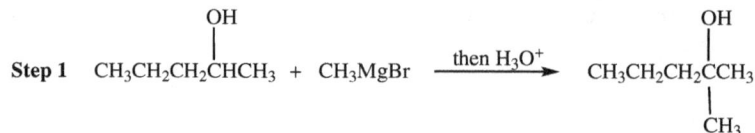

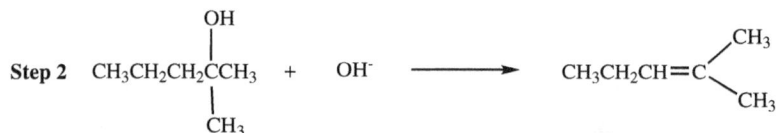

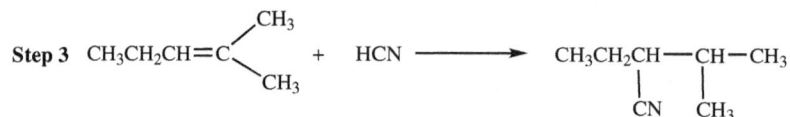

Step 1 CH$_3$CH$_2$CH$_2$CHCH$_3$ (with OH) + CH$_3$MgBr $\xrightarrow{\text{then } H_3O^+}$ CH$_3$CH$_2$CH$_2$CCH$_3$ (with OH and CH$_3$)

Step 2 CH$_3$CH$_2$CH$_2$CCH$_3$ (with OH and CH$_3$) + OH$^-$ $\longrightarrow$ CH$_3$CH$_2$CH=C(CH$_3$)(CH$_3$)

Step 3 CH$_3$CH$_2$CH=C(CH$_3$)(CH$_3$) + HCN $\longrightarrow$ CH$_3$CH$_2$CH—CH—CH$_3$ (with CN and CH$_3$)

(b) Correct the above synthesis so that product II could be obtained.

7-8. Give the IUPAC names, including any stereochemical descriptors, for each of the following.

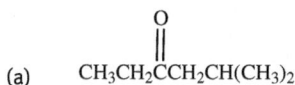

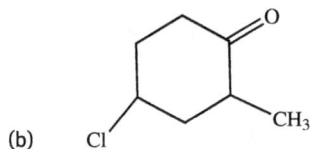

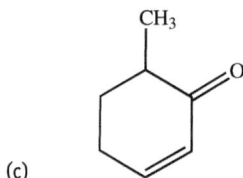

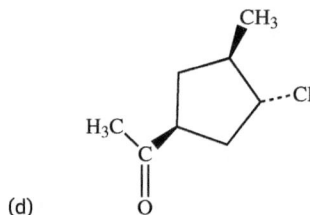

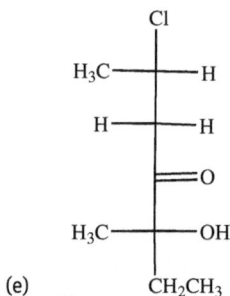

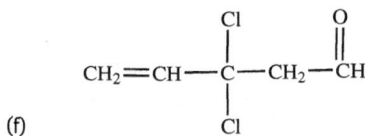

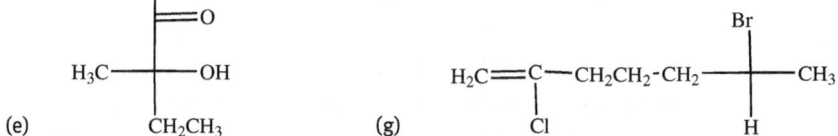

(a)   CH$_3$CH$_2$CCH$_2$CH(CH$_3$)$_2$ (with =O)

(b)

(c)

(d)

(e)

(f)   CH$_2$=CH—C—CH$_2$—CH (with CH$_3$, Cl, Cl, =O)

(g)   H$_2$C=C—CH$_2$CH$_2$-CH$_2$—CH$_3$ (with Cl, Br, H)

7-9. Draw the complete mechanism for the acid-catalyzed reaction of 3-methylbutanal with excess methanol. Make sure you show which steps are reversible.

7-10. Replace the "?" with the correct reagent or product. Show any pertinent stereo-chemical details and any required catalysts.

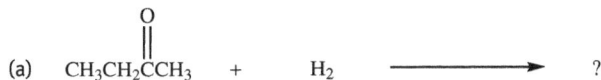

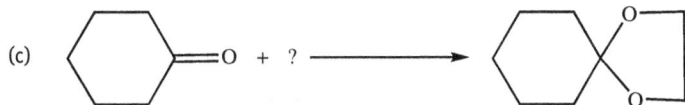

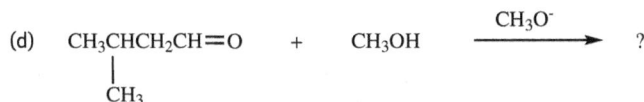

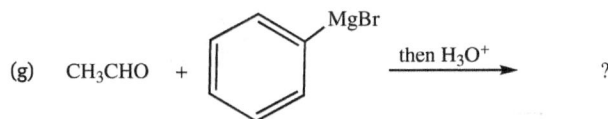

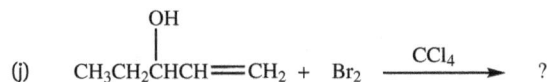

(a) $CH_3CH_2CCH_3$ (with O double bonded) + H_2 $\longrightarrow$?

(b) $CH_3CH_2CH{=}O$ + ? $\longrightarrow$ $CH_3CH_2C{-}{-}OH$ (with CH_3 up and CN down)

(c) =O + ? $\longrightarrow$

(d) $CH_3CHCH_2CH{=}O$ (with CH_3 below first carbon) + CH_3OH $\xrightarrow{CH_3O^-}$?

(e) $CH_3CH_2C{\equiv}CCH_2CH_3$ + H_2O $\longrightarrow$?

(f) $CH_3CH_2CHCH_3$ (with OH below) + ? $\longrightarrow$ $CH_3CH_2CCH_3$ (with O double bonded)

(g) CH_3CHO + (phenyl MgBr) $\xrightarrow{\text{then } H_3O^+}$?

(h) + ? $\longrightarrow$

(i) $BrCH_2CH_2CH_2CCH_3$ (with O double bonded) + CH_3O^- $\longrightarrow$?

(j) $CH_3CH_2CHCH{=}CH_2$ (with OH above third carbon) + Br_2 $\xrightarrow{CCl_4}$?

(k) $CH_3CH_2CH_2\overset{\overset{O}{\|}}{C}CH_2CH_3$ + $NaBH_4$ ⟶ ?

(l)   [structure: cyclopentane with CHO group] + ? ⟶ [structure: cyclohexane-CH(OH)-cyclopentane, with H and OH on the central carbon]

(m) $CH_3CHCH_2CH_2CHOCH_3$ + H_2O $\xrightarrow{H^+}$?
      with OCH_3 and OCH_3 substituents

(n)   [structure: (CH3)2C=C(CH3)CH3] + ? ⟶ [structure: (CH3)2C=O]

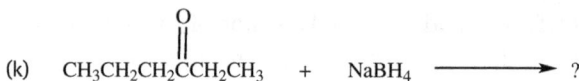

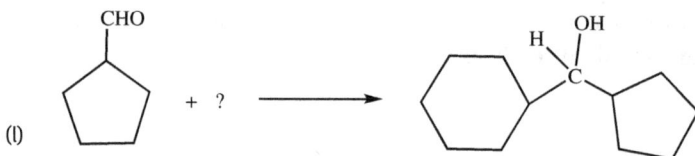

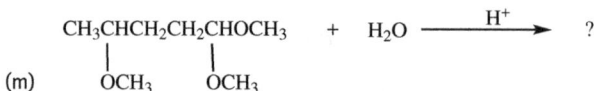

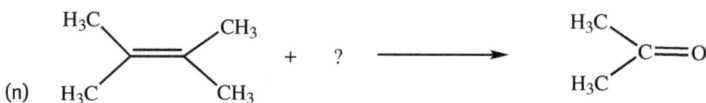

7-11. If the starting material in 7.3(h) is optically active, how many products are possible?

7-12. Suggest how you could distinguish between the following pairs of compounds. Describe what you would do and what would be observed in each case.
(a) 2-butanone and 2-butanol
(b) cyclohexanone and cyclopentanecarboxaldehyde
(c) 2-hexanol and 3-hexanol
(d) 5-hexenal and cyclohexanone

7-13. Which of the following reactions would give a product which, *as isolated*, would be optically active?

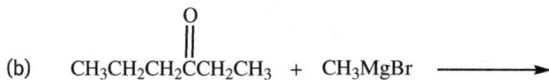

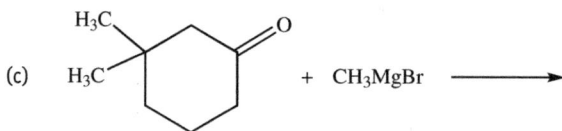

(a) CH_3CH_2CHO + CH_3MgBr ⟶

(b) $CH_3CH_2CH_2\overset{\overset{O}{\|}}{C}CH_2CH_3$ + CH_3MgBr ⟶

(c)   [structure: 4,4-dimethylcyclohexanone] + CH_3MgBr ⟶

(d) $CH_3\overset{\displaystyle O}{\overset{\displaystyle \|}{C}}CH_3$ +   [cyclopentane with two H_3C groups]—$MgBr$ ———→

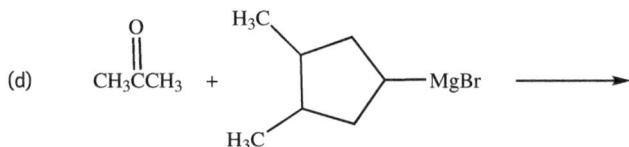

7-14. If cyclohexanone in which the oxygen atom is the isotope ^{18}O is reacted with excess ethanol under acidic conditions, will the organic product contain ^{18}O? Explain your reasoning referring to the mechanism of the reaction.

7-15. Show all possible combinations of reactants that could be used in Grignard syntheses of the following compounds.

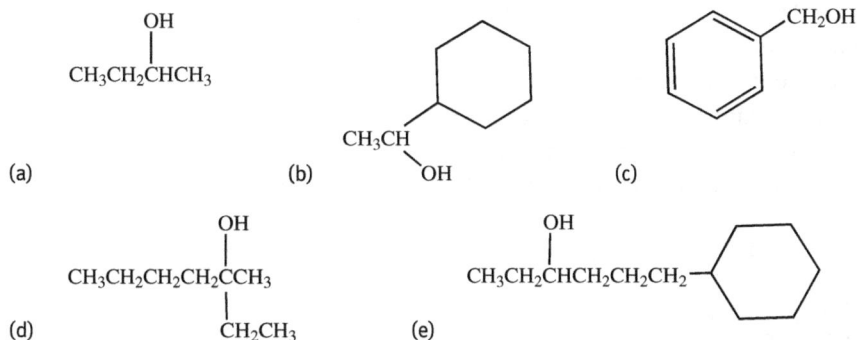

(a) $\overset{\displaystyle OH}{\overset{\displaystyle |}{CH_3CH_2CHCH_3}}$

(b) $\overset{\displaystyle CH_3CH}{\diagdown}$ OH [on cyclohexane]

(c)   [benzene ring]—CH_2OH

(d) $\overset{\displaystyle OH}{\overset{\displaystyle |}{CH_3CH_2CH_2CH_2\overset{\displaystyle |}{\underset{\displaystyle CH_2CH_3}{C}}CH_3}}$

(e) $\overset{\displaystyle OH}{\overset{\displaystyle |}{CH_3CH_2CHCH_2CH_2CH_2}}$—[cyclohexane]

7-16. Indicate which of the pairs of reagents you used in 7.8 would give optically active products.

7-17. Suggest using equations four ways each of the following might be prepared, each of which uses a *different starting material*. (Changing Br for Cl does not make a new starting material!!)

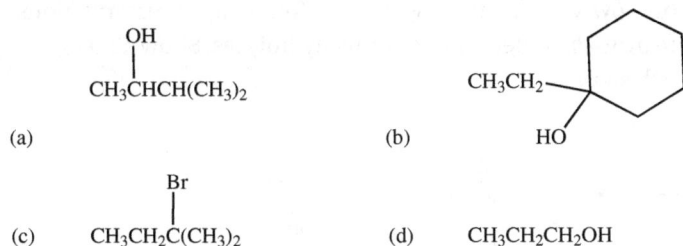

(a) $\overset{\displaystyle OH}{\overset{\displaystyle |}{CH_3CHCH(CH_3)_2}}$

(b) CH_3CH_2—[cyclohexane]—HO

(c) $\overset{\displaystyle Br}{\overset{\displaystyle |}{CH_3CH_2C(CH_3)_2}}$

(d) $CH_3CH_2CH_2OH$

7-18. Glucose can exist in two optically active hemiacetal forms called the *alpha* and the *beta* isomers (I and II). When I is dissolved in water and a catalytic amount of acid is added, the observed rotation falls until a steady reading is reached.
(a) What is the relationship between I and II?
(b) Write a mechanism for this acid-catalyzed reaction.
(c) What can be said about the relative sizes of the specific rotations of I and II?

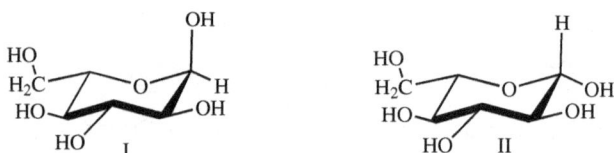

7-19. When II (question 7.18) is dissolved in excess acidic methanol, a new compound (III) can be isolated.
(a) Why did only one OH group react?
(b) Write a mechanism for this reaction.
(c) If the methanol used had ^{18}O in its structure, would III have this isotope present?
(d) If I (question 7.11) was used, what would be the structure of the product?

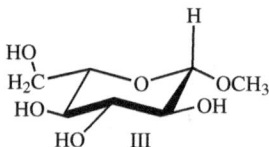

7-20. Only one of the following will react as an *acid* when treated with methoxide ion. Which is it and why?

$$CH_3CH_2CH_2Br \quad CH_3CH_2CH_3 \quad CH_3CH_2CHF_2$$

7-21. Suggest by equation how you might carry out the following transformations, each of which *may* require up to *three* steps not counting hydrolyses. Show all reagents and the products from each step.

(a) $CH_3CHO \quad \xrightarrow{?} \quad$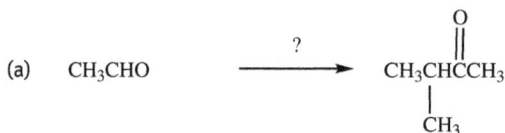

(b)  $\quad \xrightarrow{?} \quad$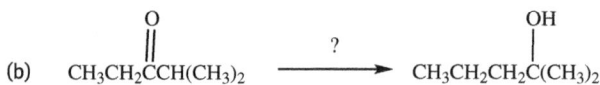

(c)

(d)

$$\underset{\underset{OCH_3}{|}}{\overset{\overset{OCH_3}{|}}{CH_3CCH(CH_3)_2}} \xrightarrow{?} CH_3CH=C(CH_3)_2$$

(e) $CH_3CH_2CH_2CH_2OH \xrightarrow{?} \underset{\overset{|}{OH}}{CH_3CH_2CHCH_3}$

(f) $(CH_3)_2CHCH_2CH_2OH \xrightarrow{?} (CH_3)_2CHCHO$

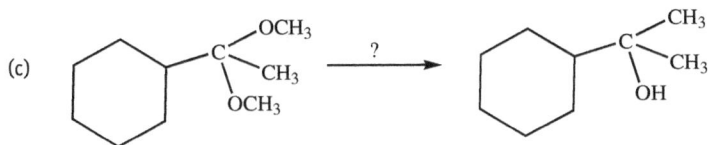

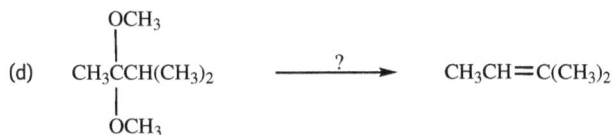

7-22. When methylmagnesium bromide reacts with the molecules shown below, how many stereoisomers will be formed and, for each compound, if more than one is formed, what is the relationship between them? In each case, is the reaction product, as formed, optically active?

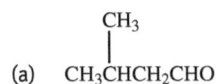

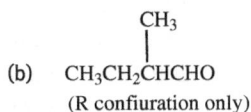

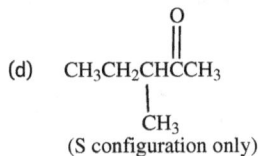

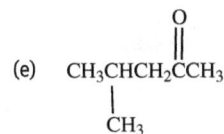

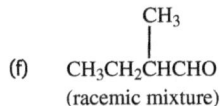

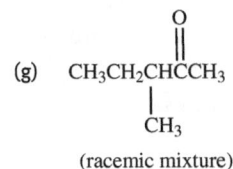

(a) $\underset{\overset{|}{CH_3}}{CH_3CHCH_2CHO}$

(b) $\underset{\overset{|}{CH_3}}{CH_3CH_2CHCHO}$
(R confiuration only)

(c) $CH_3CH_2CH_2\overset{\overset{O}{\|}}{C}CH_3$

(d) $\underset{\overset{|}{CH_3}}{CH_3CH_2CH\overset{\overset{O}{\|}}{C}CH_3}$
(S configuration only)

(e) $\underset{\overset{|}{CH_3}}{CH_3CHCH_2\overset{\overset{O}{\|}}{C}CH_3}$

(f) $\underset{\overset{|}{CH_3}}{CH_3CH_2CHCHO}$
(racemic mixture)

(g) $\underset{\overset{|}{CH_3}}{CH_3CH_2CH\overset{\overset{O}{\|}}{C}CH_3}$
(racemic mixture)

8 More Carbonyl Chemistry – Acids and Their Derivatives

8.1 Introduction

In Chapter 7, we encountered the chemistry of compounds that contained the carbonyl group attached to carbon and/or hydrogen. In this chapter, another set of functional groups will be considered that also contain a carbonyl group, but in this case, attached to one carbon and one *heteroatom* (i.e., an oxygen, nitrogen, halogen, etc.). The chemistry of these is governed by the principles learned in Chapter 7 with a few important additions. The molecular fragment RC=O is called an *acyl* group (compare with the definition of the term alkyl – Sect. 5.2.2).

8.2 Carboxylic Acids, Esters, Chlorides, Anhydrides, Amides (and Nitriles)

The functional groups to be considered here have the general formula RC(=O)X. Included also is the nitrile (−C≡N) group because its chemistry is very similar to the others. Depending on the nature of X, they are considered to be different functional groups. (Table 8.1). As usual, the nomenclature of molecules containing these functional groups will be covered in the nomenclature text. In this chapter we will start by considering the parent type of this class of functional groups – i.e., carboxylic acids – and then see how our ideas must be modified for others of the class.

8.3 Physical and Chemical Properties of Carboxylic Acids

As their name implies, carboxylic acids are acidic: i.e., they undergo ionization as

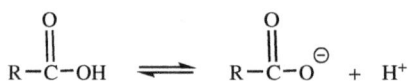

$$\underset{\text{R-C-OH}}{\overset{O}{\|}} \quad \rightleftharpoons \quad \underset{\text{R-C-O}^{\ominus}}{\overset{O}{\|}} \quad + \quad H^+$$

In the absence of special effects, the pK_as of carboxylic acids lie in the general range of 4–6. They are significantly weaker acids than mineral acids like sulfuric or hydrochloric acid. Nevertheless, they are much stronger acids than water or alcohols whose pKas are ca. 17–18. (Note what this implies about the leaving group abilities of Cl^-, RCO_2^-, and RO^-!).

https://doi.org/10.1515/9783110778311-008

The question arises as to why carboxylic acids should be more acidic than alcohols. This question was fully discussed in Sect. 4.10 and you should re-read that section at this point.

A principle that should be familiar to you from previous chemistry courses is "like dissolves like": that is highly polar molecules (and particularly ionic materials) dissolve best in polar solvents (like water), whereas nonpolar molecules (covalent organic molecules) dissolve very well in nonpolar solvents like ether. Conversely, ionic materials are essentially insoluble in nonpolar solvents, and organic materials are insoluble in solvents like water. The alkali-metal salts of acids are generally ionic and therefore quite water soluble and ether insoluble. Conversely, heavier metal (even Ca^{+2}) salts are much more insoluble in water (see Sect. 8.6.1). The un-ionized carboxylic acids are insoluble in water above about C5 (when their hydrocarbon bulk overcomes the ability of the carboxyl group to hydrogen bond to water and make them soluble). (Can you suggest a practical method for the separation of an acid from a neutral compound like an alcohol?) The acids have high boiling points for their molecular weights. Look again at Table 7.1. For comparison, ethanoic acid has a molecular weight of 60, but a boiling point of 101 °C. They can form hydrogen bonds both as donors and acceptors and even in the vapor state, exist largely as dimers. Many of them have very pungent odors. For example, ethanoic acid (commonly called acetic acid) is responsible for the odor of vinegar and butanoic acid is the principle malodorous compound in rancid fat.

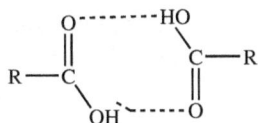

Tab. 8.1: Acids and their Derivatives [R(C=O)X].

X	class name
OH	carboxylic acids
Cl	(carboxylic) acid chlorides or *acyl* chlorides
–O–C–	(carboxylic) esters
–NR$_2$	amides
–O–C–R anhydrides	
R–C≡N	nitrile

8.4 Preparations of Carboxylic Acids

To date, we have seen the following methods of preparation of acids (Fig. 8.1). In this chapter, we will see several others at various points.

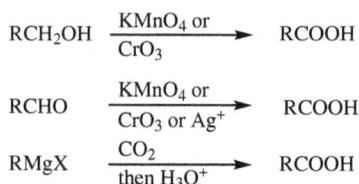

$$RCH_2OH \xrightarrow[\text{CrO}_3]{\text{KMnO}_4 \text{ or}} RCOOH$$

$$RCHO \xrightarrow[\text{CrO}_3 \text{ or } Ag^+]{\text{KMnO}_4 \text{ or}} RCOOH$$

$$RMgX \xrightarrow[\text{then } H_3O^+]{CO_2} RCOOH$$

Fig. 8.1

8.5 Reactions of Carboxylic Acids – Esterification

The carboxylic acid group contains a carbonyl and one might expect that the mechanisms we learned in Chapter 7 might apply. In some ways this is true but there are some complicating factors. Recall that all nucleophiles are also potentially *basic*. Since carboxylic acids are acidic, the use of sufficiently basic nucleophiles will lead to the formation of the carboxylate ion (Fig. 8.2). This ion, being negatively charged will resist attack by nucleophiles and is therefore unreactive. In fact, it is itself nucleophilic (see Table 6.2). For this reason, *base-catalyzed reactions of acids with nucleophiles are not feasible.*

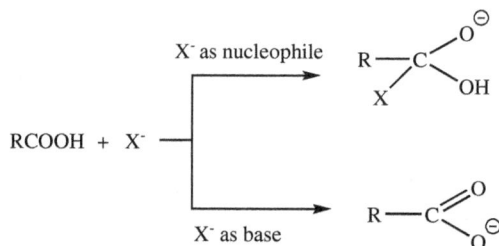

Fig. 8.2

Conversely, acids can be protonated on the carbonyl oxygen by strong (mineral) acids (Fig. 8.3). The protonated form has increased positive character on the carbonyl carbon and the attack of nucleophiles can occur just as it did with aldehydes and ketones. This is shown in Fig. 8.3 using an alcohol as the nucleophilic species. The intermediate

A is exactly analogous to the corresponding one using aldehydes and ketones (Fig. 7.5) and intermediate B is comparable to a hemiacetal or hemiketal.

As in the acid-catalyzed formation of acetals, further reaction of B is possible. C and D are formed by the same processes as found in the formation of acetals (Fig. 7.13). However, a comparison of D with the analogous intermediate in Fig. 7.13 indicates a significant difference (Fig. 8.4). Both have intermediates that are positively charged. To achieve neutrality, either something negative must be *added* or something positive must be removed.

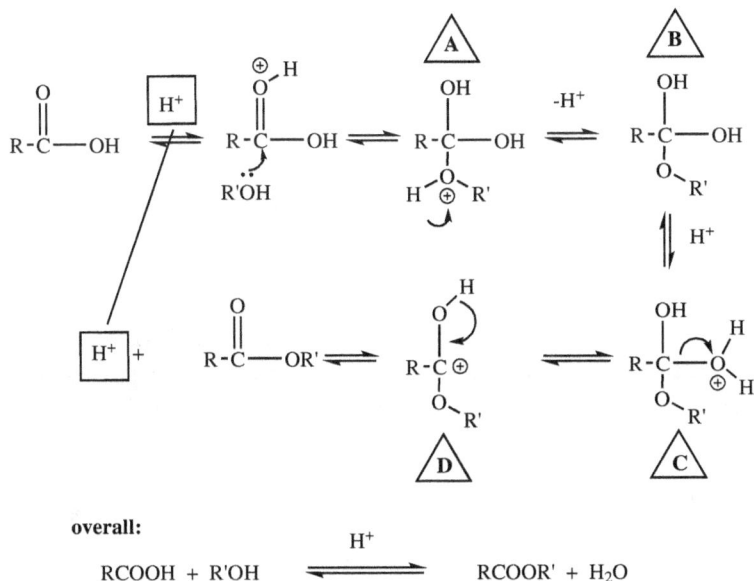

overall:

$$RCOOH + R'OH \;\underset{}{\overset{H^+}{\rightleftharpoons}}\; RCOOR' + H_2O$$

Fig. 8.3

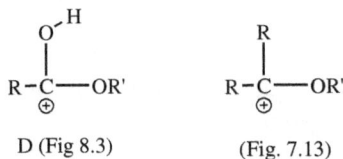

D (Fig 8.3)   (Fig. 7.13)

Fig. 8.4

The intermediate in the acetal formation has no choice but to neutralize its positive charge by reaction with a nucleophile since no stable, positively charged species can be lost. However, in D (Fig. 8.3), loss of a proton is a very facile reaction. This occurs to the exclusion of other reactions and the overall result is a *substitution* of the OH group of the acid by the nucleophile.

Reactions of carboxylic acids generally occur under acid-catalyzed conditions, require neutral nucleophiles, and result in *substitution*.

The overall result of the mechanism shown in Fig. 8.3 is the formation of an *ester* (Sect. 8.2). This process is often called a *Fisher esterification*. The whole process is *reversible* and Le Chatlier's principle applies. By control of the concentrations of water and alcohol, acids may be *esterified* or esters may be *hydrolyzed* under acid catalysis. In the hydrolysis reaction, the nucleophile is water (Fig. 8.3).

The intermediate B in Fig. 8.3 is frequently referred to as the "tetrahedral intermediate." Both the starting material and the product of the reaction contain a sp^2-hybridized carbonyl carbon, but the intermediate has this carbon in the sp^3 state and this difference is denoted by the name.

An often-cited dictum in general chemistry states that "An acid plus a base gives a salt plus water." The organic chemist's version is "An acid plus an alcohol gives an ester plus water." Note that OH^- is a poor leaving group. Therefore, the loss of OH^- from intermediate B in Fig. 8.3 is unlikely and the protonation to form intermediate C first is most likely.

What nucleophiles will work in this process? They must be molecules that are *non-basic but still nucleophilic*. For example, HCN will not work, nor will NH_3 (because it is basic!). Halide ions are not applicable either since they undergo the reverse reaction too readily. In fact, only alcohols and water are really useful.

8.6 Reactions of Esters

Esters differ from acids only in the replacement of the acidic hydrogen atom by an alkyl group. This seemingly minor change has a major effect on their reactions. Since esters are not acidic materials, they can react with basic nucleophiles as well as neutral ones. In this regard, their chemistry is similar to that of aldehydes and ketones.

The reaction shown in Fig. 8.3 is completely reversible. The *acid-catalysis* of the reaction of esters with water (or any other nucleophile) is simply that same reaction but where the equilibrium is being pushed in the opposite direction, usually by manipulation of reactant concentrations.

Now consider the reaction of an ester with hydroxide ion (Fig. 8.5). Nucleophilic attack at the carbonyl group generates an intermediate (A) which is analogous to that formed by attack on an aldehyde or ketone. This can decompose in one of two ways: by expulsion of hydroxide ion which regenerates the ester *or* by loss of alkoxide which generates B. Which pathway is more favorable? We know that H_2O is more acidic than an alcohol. Thus, OH^- should be a better leaving group than $C-O^-$. However, the alkoxide does depart. Why this apparent anomaly?

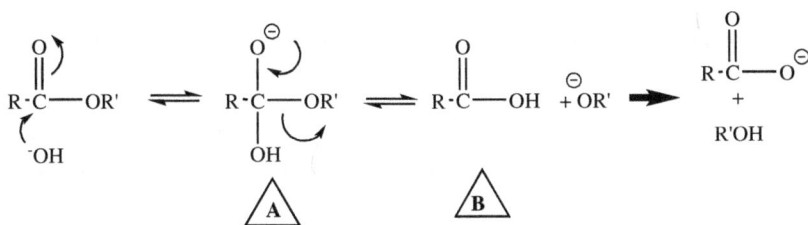

Fig. 8.5

You must realize that both the loss of OH^- and $C-O^-$ are *equilibria*. The fact that OH^- is a better leaving group is reflected in a *larger* K_{eq} than for the loss of alkoxide. However, there will always be a *finite* amount of B present since this $K_{eq} > 0$. Now look at the species that are formed by this loss of alkoxide. One is an *acid* and the other is a *base*. These will always react to form the salt of the acid (Fig. 8.5). Because there is a huge difference in the acidities of acids and alcohols, this reaction has an enormous K_{eq}. The result of this is that the concentration of acid is essentially *zero* at all times. If this is so, the K_{eq} for $A \rightleftharpoons B$ which is of the form

$$K_{eq} = [B]/[A]$$

can never be satisfied. Mother Nature will keep trying to produce a finite amount of B in order to satisfy the K_{eq}, but the acid is continuously being removed. The overall effect is that, even though the K_{eq} for loss of hydroxide is larger, the whole reaction becomes *irreversible* for the loss of alkoxide.

If, after the reaction is complete, aqueous mineral acid is added, the carboxylate ion, being the salt of a weak acid, will be protonated forming the free carboxylic acid. The overall reaction is then a *basic hydrolysis* of an ester to an acid. *It is enormously important that you understand the various equilibria involved in these processes and why they lie where they do!* The competition between an equilibrium and an irreversible sequence is very common in organic chemistry. In such a situation, the irreversible sequence will *always* win.

The scheme shown in Fig. 8.5 also shows again why base-catalyzed reactions of acids is unproductive. This, if it occurred, would be the reverse of the sequence shown below. Since the first step is then an acid-base reaction, again the concentration of acid approaches zero and the $A \rightleftharpoons B$ equilibrium will not generate any A.

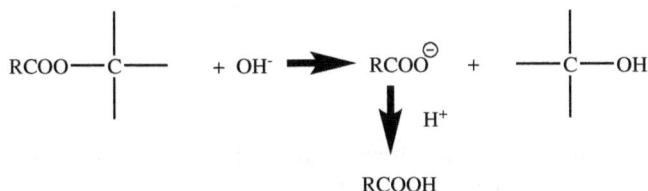

Esters may be converted into acids (hydrolyzed) under either acidic (Fig. 8.3) or basic (Fig. 8.5) conditions, but acids may be converted into esters (esterified) *only* under acidic conditions.

Since esters are not acidic, they can be attacked by basic nucleophiles. Figure 8.5 can be rewritten as in Fig. 8.6. If Nu has no further acidic hydrogens, no irreversible step can occur and the position of the overall equilibrium will be governed partially by the relative nucleophilicities and leaving group abilities of Nu⁻ and C–O⁻ and also, very importantly, by the relative *concentrations* of Nu⁻ and alkoxide.

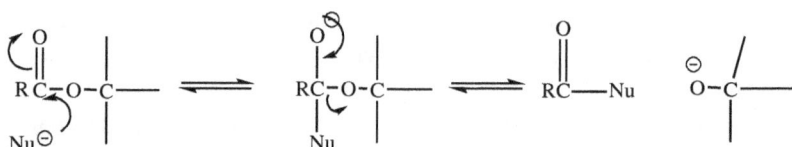

Fig. 8.6

Consider the reaction shown below. In this reaction we have an ester undergoing reaction with an alcohol to give a new ester. This can happen under either acidic or basic conditions. Such a process is called a *transesterification*. The position of the equilibrium is determined by steric and concentration effects. *You must be able to write mechanisms for both the acid- and base-catalyzed reactions.*

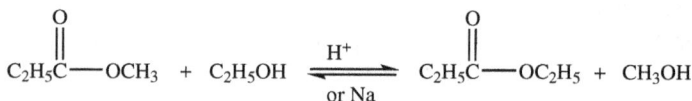

Consider the reactions shown in Fig. 8.7. In the first two reactions, the nucleophilic species on the left is a poorer nucleophile than that on the right and the leaving group ability of the group attached to the carbonyl on the right is higher than on the left. *Both* effects predict that these equilibria lie heavily to the left (i.e., $K_{eq} \ll 1$). For the third equation, an irreversible step is present, which can draw this equilibrium to the right (NH_4^+ is acidic). Thus, it is predicted and found that acyl chlorides or anhydrides react with methoxide ion to form esters, but not the reverse, whereas ammonia reacts with esters to form amides but not the reverse.

8.6.1 Detergents

The reaction of esters with alkali is one of the oldest organic chemistry procedures. Animal fats are esters of 1,2,3-propanetriol (glycerol) and long-chain (C_{14}–C_{18}) acids. Treatment of these with lye (=hydroxide) causes hydrolysis of the esters and leaves

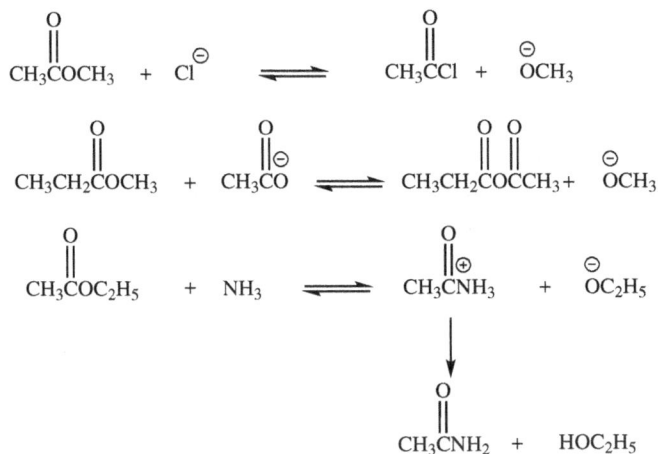

$$CH_3\overset{\displaystyle O}{\overset{\|}{C}}OCH_3 \ + \ \overset{\ominus}{Cl} \ \rightleftharpoons \ CH_3\overset{\displaystyle O}{\overset{\|}{C}}Cl \ + \ \overset{\ominus}{O}CH_3$$

$$CH_3CH_2\overset{\displaystyle O}{\overset{\|}{C}}OCH_3 \ + \ CH_3\overset{\displaystyle O}{\overset{\|}{C}}\overset{\ominus}{O} \ \rightleftharpoons \ CH_3CH_2\overset{\displaystyle O}{\overset{\|}{C}}O\overset{\displaystyle O}{\overset{\|}{C}}CH_3 + \ \overset{\ominus}{O}CH_3$$

$$CH_3\overset{\displaystyle O}{\overset{\|}{C}}OC_2H_5 \ + \ NH_3 \ \rightleftharpoons \ CH_3\overset{\displaystyle O}{\overset{\|}{C}}\overset{\oplus}{N}H_3 \ + \ \overset{\ominus}{O}C_2H_5$$

$$\downarrow$$

$$CH_3\overset{\displaystyle O}{\overset{\|}{C}}NH_2 \ + \ HOC_2H_5$$

Fig. 8.7

the salts of the acids. These materials are commonly known as *soap*. Their action is based on the following generalized principle. The long hydrocarbon chain is hydrophobic (water-hating) while the ionic carboxylate end of the chain is hydrophilic (water-loving). Much dirt is organic in nature (but of course, not all organic matter is dirt!). The hydrocarbon end of the chain will dissolve in this while the hydrophilic end remains firmly attached to the water. When the water is removed, the soap molecules are carried away with the grease still attached to the other end of the molecule. As previously pointed out, calcium salts of carboxylic acids are not soluble in water. Therefore, the salts formed in hard water (rich in Ca^{+2} ions) precipitate as a scum. Modern detergents avoid this problem by using salts of sulfonic (SO_3H) acids whose calcium salts are water soluble.

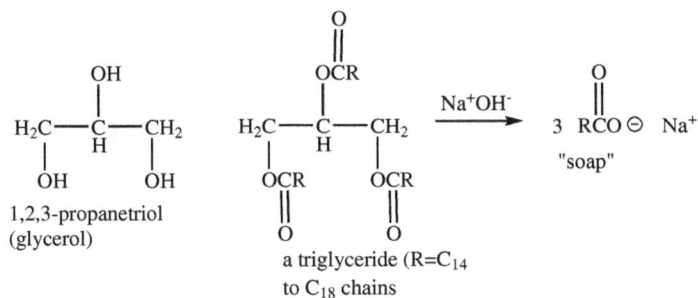

$$
\begin{array}{ccc}
 & OH & \\
 & | & \\
H_2C\!-\!\!\!\!\overset{}{\underset{H}{C}}\!\!\!\!-\!CH_2 & & \\
 & | \qquad\ | & \\
 & OH \quad\ OH &
\end{array}
$$

1,2,3-propanetriol
(glycerol)

$$
\begin{array}{ccc}
 & O\!\!=\!\!\overset{\|}{C}R & \\
 & OCR & \\
 & | & \\
H_2C\!-\!\!\!\!\overset{}{\underset{H}{C}}\!\!\!\!-\!CH_2 & \xrightarrow{\ Na^+OH^-\ } & 3\ \ RCO\ominus\ \ Na^+ \\
 & | \qquad\quad\ | & \\
 & OCR \quad\ OCR & \text{"soap"} \\
 & \| \qquad\quad\ \| & \\
 & O \qquad\quad\ O &
\end{array}
$$

a triglyceride (R=C$_{14}$
to C$_{18}$ chains

8.7 Acyl Halides

As suggested by the first equation in Fig. 8.7, acyl halides cannot be prepared by reaction of esters or acids with halide ion. These compounds are usually prepared by the reaction shown. This reaction is highly exothermic and irreversible since both SO_2 and HCl, being gases, are removed from the reaction as it proceeds. Acyl bromides and iodides are so reactive they are usually not used.

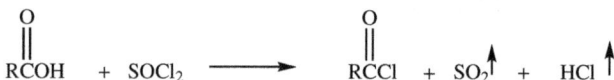

$$\underset{RCOH}{\overset{O}{\|}} + SOCl_2 \longrightarrow \underset{RCCl}{\overset{O}{\|}} + SO_2\uparrow + HCl\uparrow$$

Consideration of the expected properties of acyl chlorides leads to the following conclusions. Since chlorine is strongly electronegative, its presence should increase the positive character of the carbonyl group and therefore increase the ease of attack of nucleophiles compared to that found in aldehydes, ketones, acids, or esters. The chloride ion, being a good leaving group and poor nucleophile should be easily displaced and the reactions should be essentially irreversible. These predictions are confirmed in practice. Acyl halides react with even weakly nucleophilic neutral nucleophiles to give products of substitution. Some examples are shown in Fig. 8.8. The formation of amides is best carried out using acyl halides.

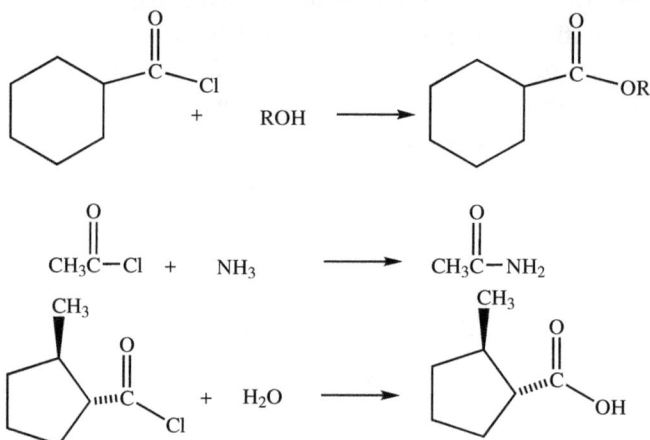

Fig. 8.8

8.8 Anhydrides

Anhydrides have the structure shown in Table 8.1. They are formally derived by the condensation of two molecules of a carboxylic acid with the loss of water. They have no acidic protons. Anhydrides can be written as RCOX where X = OCOR. The protonated form of this X group is a carboxylic acid which is more acidic than water or alcohol; it should be a better leaving group than OH⁻ or OR⁻ but not as good as Cl⁻. Therefore, anhydrides will react with many nucleophiles to give substitution reactions. Many esters are prepared by reaction of anhydrides with alcohols, frequently in the presence of pyridine, a basic catalyst (Fig. 8.9).

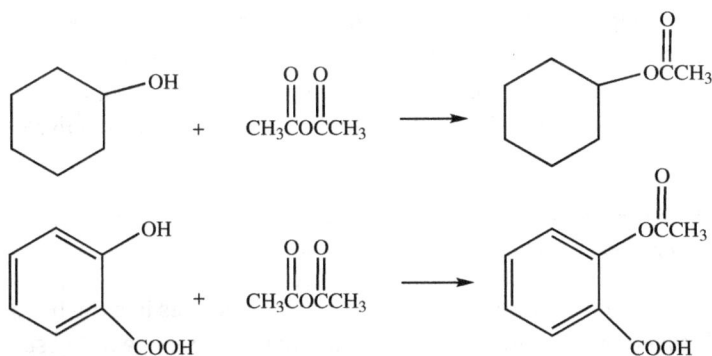

Fig. 8.9

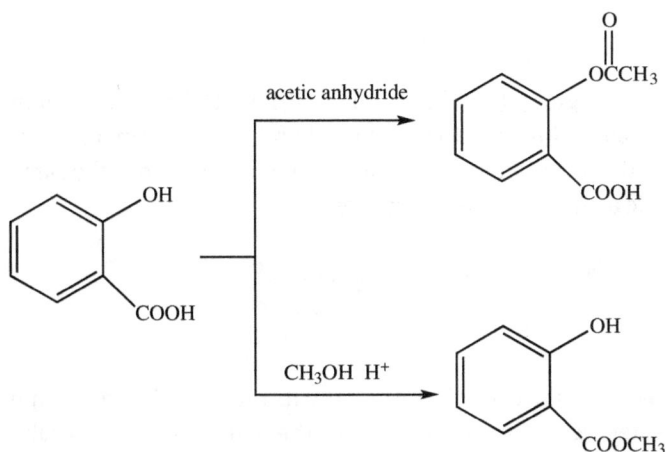

Fig. 8.10

The second example in Fig. 8.10 is an interesting one. Salicylic acid has both an alcohol and a carboxylic acid group. It has the potential to form esters at two sites – one by reaction (at the alcohol) with an acid derivative and the other by reaction (at the acid) with an alcohol. The products are vastly different materials! One, the ester of the carboxyl group of salicylic acid with methanol, is oil of wintergreen while the other, the ester of the alcohol of salicylic acid with acetic acid is commonly called aspirin.

8.9 Amides

Amides have the structure shown in Table 8.1 in which a trivalent nitrogen atom is attached to the carbonyl group. They are best prepared by reaction of acyl halides with ammonia or ammonia in which one or two of the hydrogens have been replaced by alkyl groups. Such alkylated ammonia molecules are called *amines* (see Chapter 10).

Amides can be hydrolyzed to acids with water under either acid- or base-catalyzed conditions.

> Write a mechanism for these two processes and decide which you think would be the better procedure. Apply the type of logic used in the section on esterification.

The amide bond is one of the most important bonds in biochemistry as it is the bond that joins amino acids together in proteins. More of their chemistry will be discussed in Chapter 10.

8.10 Nitriles

Molecules containing the cyano group ($C{\equiv}N$) are called nitriles. The carbon atom in nitriles is sp-hybridized. Because nitrogen can form three bonds, but oxygen only two, one nitrogen atom can take the place of two oxygen atoms and nitriles are at the same oxidation level as carboxylic acids and their derivatives.

$$\underset{RC-O-}{\overset{\overset{\textstyle O}{\|}}{}} \qquad RC{\equiv}N$$

The carbon atom of nitriles is positively polarized and will undergo the same type of reaction as acids and esters, etc. The mechanism for the acid-catalyzed hydrolysis of nitriles is shown in Fig. 8.11. The relationship between A and B in Fig. 8.11 has been seen before. In the addition of water to alkynes, the initial product had the partial structure C=C–OH (Sect. 5.11.2). This was unstable and immediately changed into CH–C=O. Such isomers derived by the migration of one hydrogen atom and one π bond are called *tautomers*. The same relationship exists between A and B in Fig. 8.11.

Tautomers are isomers that can be interconverted by moving a proton from one site to another with the concomitant change of position of a pi bond. The *process* is called *tautomerism.*

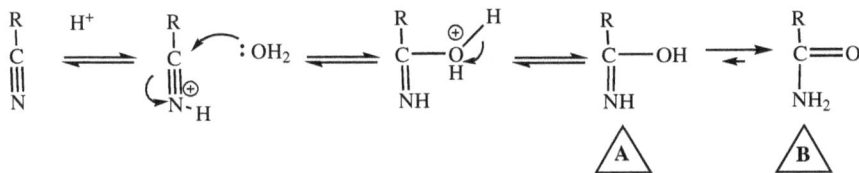

Fig. 8.11

B is an amide and A is a tautomer of an amide. These are in equilibrium with each other, but B is much more stable than A and the amide is by far the predominant species. The base-catalyzed hydrolysis process is also possible (Fig. 8.12).

Fig. 8.12

8.11 Other Reactions of Acids and Their Derivatives

8.11.1 Grignard Reactions

Grignard reagents (RMgX) are basic and *cannot* be used with acidic materials (cf. Chapter 7). However, esters, acyl halides and nitriles will all react to give, as the product of the first step, a species analogous to that formed with aldehydes and ketones (Sect. 7.6). The fate of the addition product depends on its particular structure (Fig. 8.13).

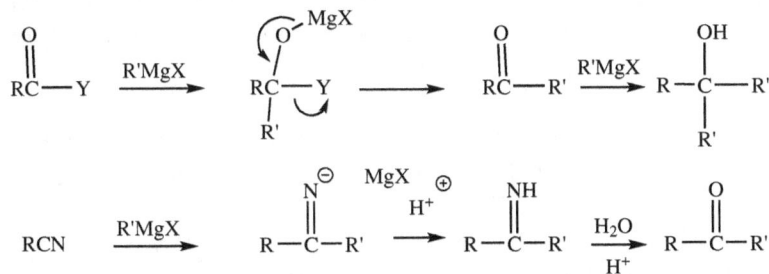

Fig. 8.13

When X is a leaving group (Cl or OR, the same fate befalls it as we saw in the base-catalyzed hydrolysis of these molecules. The product of this is a ketone which itself reacts with *another* molecule of Grignard reagent to give, as the final product after the usual addition of acid, a tertiary alcohol which has two identical alkyl groups attached to the carbinol carbon atom. *This reaction cannot be stopped at the ketone stage!* In the case of the nitrile, the intermediate cannot eliminate anything and since it is negatively charged, it resists further reaction with the nucleophilic Grignard reagent. When aqueous acid is added to this, the compound with the C=NH is formed. In the presence of water and acid this is converted into the ketone which is the final product. (*Write a mechanism for this process.*)

8.11.2 Reduction

Lithium Aluminum Hydride (LAH) (Sect. 7.8) but *not* sodium borohydride reacts with esters, nitriles, and even acids in the same way as Grignard reagents (Fig. 8.14). You should be familiar enough by now with reaction mechanisms to be able to design mechanisms that account for these results. It is apparent that H^- ion is a potent enough nucleophile to attack even the negatively charged intermediates.

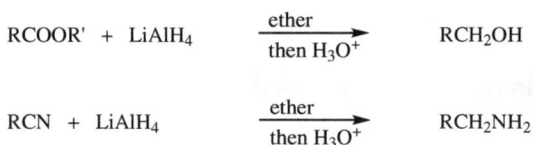

$$RCOOR' + LiAlH_4 \xrightarrow[\text{then } H_3O^+]{\text{ether}} RCH_2OH$$

$$RCN + LiAlH_4 \xrightarrow[\text{then } H_3O^+]{\text{ether}} RCH_2NH_2$$

Fig. 8.14

8.12 Problems

8-1. Write the mechanisms for both the acid and base-catalyzed reactions of ethyl propanoate with water. Explain why the base-catalyzed process is a synthetically more useful one.

8-2. When pentanoic acid is esterified with methanol that contains ^{18}O in the hydroxyl group, does the product ester contain the ^{18}O atom? Explain your reasoning.

8-3. When acetylsalicylic acid is treated with methanol under acid catalysis, the product is methyl salicylate. Write a mechanism for this transformation.

8-4. Write the mechanism for the hydrolysis of N,N-dimethyl butanamide under both acid and base catalysis.

8-5. When succinic anhydride (I) is reacted with excess methanol, the following facts are noted.
1. If *no* catalyst is added, the product is monomethyl succinate (II).
2. If an acid catalyst is added, the product is the dimethyl ester (III).

(a) Explain why there is a reaction in the *absence* of a catalyst.
(b) Explain why the products are different in the two reactions.
(c) Predict what product would be formed if a catalytic amount of sodium methoxide was added to the solution.

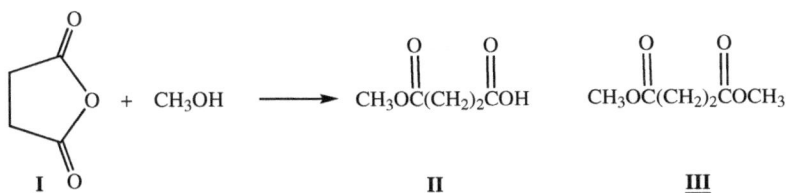

I $+$ CH_3OH ⟶ $CH_3OC(CH_2)_2COH$ $CH_3OC(CH_2)_2COCH_3$

I   II   III

8-6. Replace the "?" in the following equations with the correct reagent or product. Show any necessary catalysts and stereochemical features of the reactions.

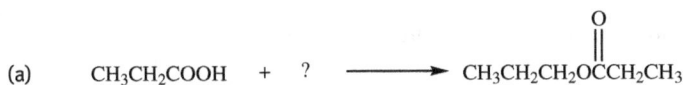

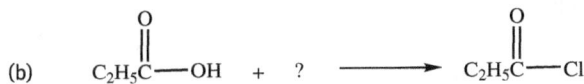

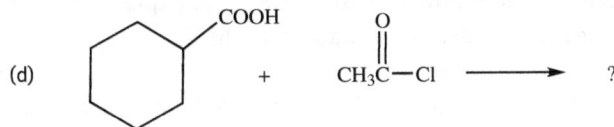

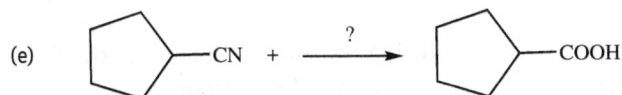

(a) CH_3CH_2COOH $+$?   ⟶ $CH_3CH_2CH_2OCCH_2CH_3$

(b) C_2H_5C-OH $+$?   ⟶ C_2H_5C-Cl

(c) $(CH_3)_2CHCOCH_3$ $+$ C_2H_5OH ⟶   ?

(d)   [cyclohexane-COOH] $+$ CH_3C-Cl ⟶   ?

(e)   [cyclopentane-CN] $+$?   ⟶   [cyclopentane-COOH]

(f) $CH_3\overset{\overset{\displaystyle O}{\|}}{C}OCH_3$ + ? ⟶ $CH_3\overset{\overset{\displaystyle OH}{|}}{\underset{\underset{\displaystyle CH_2CH_3}{|}}{C}}CH_2CH_3$

(g) $CH_3\overset{\overset{\displaystyle O}{\|}}{C}CH_2CH_2COOH_2$ + ? ⟶ $HOOCCH_2CH_2COOH$

(h) $CH_3CH_2\overset{\overset{\displaystyle O}{\|}}{C}-O^{\ominus}$ + $\overset{\overset{\displaystyle CH_2CH_3}{|}}{\underset{\underset{\displaystyle CH_2CHBr}{}}{}}$ ⟶ ?

(i) $CH_3CH_2\overset{\overset{\displaystyle O}{\|}}{C}-Cl$ + ⬡NH_2 ⟶ ?

(j) $(CH_3)_2CH\overset{\overset{\displaystyle O}{\|}}{C}-Cl$ + ? ⟶ $(CH_3)_2CHO\overset{\overset{\displaystyle O}{\|}}{C}CH(CH_3)_2$

(k) $CH_3CH_2COCH_3$ + ? ⟶ $CH_3CH_2CH_2OH$

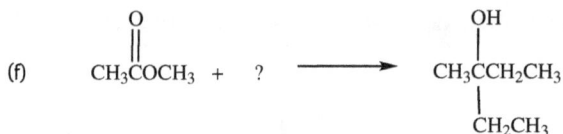

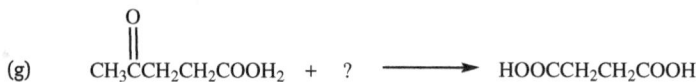

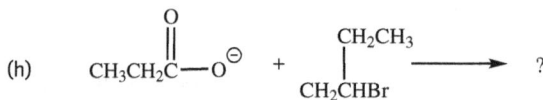

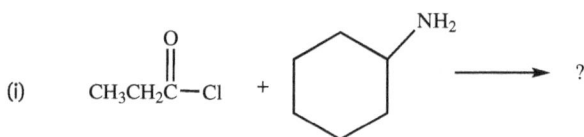

8-7. Carbamic acids are adducts of carbon dioxide and amines.

R_2NH + CO_2 ⇌ $R_2N\overset{\overset{\displaystyle O}{\|}}{C}-OH$

They are difficult to isolate and rapidly decompose in neutral or acidic solution. However, in *basic* solution they are stable. Explain why this might be so.

8-8. A standard method for the formation of certain types of esters is the reaction of acid chlorides with alcohols in aqueous sodium hydroxide solution. Explain why the yields of esters are best when the ester formed is not water soluble.

8-9. The reaction of nitriles with alcohols under acid catalysis leads to *esters* after treatment with water. Write a mechanism for this reaction.

8-10. Show how you would carry out the following transformations, each of which may require more than one step. Show all reagents and isolable products.

(a) $CH_3CH_2CH_2OH$ ⟶ $CH_3CH_2\overset{\overset{\displaystyle O}{||}}{C}-OCH_3$

(b) $(CH_3)_3C\overset{\overset{\displaystyle O}{||}}{C}-OH$ ⟶ $(CH_3)_3C\overset{\overset{\displaystyle O}{||}}{C}-NH_2$

(c)

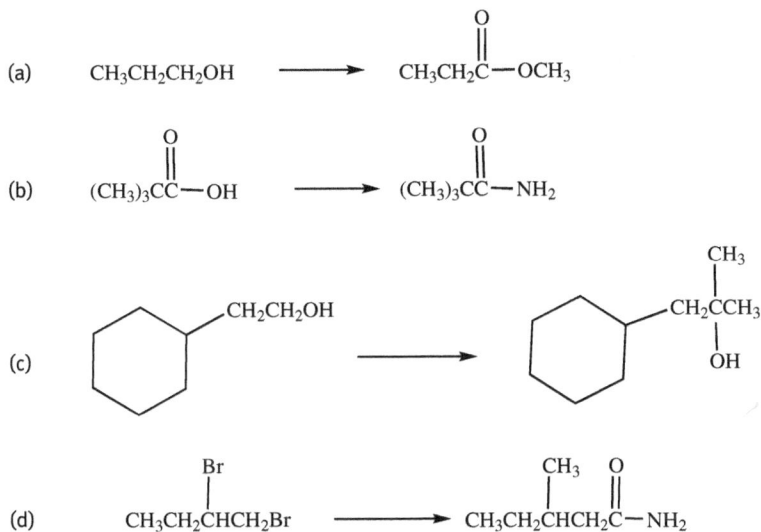

(d) $CH_3CH_2\overset{\overset{\displaystyle Br}{|}}{C}HCH_2Br$ ⟶ $CH_3CH_2\overset{\overset{\displaystyle CH_3}{|}}{C}HCH_2\overset{\overset{\displaystyle O}{||}}{C}-NH_2$

8-11. Lactones are cyclic esters formed by the intramolecular reaction of a molecule which contains both an alcohol and a carboxylic acid.
(a) Write the structures of the hydroxy acids that lead to the following lactones.

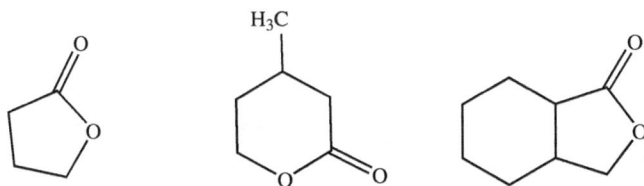

(b) Show the product that would be obtained when each of these lactones are treated with LiAlH₄ and then dilute aqueous acid.

8-12. Heating phthalic anhydride (I) and 1,2-ethanediol (ethylene glycol) together leads to a polymer called a polyester. Show the structure of a section of this polymer.

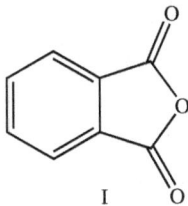

I

8-13. Outline a simple chemical test that would distinguish between:
(a) Propanoic acid and methyl propanoate
(b) An acid and an acid chloride
(c) An acid and an alcohol

8-14. Show how pentanoic acid can be prepared from:
(a) 1-pentanol
(b) 1-bromobutane (two ways)
(c) 1-hexene
(d) 5-decene
(e) 2-hexanone

8-15. Phosgene ($COCl_2$) is the diacid chloride of carbonic acid (H_2CO_3). How could you prepare diethyl carbonate from phosgene?

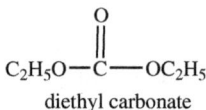

$$\underset{\text{diethyl carbonate}}{C_2H_5O-\overset{\overset{\displaystyle O}{\|}}{C}-OC_2H_5}$$

8-16. Write the products from the reaction of (i) CH_3MgBr and (ii) $LiAlH_4$ with each of the following compounds.

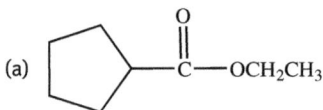

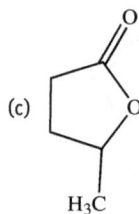

(a)

(b) $CH_3CH_2\overset{\overset{\displaystyle O}{\|}}{C}CH_2CH_2\overset{\overset{\displaystyle O}{\|}}{C}-OH$

(c)

9 Aromatic Compounds – Benzene and Its Derivatives. Resonance as a Force Majeure

9.1 Introduction

To this point, we have been considering reactions of molecules that have been, for the most part, made up of carbon in the sp^3-hybridized state. Isolated sp^2-carbons in carbonyl groups and alkenes have also been examined and we saw that, when more than two sp^2-hybridized carbons were present in a contiguous chain, resonance was a major factor in determining the molecules' reactivities (Chapter 4).

In this chapter, we will consider a group of molecules which, as a class, are called *aromatic* molecules. We will see that these molecules contain cyclic structures with all the atoms in the ring being sp^2-hybridized. This situation has been briefly presented previously (Sect. 4.10). These molecules have the six π electrons completely delocalized around the ring and are thus highly resonance-stabilized. Such molecules were originally called aromatic for historical reasons. Their distinctive odor was responsible for this name, but in modern times the name has been retained to distinguish them from aliphatic molecules in which the possibilities for resonance are much more limited.

9.2 Aromatic Molecules, Aromaticity, and Hückel's Rule

Many molecules are aromatic. Since they were some of the earliest identified organic molecules, trivial names abound and many are in very common use. The systematic IUPAC naming of these molecules will be covered in your nomenclature text. The structures of some very common aromatics and their trivial names are shown in Fig. 9.1. Note the difference between the words phenol (hydroxybenzene) and phenyl which is the word to describe benzene as a substituent (note the "yl" ending).

It was noted very early that molecules having a six-membered ring with six π electrons did not behave in the manner expected in analogy with cyclohexene. Compare for example the reactions of benzene and cyclohexene shown in Fig. 9.2. Clearly there is an unusual stability associated with this particular arrangement of atoms. A prediction as to whether a molecule will have this stability can be made on the basis of Hückel's Rule.

If a molecule has $4n + 2$ $(n \geq 1)$ π electrons in cyclic conjugation, the molecule will be more stable than expected on the basis of simple aliphatic analogs.

https://doi.org/10.1515/9783110778311-009

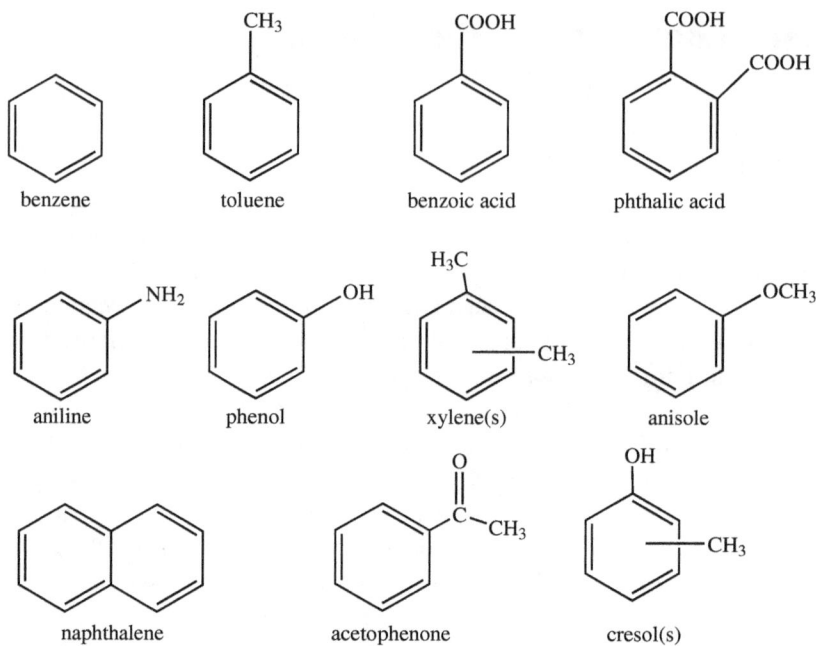

benzene   toluene   benzoic acid   phthalic acid

aniline   phenol   xylene(s)   anisole

naphthalene   acetophenone   cresol(s)

Fig. 9.1

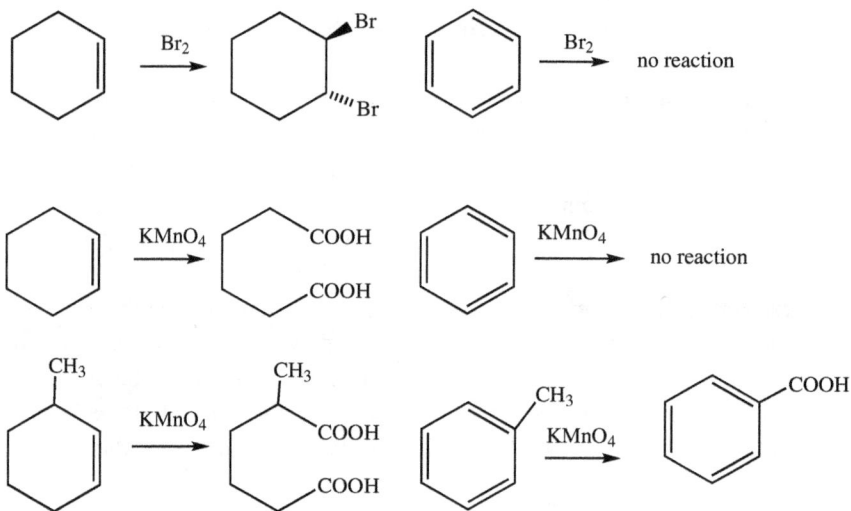

Br$_2$

Br$_2$ → no reaction

KMnO$_4$ → COOH, COOH

KMnO$_4$ → no reaction

KMnO$_4$ → CH$_3$, COOH, COOH

KMnO$_4$ → COOH

Fig. 9.2

Molecules with this arrangement of π electrons are called aromatic and the amount of energy by which they are stabilized is the *Resonance Stabilization Energy* (Sect. 4.10) or *Aromaticity*. Let us see how this might be calculated.

When cyclohexene is hydrogenated to cyclohexane, the exothermic reaction, which involves breaking one π bond and forming two C–H bonds, releases 28.6 Kcal/mole. Therefore, if benzene was simply three double bonds in a ring, hydrogenation to cyclohexane would be expected to release $3 \times 28.6 = 85.8$ Kcal/mole. The actual amount of energy released is actually 49.8 Kcal/mole. Since cyclohexane, the common product of both reactions, must have the same energy regardless of how it is formed, this result means that the actual energy of benzene was 36.0 Kcal/mole *less* than that calculated on the cyclohexatriene model. Thus, benzene is aromatic with a Resonance Stabilization Energy of 36 Kcal/mole. This situation is diagramed in Fig. 9.3.

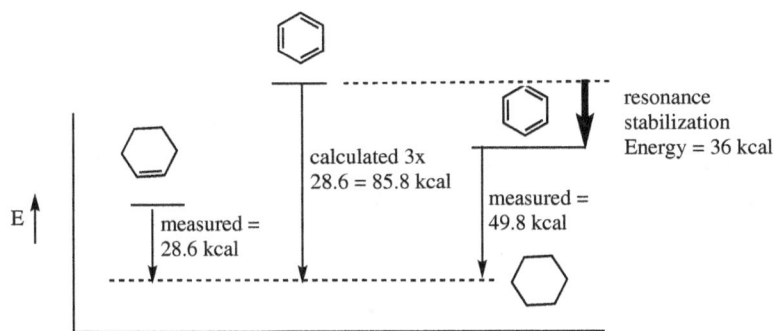

Fig. 9.3

9.3 Benzene – The Prototype of Aromatic Molecules

Benzene (C_6H_6) is the simplest molecule showing aromatic stabilization. All the carbons are sp^2-hybridized and therefore the molecule must be flat with the single hydrogen atom attached to each carbon also in the plane of the ring atoms (Fig. 9.4). (Contrast this with the structure of cyclohexane. Do the terms axial, equatorial, *cis*, or *trans* have application in the structure of benzene?) Each ring atom has a p-orbital that contains one electron. These p-orbitals are all coplanar and the π electrons can move around this "electrical circuit" in an endless fashion (Fig. 9.4). Each of the π electrons can become associated with a total of six atoms via the resonance process (recall the meaning of the symbol ↔) and the increased stability is predicted by the resonance concept.

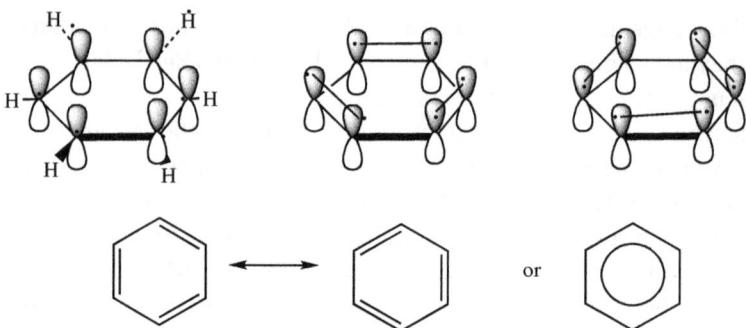

Fig. 9.4

9.4 Reactions of Aromatic Molecules

9.4.1 General Considerations

Consideration of the structure of aromatic molecules leads one to conclude that, like alkenes, they are electron-rich and therefore should be attacked by electrophiles. (Review the electrophilic addition mechanism in Chapter 5). Let us compare this possibility with the usual electrophilic addition mechanism shown again in Fig. 9.5.

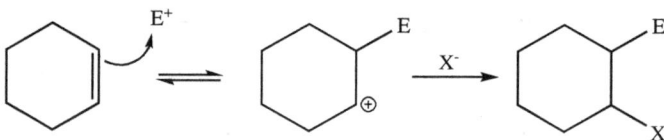

Fig. 9.5

If a set of steps occurs that is exactly analogous to the reaction of alkenes (e.g., cyclohexene), the addition product that is formed has *lost its aromatic resonance stabilization* (Fig. 9.6). This is a *very unfavorable* process (i.e., $K_{eq} \ll 1$). Therefore, the reverse reaction (i.e., the loss of E^+), which regenerates the aromatic material, is much more favorable. However, remember that H^+ is an excellent electrophile and could also be lost. If $E^+ = H^+$, H^+ can be lost to regenerate the aromatic system and the overall result is an *electrophilic substitution* reaction. The mechanism leading to a substitution product is termed an S_E2 reaction (Substitution, Electrophilic, Bimolecular).

What is the nature of E^+ in Fig. 9.6? In this chapter we will look at four specific examples.

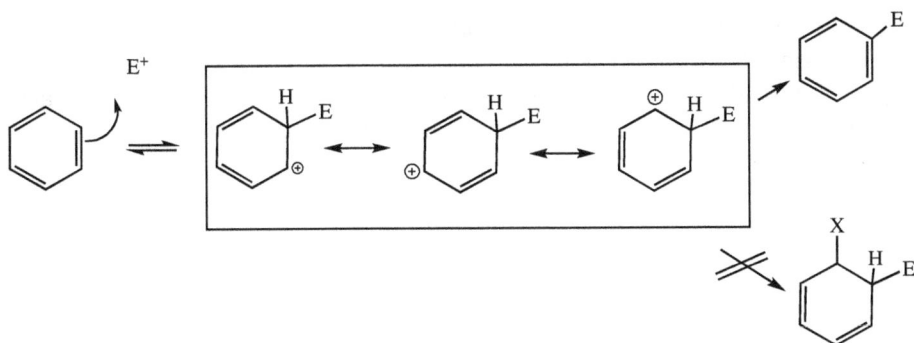

Fig. 9.6

9.4.2 Nitration

The reaction of nitric acid (HNO_3) with strong acid establishes the following equilibrium.

$$HNO_3 + H^+ \rightleftharpoons H_2O + NO_2^+$$

Frequently, concentrated sulfuric acid is used to generate a reasonable concentration of the electrophilic nitronium ion NO_2^+. This can then react in the manner suggested in the figure below.

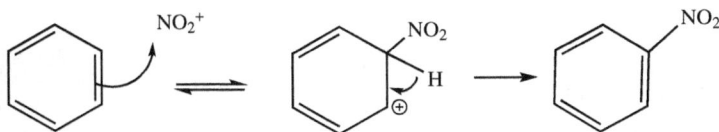

9.4.3 Halogenation

The reaction of halogens (usually Cl_2 or Br_2) with the Lewis acid $FeCl_3$ (or $FeBr_3$) generates a complex that can be written as shown.

$$Cl_2 + FeCl_3 \rightleftharpoons FeCl_4^+ \; Cl^-$$

$$Br_2 + FeBr_3 \rightleftharpoons FeBr_4^+ \; Br^-$$

The chloronium or bromonium ions so generated can undergo the S_E2 reaction. The iron salt is catalytic since the HX generated is liberated from the reaction mixture and the overall reaction can then be written as shown below. It is important to choose the correct iron halide when doing an aromatic halogenation. The iron must be in the

+3 oxidation state *and* the halogen atom must correspond to that being introduced. Therefore, use $FeCl_3$ with Cl_2, $FeBr_3$ with Br_2, etc. In some cases you may see the use of iron metal (oxidation state 0) used as catalyst. In these cases the iron first reacts with the halogen to form FeX_3 which is the real catalyst.

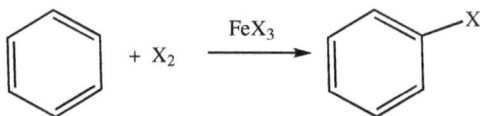

9.4.4 Acylation

The reactions of acids, esters, and acyl halides with an appropriate catalyst generate electrophilic species called *acylium ions* (RCO^+). Strong *protonic* acids (e.g., sulfuric, phosphoric, HF) are used with *carboxylic acids or esters* and *Lewis* acids (usually $AlCl_3$) are used with *acyl chlorides*. The acylium ion, being electrophilic, undergoes the S_E2 reaction, which in this case is referred to as a *Friedel–Crafts acylation* reaction. Note that the product from such a reaction is a ketone in which one of the substituents attached to the carbonyl carbon is an aromatic group.

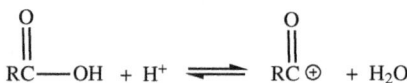

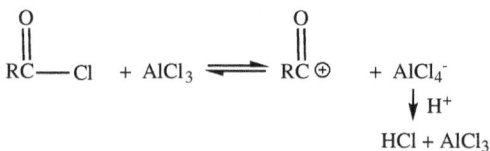

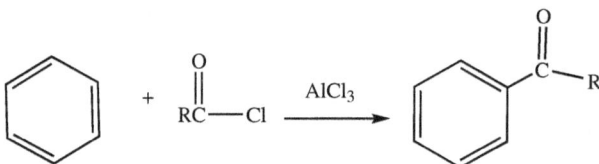

9.4.5 Alkylation

Carbocations are electrophiles. We have already seen several examples of the generation of these positively charged carbon ions. For an example, see Fig. 9.7. Also shown in this figure is another very common method of generation of carbocations: i.e., the reaction of alkyl halides with strong Lewis acids like $AlCl_3$. Regardless of how they are generated, these positively charged species can undergo the S_E2 reaction to generate alkylated aromatic compounds. The reaction is usually referred to as a *Friedel–Crafts alkylation* reaction since it introduces an alkyl group.

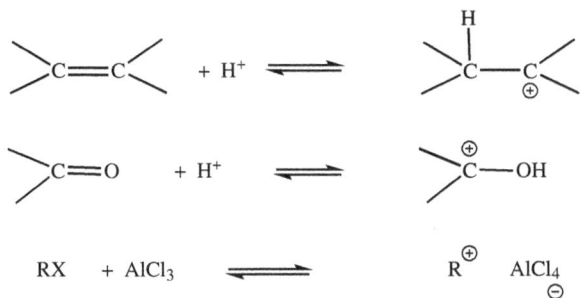

Fig. 9.7

9.5 Reversibility

As has been pointed out several times, all reactions are potentially reversible. As usually carried out, aromatic substitution reactions proceed under *kinetic control* (Chapter 4); that is *activation energy and not product stability govern the nature of the product formed*. This will become very important in subsequent sections. For our purposes, we will simply say that the reverse reactions of nitration and Friedel–Crafts (F/C) acylation are not observed (i.e., they are irreversible), while it is possible (but very difficult) to reverse the processes of halogenation and F/C alkylation.

9.6 Reactions of Substituted Benzenes

9.6.1 General Considerations

Since benzene is symmetrical, reaction of an electrophile at anyone of the six carbons atoms will lead to the same product. However, if a substituent is already present in the reacting molecule, three isomeric products are possible (Fig. 9.8). Note that, since there are *two ortho, two meta, and only one para position*, a completely statistical distribution of products would be 2:2:1. In practice, this is almost never observed. What factors could influence where the incoming electrophile will attack? The two most obvious possibilities are the nature of the electrophile and that of the substituent already in place. Consider the information shown in Fig. 9.9.

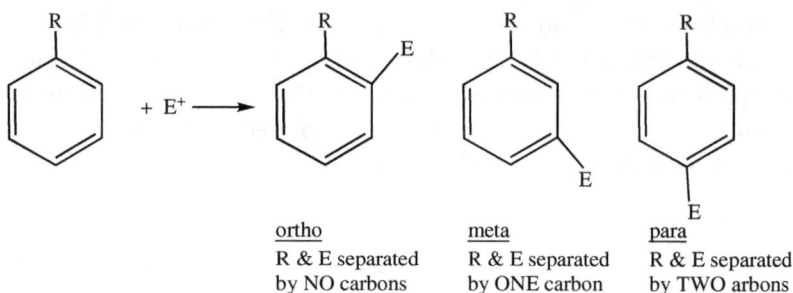

ortho
R & E separated
by NO carbons

meta
R & E separated
by ONE carbon

para
R & E separated
by TWO arbons

Fig. 9.8

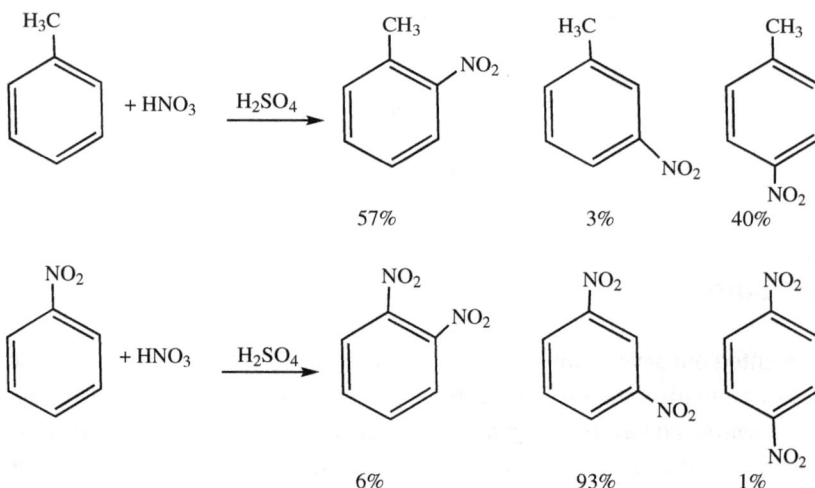

57%   3%   40%

6%   93%   1%

Fig. 9.9

From these two reactions it is clear that the substituent already in place has a major effect on the structure of the product formed. From many other examples, it is equally clear that the nature of the electrophile has little effect.

The isomeric composition of products of S_E2 reactions on substituted benzenes is controlled by the substituent(s) already in place.

Another factor of immense importance is the difference in reactivity (i.e., the rates of reaction) of the two substrates in Fig. 9.9. The presence of the methyl substituent makes the reaction of toluene *faster* than that of benzene, while the nitro substituent in nitrobenzene makes its substitution reactions much slower than those of benzene. *These observations are true regardless of the nature of the electrophile.*

The *rate of reaction* of a substituted aromatic ring in S_E2 reactions is determined by the substituents already in place on the ring.

9.6.2 Reactivities

How can we account for these orientation and reactivity effects in terms of the mechanism we have developed (Fig. 9.6)? Since the rate of reaction of a compound is determined by the energy difference between the starting material and the highest point on the energy curve (i.e., the activation energy) and since the first step in an S_E2 reaction is the slow step, effects that alter rates of reaction must do so by some effect on this step. This is illustrated in Fig. 9.10 for the example in Fig. 9.9.

In the first step of the reaction, the electrons of the aromatic ring must be released by the ring to form a bond to the electrophile. Therefore, it is reasonable that *anything that increases the electron-density* on the ring (i.e., an electron-donating substituent) *would make this bond-forming process easier* and, conversely, anything that withdraws electrons from the ring would reduce the electron density and make the bond-forming process more difficult (Case C, Fig. 9.10). What effects can cause electron-density increases or decreases? Clearly both resonance and inductive effects are possibilities. Consider the molecule chlorobenzene (Fig. 9.11). Since chlorine is electronegative relative to carbon, its presence will *withdraw electrons* from the ring by an *inductive* effect. Substituents are designated as (-I) substituents (Sect. 4.8), while those that donate electrons to a ring by this effect are designated as (+I) substituents. The best example of a +I substituent is provided by an alkyl group (e.g., CH_3). (-I) substituents are ring-deactivating (relative to benzene), while (+I) substituents are ring-activating (again relative to benzene).

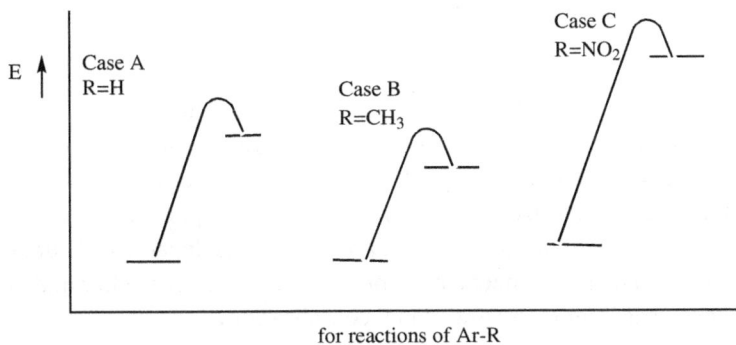

for reactions of Ar-R

Fig. 9.10

Now consider the molecule acetophenone (Fig. 9.11) Since the carbonyl atom has a p-orbital and is adjacent to the aromatic ring, it can participate in resonance as indicated. Therefore, the carbonyl group withdraws electrons from the ring by a *resonance* effect.

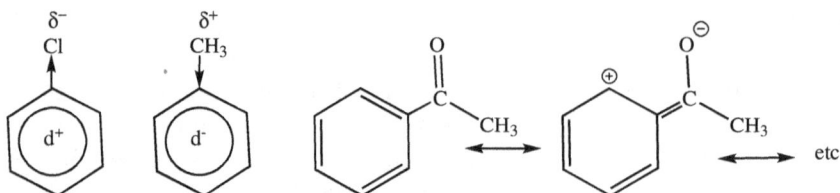

Fig. 9.11

Such substituents are termed (-M) substituents (electron-withdrawing by a reso-nance – or mesomerism – effect). In the same way, since oxygen has nonbonding electrons in p-orbitals, when it is attached to an aromatic ring, it may participate in res-onance by donating these electrons (Fig. 9.12). This increases the electron-density in the ring by a resonance effect and such substituents are designated as (+M). (Note that oxygen cannot withdraw electrons by a resonance effect since it has no vacant orbitals into which the electrons can go. It can only withdraw electrons by an *inductive* effect.)

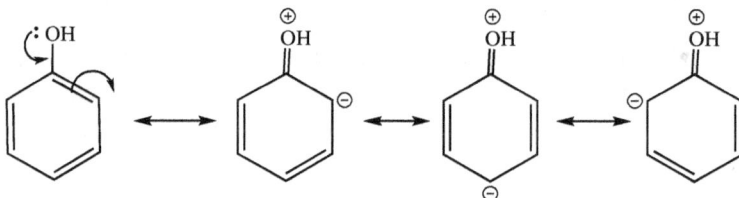

Fig. 9.12

One factor must be noted. Some substituents appear to be (-I) but also (+M). For ex-ample, oxygen, nitrogen, and halogens are all more electronegative than carbon and should be -I, but all also have electrons in p-orbitals and are thus capable of being (+M). In such cases, it is found that oxygen and nitrogen-substituted compounds re-act faster than benzene and therefore the +M effect dominates. Halogen-substituted compounds react more slowly than benzene and therefore the -I effect predominates in these cases. (However, see later for an additional complication).

To summarize: (-M) and (-I) substituents slow down S_E2 reactions while (+M) and (+I) substituents speed up S_E2 reactions.

9.6.3 Orientations

When an electrophile reacts with a monosubstituted benzene, there are three possible intermediates (Fig. 9.13).

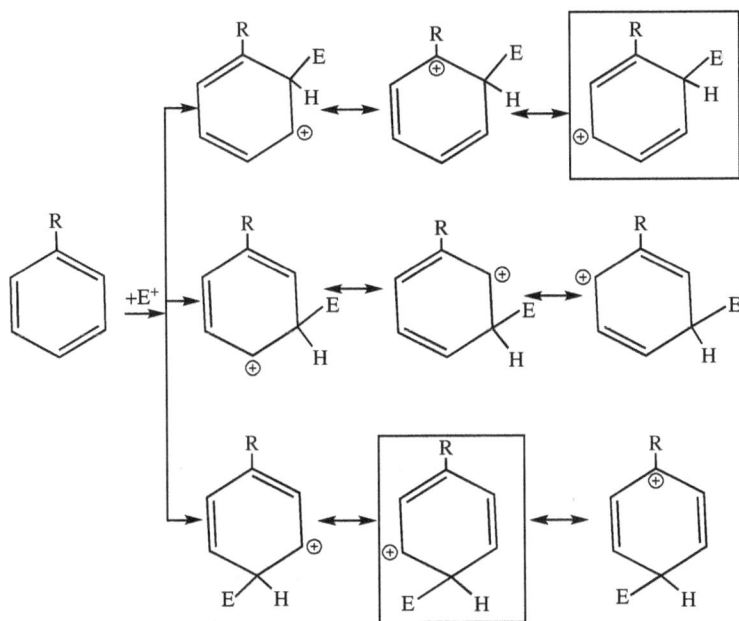

Fig. 9.13

These three intermediates can be differentiated in one very important way. All can distribute the resulting positive charge around the rest of the ring atoms by resonance. However, *only in those derived from attack of the electrophile at the ortho or para positions* does the intermediate allow localization of the positive charge on the same atom which bears the substituent (boxed structures on Fig. 9.13). If R is a substituent that *donates* electrons (i.e., either a (+I) or (+M) substituent), the positive charge can be further delocalized by an inductive or resonance effect. This spreads out the charge even more, increasing the stability of the intermediate and reducing the activation energy of the reaction. In the case of the *meta* isomer, this direct interaction cannot take place and stabilization is much less efficient. This results in *preferential formation of the ortho and para isomers when R is an electron-donating substituent. Note that these are also the same types of substituents that are ring-activating.* In terms of activation energies, since the ortho and para intermediates are more highly stabilized, they have lower activation energies leading to them (Fig. 9.14)

In the case of the (-I) or (-M) substituents, the electron-withdrawing effect would be expected to *destabilize* the ortho and para intermediates relative to the meta. (Any effect that draws electrons away from something already positive, will make it more positive and therefore less stable.) Therefore, the groups we saw deactivating rings also direct substitution to the meta position.

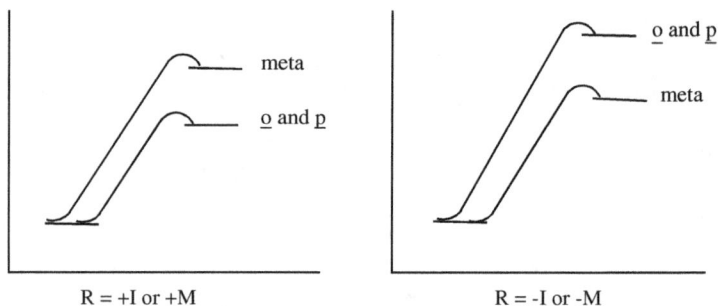

Fig. 9.14

One apparent exception to these rules is found in the halogen substituents chlorine, bromine, and iodine. All of these are ring-deactivating as we have already seen. However, they are also found to be ortho, para directing. This apparent contradiction can be explained in the following way. Halogen does reduce the density of electrons on the ring and makes the attack of an electrophile on the ring more difficult. However, if attack does occur, a positively charged ring is formed which is now more electron-attracting than the halogen atom and, if the attack had occurred in the ortho or para position the nonbonding electrons on the substituent can be used to stabilize the intermediate by resonance. In effect, it becomes a +M substituent.

9.6.4 Summary

The results of Sects. 9.6.1 to 9.6.3 can be summarized in the way shown in Table 9.1. As can be seen from that Table, those substituents that are above hydrogen in reactivity, and also the halogens, are ortho-para directing, while those below hydrogen are meta directing. In fact, *rings containing a nitro group are so deactivated that neither F/C alkylation nor acylation will occur.*

9.6.5 Which Is It – Ortho or Para?

There are twice as many ortho positions as para positions. However, the ratio of ortho to para products is rarely 2:1. In actual fact, it is dependent on the size of the substituent in place and the nature of the electrophile. This is a complex situation which is beyond the scope of this book. The major factor is a steric one and all but the smallest groups give mainly para substitution.

Tab. 9.1: Summary of reactivities and orientations in S_E2 reactions.

Substituent	Typical orientation (%)		
	Ortho	Meta	Para
NH_2	17	0	83
OH	10	0	90
NHCOR	19	1	80
alkyl	10–60*	5	35–65*
H			
halogen	35	1	64
C=O	19	80	1
R_3N^+	0	89	11
NO_2	6	93	1

*depends on steric and other factors

9.7 Reactions of Disubstituted Compounds

When *two* substituents are already in place on a ring, three different situations can exist.

1. *Both substituents are deactivating.* A doubly deactivated ring will rarely undergo further reaction.

2. Both substituents are activating. The substituent closer to the top in Table 9.1 (i.e., the most strongly activating group) will control the reaction site.

3. *One activating and one deactivating group are present.* The activating group will control the reaction. The presence of one activating group can outweigh several strong deactivators as is evidenced by the last reaction shown in Fig. 9.15.

9.8 Some Special Functional Group Transformations

Two reactions, both of which are hydrogenations, are extremely useful in the type of chemistry we will be discussing in the next sections. The common result of these reactions is to convert one, easily introduced substituent which is meta directing into another which is ortho-para directing.

The nitro group is very easily introduced into an aromatic ring. It is a strongly deactivating and meta-directing group. Nitrobenzene will not undergo F/C alkylation or

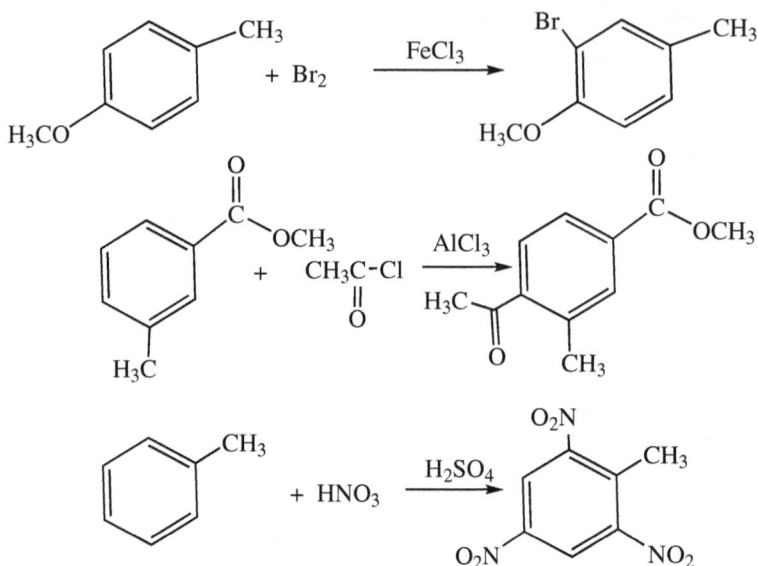

Fig. 9.15

acylation, but can be halogenated. *Catalytic hydrogenation* using a nickel, palladium, or platinum catalyst very rapidly changes the nitro group into an amino (NH_2) group (see next chapter). The amino group is strongly *activating* and ortho-para directing. We will see that some problems can occur when electrophilic substitution reactions are attempted on rings containing this group.

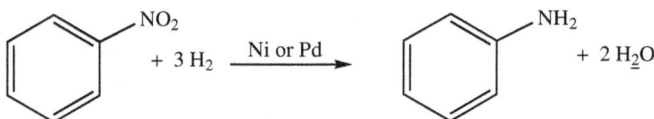

The other reaction introduced here involves the simple conversion of an acyl group to an alkyl group. In this way, the product of an F/C acylation, which is a ketone and therefore meta directing, can be changed into the product of an F/C alkylation which contains an alkyl substituent and is therefore ortho-para directing. We saw in Chapter 7 that ketones can be reduced using hydrogen and a catalyst to the alcohol. In the special case where the ketone is immediately adjacent to an aromatic ring, the alcohol formed by this reaction reacts further to produce the hydrocarbon (Fig. 9.16). This reaction is more difficult than the hydrogenation of a nitro group. Therefore, in a molecule that contains both a nitro group and a ketone, hydrogenation of the nitro group will occur first.

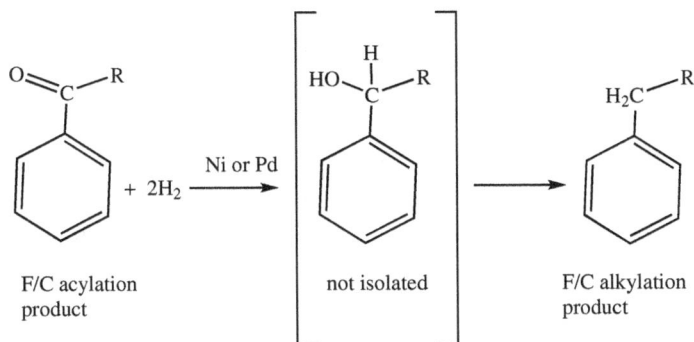

F/C acylation product

not isolated

F/C alkylation product

Fig. 9.16

9.9 Control of Reactivity

Let us examine a simple F/C alkylation (Fig. 9.17). Since alkyl groups are activating, the product of the reaction is *more reactive than the starting material*. If one mole of benzene is reacted with one mole of ethyl chloride, at the point where the reaction has reached only 20% of completion, a significant amount of ethylbenzene is already present. This will react preferentially (since it is more activated) and some of the product will be diethylbenzene. This, in turn, is still more reactive and alkylates again. Therefore, unless some precautions are taken, *polyalkylation* will occur. The product of the reaction will be a mixture of benzene, ethylbenzene, diethylbenzene, triethylbenzene, etc. Two solutions to this problem are available.

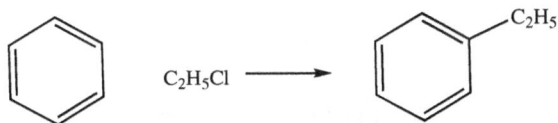

Fig. 9.17

The alkylation of both benzene and ethylbenzene are S_E2 reactions. Therefore, the rate equations for their reactions with ethyl chloride will be:

$$rate_{(benzene)} = k[benzene][ethyl\ chloride]$$

$$rate_{(ethylbenzene)} = k'[ethylbenzene][ethyl\ chloride]$$

The difference in *reactivities* of benzene and ethylbenzene is reflected in the rate constants k and k'. From these equations, it should be clear that if the concentration of benzene is much greater than that of ethylbenzene, the rate of reaction of benzene will be greater, regardless of the difference in k and k'. Running the reaction using a huge excess of benzene (typically 100-fold excess) will ensure monoalkylation. Another way of saying this is that when such an excess of benzene is present, the chances of an ethyl chloride molecule finding an ethylbenzene to react with is very small. The drawback is that 99% of the benzene is not converted to product and therefore this solution to the polyalkylation problem is only practical when very cheap and abundant starting materials are available.

An alternate solution to the problem is to introduce a *deactivating* group first and then convert it to the desired group in a subsequent step. If a deactivating group is introduced, the product of the reaction is *less reactive than the starting material* and the problem of polyalkylation does not arise. It is in this context that the hydrogenation reactions mentioned in Sect. 9.8 are so useful. Both the nitro and acyl groups are deactivating and therefore easily introduced. The amino and alkyl groups which are the hydrogenation products are strongly activating.

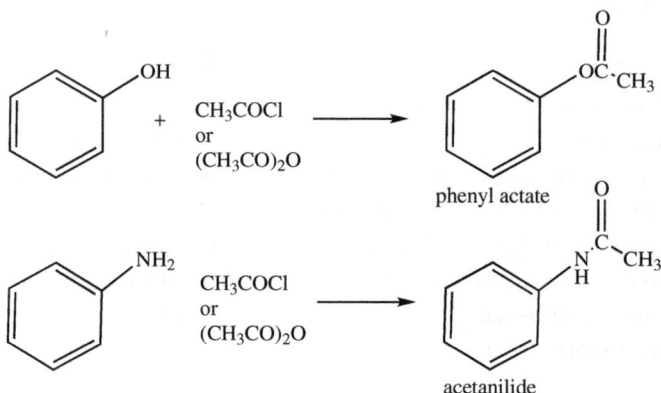

phenyl actate

acetanilide

Fig. 9.18

Both the OH and NH$_2$ groups activate aromatic rings to the point where control of reactions becomes a real problem. This effect can be attenuated by changing these groups into their acyl derivatives (Fig. 9.18). The effect of these reactions is to change a strongly activating group into one that is weakly activating (Table 9.1) and ortho-para directing, but whose reactivity is more easily controlled.

9.10 Synthetic Problems

Let us see how all the foregoing information can be used in designing methods for the preparation of specific compounds. Consider, for example, how you might convert benzene into *para*-nitrochlorobenzene (Fig. 9.19). Since two functional groups must be introduced, we could either do a nitration followed by a halogenation, or the reverse. However, when you consider the second step, you will see that only the chlorobenzene (derived from an initial chlorination of benzene) will give the desired para isomer of the final product. *The order in which the groups are introduced is very important!*

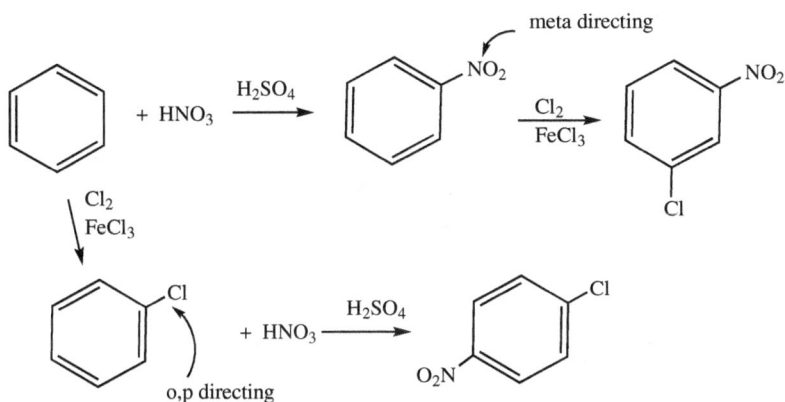

Fig. 9.19

Another example is shown in Fig. 9.20. Here, if one introduces the acetyl group first, the product of F/C alkylation is the meta-disubstituted isomer, while reversing the steps leads primarily to the para isomer.

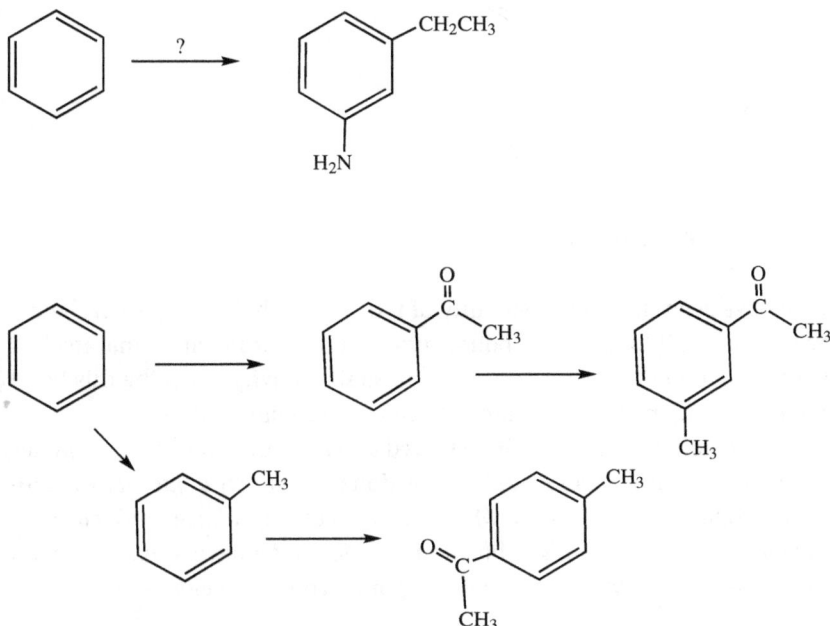

Fig. 9.20

Now consider how you might prepare *meta*-aminoethylbenzene.

Note that both groups are *ortho-para* directing. Therefore, to get the meta isomer, we must introduce a group which is meta directing, add the second group, and then change the first into the desired form. It will be recalled that nitro groups can be very easily transformed into amino groups and acyl groups into alkyl groups by hydrogenation. We could introduce either a nitro or acyl group first, use its meta-directing effect to introduce the second group in the proper place and then reduce (Fig. 9.21). However, recall that nitro groups deactivate the ring so severely that F/C reactions will not take place. Therefore, only the reaction sequence starting with acylation will be successful. More problems of this type are included at the end of this chapter.

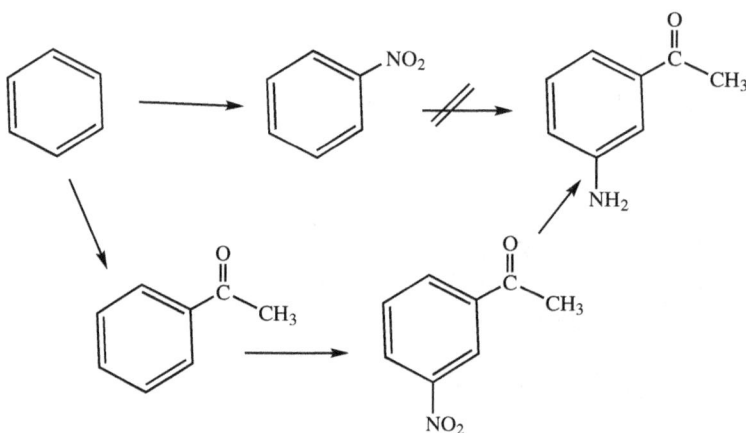

Fig. 9.21

9.11 The Benzylic Position

Previously, we have seen the special stability of ions in which the charge can be delocalized by resonance. It follows that bonds attached to a carbon atom that itself is attached to a double bond might possess some special reactivity. Breaking this bond to give either a carbocation or a carbanion will lead to a *resonance-stabilized intermediate*. This special reactivity is frequently observed and has resulted in special names being given to this position in a molecule. The position *immediately adjacent to an aromatic ring* is called the *benzylic* position while that *immediately adjacent to a carbon–carbon double bond* is called an *allylic* position (Fig. 9.22). We have already seen an example of this special reactivity in which it was pointed out that ketones next to aromatic rings are hydrogenated via the *benzylic* alcohol to the hydrocarbon, a reaction that does not occur when the aromatic ring is absent. Other reactions showing this special reactivity are known. Perhaps the most important is the oxidation of alkyl-substituted benzenes to benzoic acid. This reaction gives the same acid regardless of the length of the alkyl chain. This is useful because the direct introduction of a carboxylic acid on a benzene ring is not possible using an F/C acylation (Cl–CO_2H is not a stable molecule).

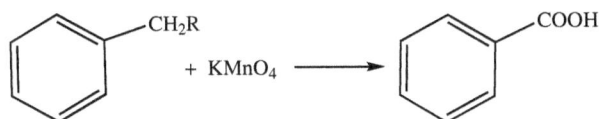

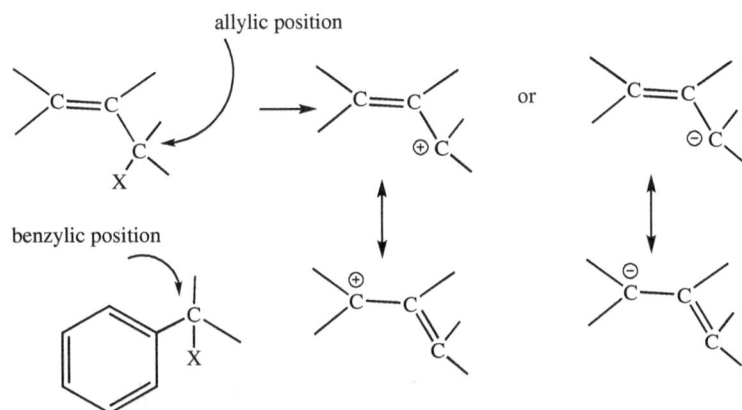

Fig. 9.22

9.12 Final Words On Reactions

It is important to note the difference in typical reaction conditions that cause substitutions on aromatic rings and aliphatic chains. All S_E2 reactions involve positive ions, usually derived by the use of some type of acid catalyst. In contrast, reactions involving aliphatic chains very frequently involve nucleophilic attack of anions. In molecules that contain both kinds of carbon (aromatic and aliphatic), it is usually possible to carry out reactions at one place or the other selectively. Some examples are shown in Fig. 9.23.

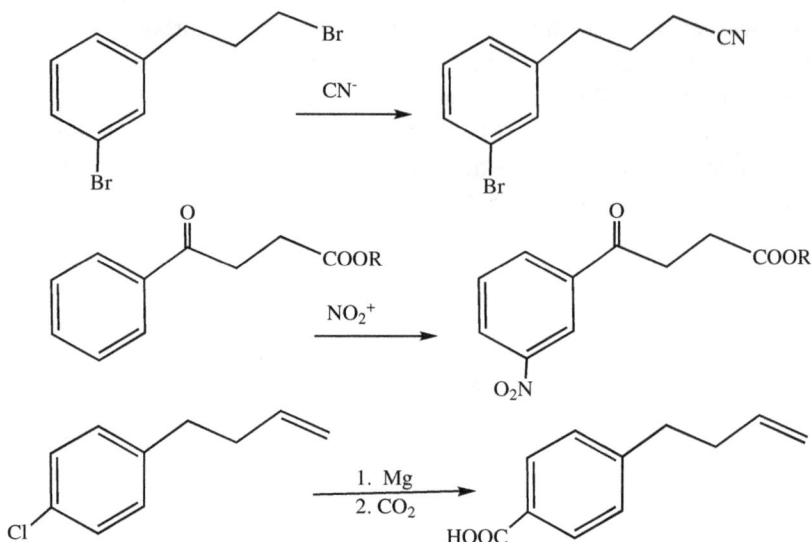

Fig. 9.23

9.13 Phenols and Anilines

9.13.1 Acidity of Phenol

The structures of these two molecules are shown below. It appears that these are examples of an alcohol and amine, but their properties are not in accord with simple aliphatic analogs. We will postpone consideration of aniline to Chapter 10, but take a brief look at phenol here.

The most outstanding difference between simple aliphatic alcohols and phenol is found in their ionization constants. From the equations shown below, it is clear that phenol is a much stronger acid with a K_a of the order of a carboxylic acid. Phenol's trivial name is carbolic acid. Why is it so acidic? Consider the resonance forms available to both phenol and its anion (phenoxide ion) (Fig. 9.24). Exactly the same situation applies here as was discussed in Chapter 4 when the reasons for the acidity of carboxylic acid were considered. The added resonance stability of the phenoxide provides a "driving force" for ionization that is not present in simple alcohols.

9.13.2 Diazonium Salts

Aniline and other aromatic amines undergo reaction with *nitrous* acid (HNO_2) [not nitric acid HNO_3] as shown below. The products are *diazonium salts* which participate in a number of useful reactions. Some of these are shown in Fig. 9.25. These reactions are *not* S_E2 reactions. Note that the second reaction in Fig. 9.25 provides a way of removing

$$CH_3OH \rightleftharpoons CH_3O^- + H^+ \qquad Keq = 10^{-18}$$

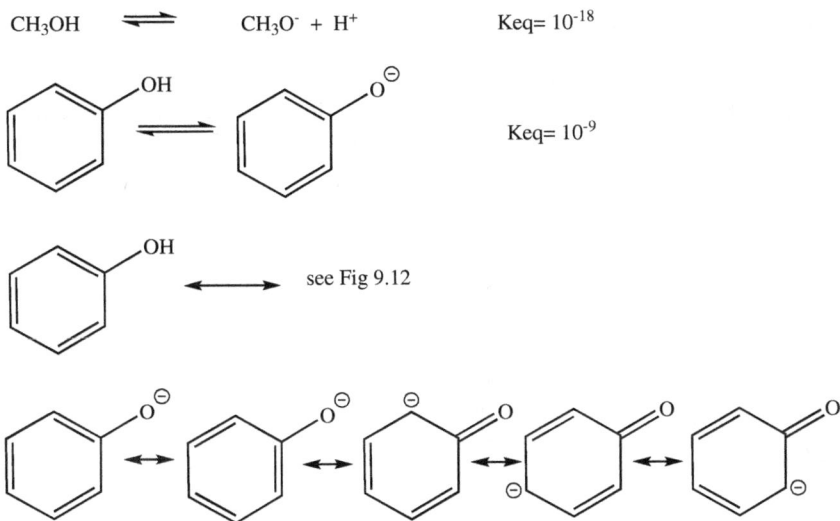

$$Keq = 10^{-9}$$

see Fig 9.12

Fig. 9.24

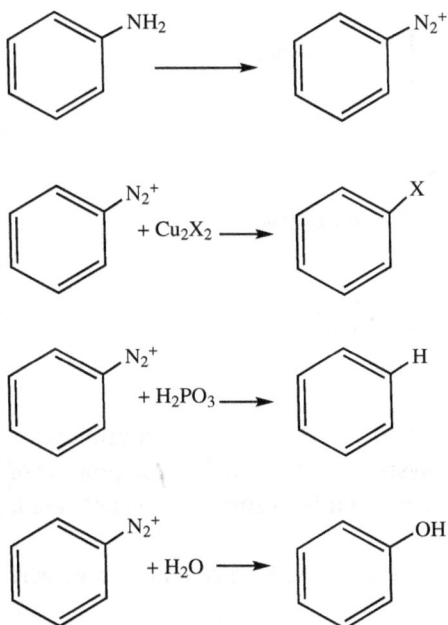

Fig. 9.25

an amino group (and therefore a nitro) group from an aromatic ring. Reactions of the first type shown using diazonium salts and Copper (I) salts are called "Sandmeyer Re-actions." Other reactions of diazonium salts are known but will not be discussed here.

9.14 Problems

9-1. Which of the following compounds will be aromatic? Remember that to be aromatic, Hückel's Rule must be satisfied and all the p-orbitals must be coplanar.

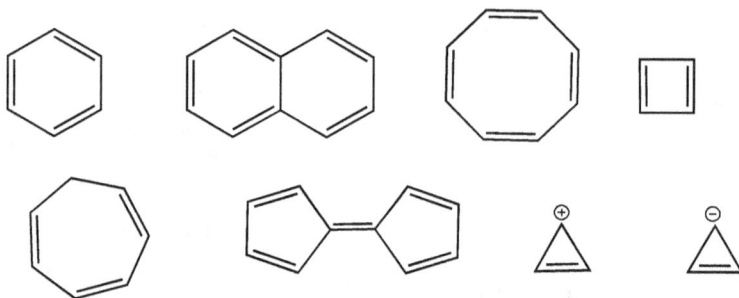

9-2. The following heats of hydrogenation have been obtained:

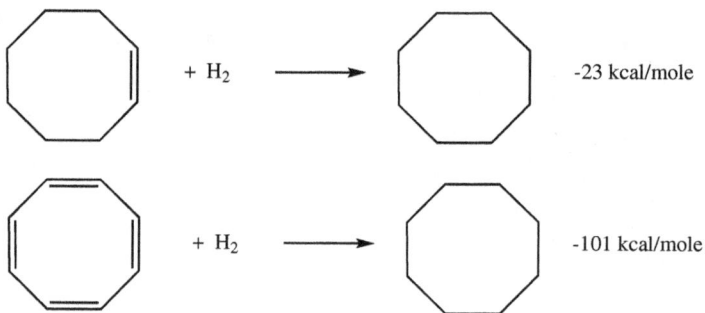

$+ H_2$ ⟶ -23 kcal/mole

$+ H_2$ ⟶ -101 kcal/mole

(a) What is the resonance stabilization energy of cyclooctatetraene?
(b) What is the significance of this number when compared to that of benzene?

9-3. Aniline is a very difficult molecule to handle in electrophilic substitution reactions. Because it is *basic*, it reacts with the catalysts (e.g., H^+ or $AlCl_3$). The product of this reaction is not reactive in $S_E 2$ reactions. Because it is so strongly ring activated, substitution reactions are difficult to control.
(a) Write the structure of the reaction product of aniline with an acid and show why it is unreactive in $S_E 2$ reactions.
(b) Write the resonance structures that indicate why aniline is so reactive.
(c) Under very forcing conditions aniline will undergo $S_E 2$ reactions (e.g., nitration). What product would be expected?

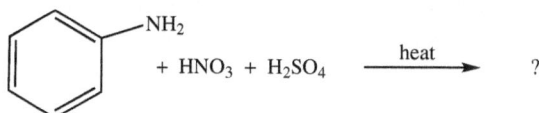

$-NH_2$

$+ HNO_3 + H_2SO_4$ $\xrightarrow{\text{heat}}$?

9-4. Suggest a synthetic sequence that could be used to convert benzene to each of the following.

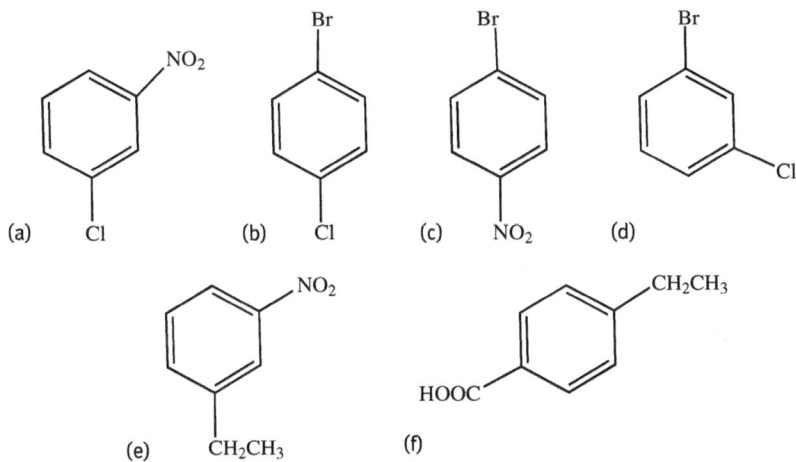

(a)   (b)   (c)   (d)

(e)   (f)

9-5. Replace the "?" with the correct reagent or product.

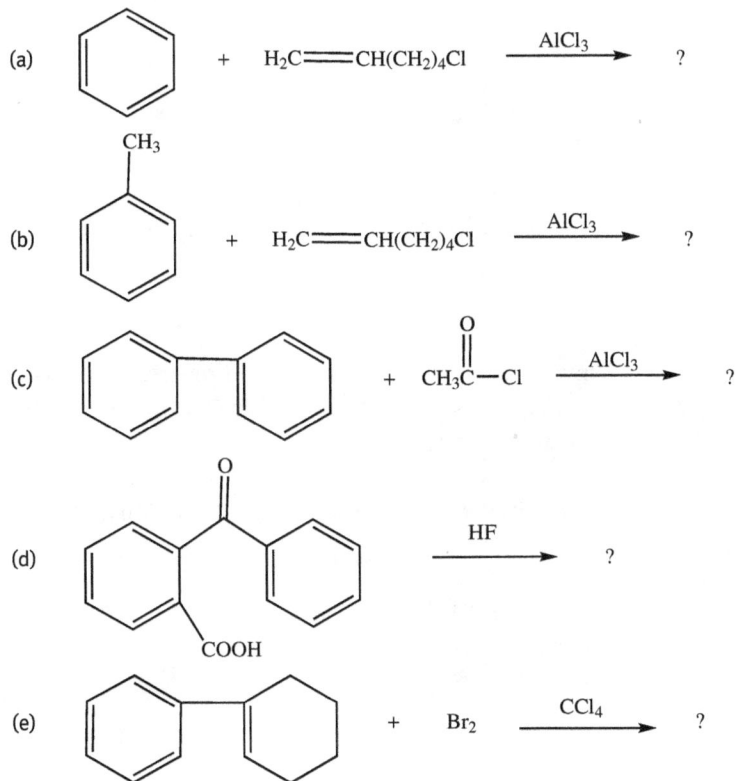

(a)   + $H_2C=CH(CH_2)_4Cl$ $\xrightarrow{AlCl_3}$?

(b)   + $H_2C=CH(CH_2)_4Cl$ $\xrightarrow{AlCl_3}$?

(c)   + CH_3C-Cl $\xrightarrow{AlCl_3}$?

(d) $\xrightarrow{HF}$?

(e)   + Br_2 $\xrightarrow{CCl_4}$?

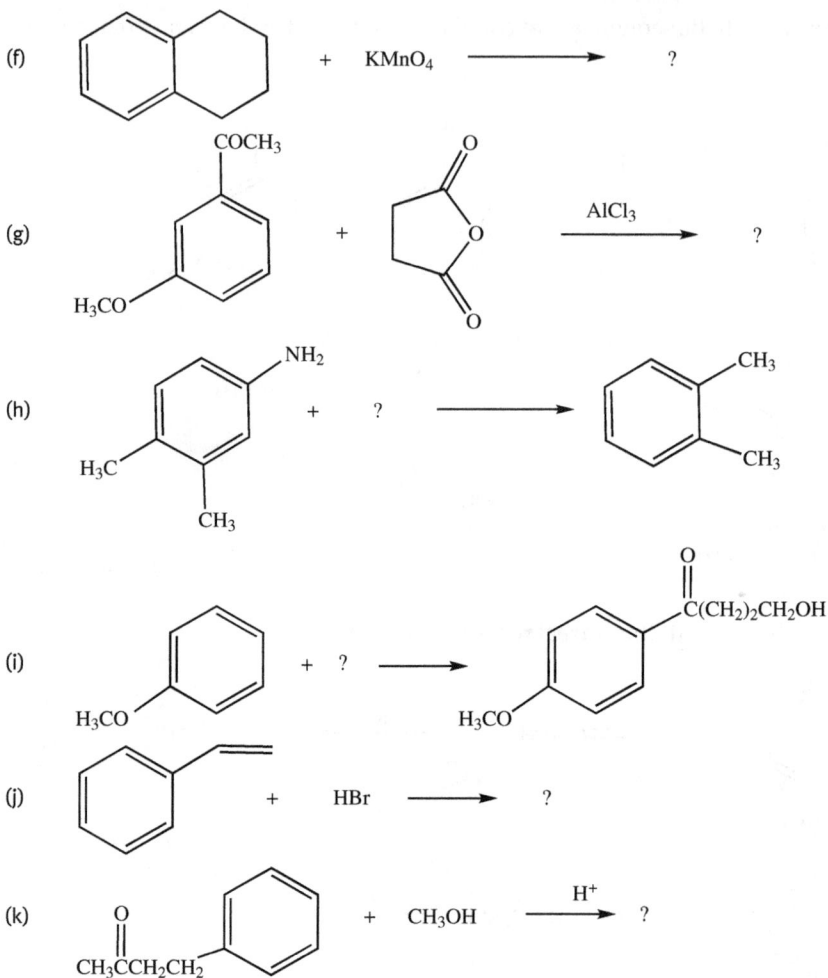

(f)

+ KMnO₄ ⟶ ?

(g)

+

$\xrightarrow{AlCl_3}$?

(h)

+ ? ⟶

(i)

+ ? ⟶

(j)

+ HBr ⟶ ?

(k)

+ CH₃OH $\xrightarrow{H^+}$?

9-6. A compound A ($C_9H_{12}O$) gives CH_3I and a carboxylic acid on treatment with $NaOH/I_2$ followed by acidification. Vigorous oxidation of A with $KMnO_4$ leads to phthalic acid (Fig. 9.1). What is the structure of A?

9-7. Show by equation how you could effect the following transformations, each of which may require more than one step.

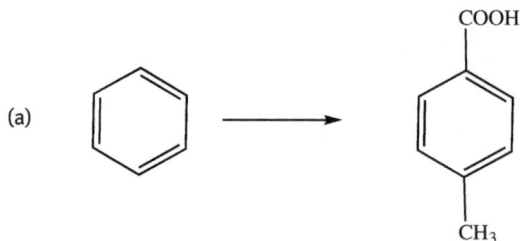

(a)

(b)

(c)

(d)

(e)

(f)

(g)

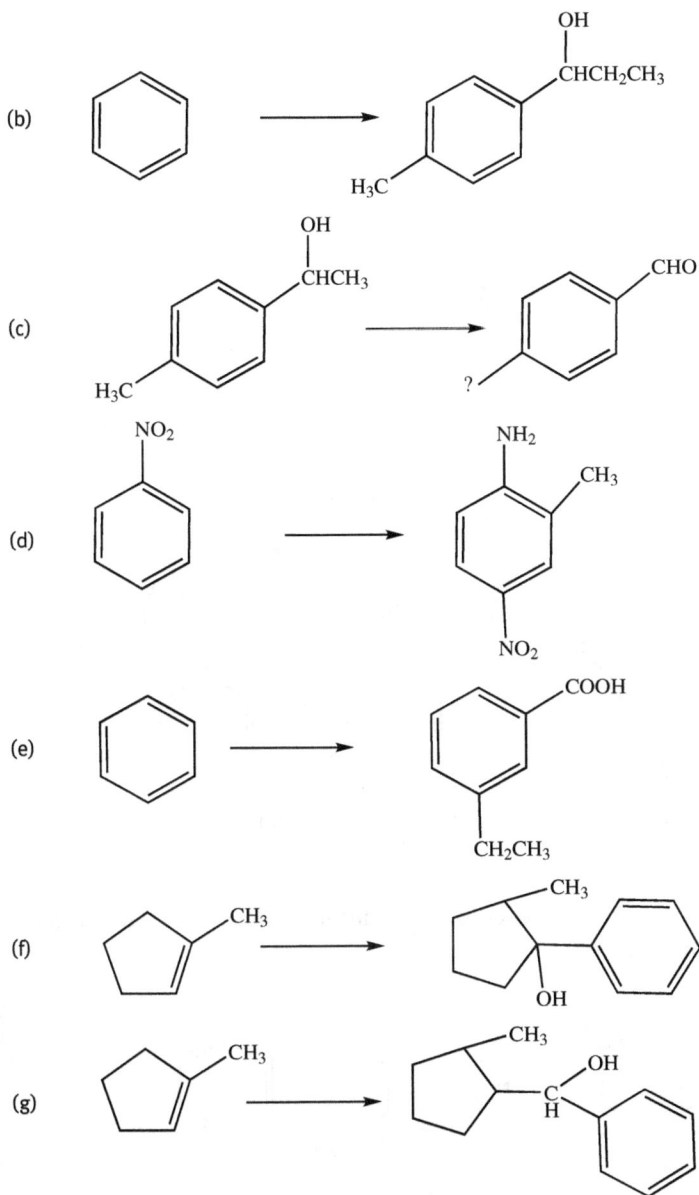

9-8. Naphthalene (Fig. 9.1) undergoes S_E2 reactions at the 1-position much more rapidly than at the 2-position. Explain this observation.

9-9. Anthracene (I) can be prepared by the following series of reactions.

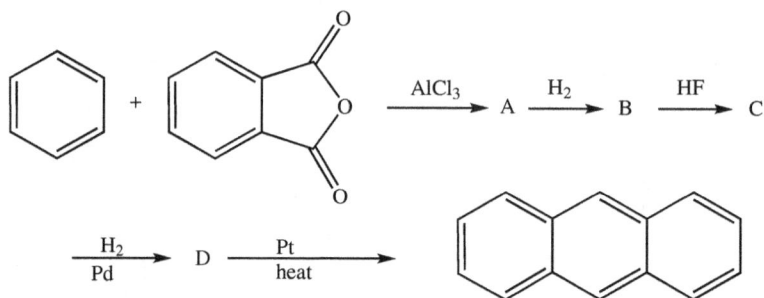

Give structures for the lettered compounds.

9-10. Predict the major product when each of the following is nitrated.

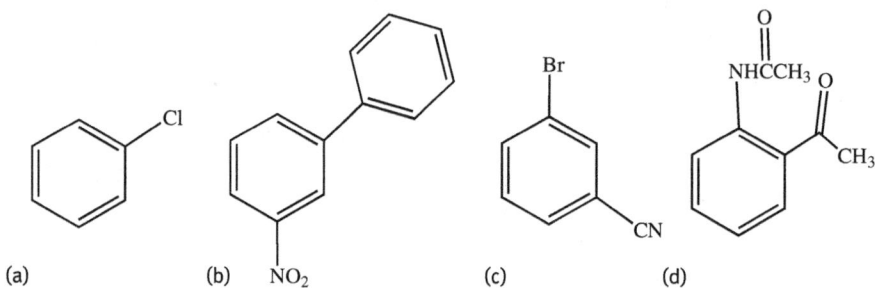

(a)   (b)   NO$_2$ (c)   (d)

9-11. Diazonium salts undergo other reactions in addition to those shown in Fig. 9.25. One of these is shown below.

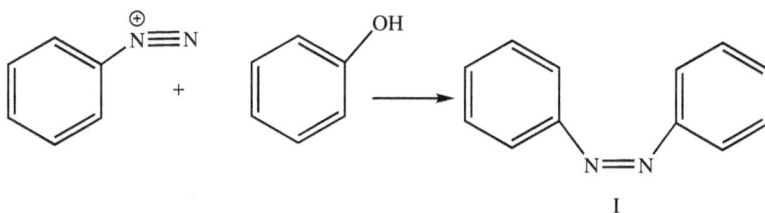

I

When 2,4,6-trimethylphenol is used, the reaction leads to a product II with the formula $C_{15}H_{16}N_2O$. This product, unlike I is not acidic.
a) Write a mechanism for the formation of I.
b) Write the structure of II and show how it is formed.

10 Organic Nitrogen Compounds – Amines and Amides

10.1 Introduction

From time to time in the course of other discussions reference to compounds called amines and amides has appeared. These molecules have the general structures:

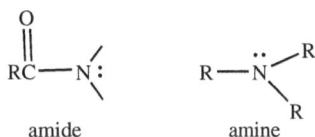

amide   amine

Amines are derivatives of ammonia (NH_3) where one, two, or all three hydrogens have been replaced by alkyl or aromatic groups. Amides are derived from ammonia by replacement of one hydrogen with an acyl group.

10.2 Amines

10.2.1 Nomenclature, Physical and Chemical Properties

As noted above, amines are related to ammonia in the same way alcohols and ethers are related to water. The IUPAC method for the nomenclature of amines is covered in your nomenclature work book and will not be covered here. However, one point must be emphasized. The terms *primary*, *secondary*, and *tertiary* as applied to alcohols were learned previously. In that case, these terms referred to the number of hydrogen atoms on the *carbon* bonded to oxygen. Unfortunately, the same terminology when applied to amines has a different connotation. Here these terms refer to the number of hydrogens on *nitrogen* as is shown below.

Amines, like ammonia, can form strong intermolecular hydrogen bonds (tertiary amines that have no hydrogens on nitrogen cannot do this, but can form hydrogen

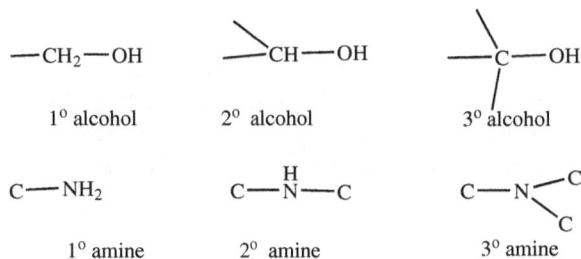

1° alcohol   2° alcohol   3° alcohol

1° amine   2° amine   3° amine

https://doi.org/10.1515/9783110778311-010

bonds with, for example, water). Therefore, they are quite soluble in water. Amines also have a characteristic odor that is usually described as "fishy." Trivalent nitrogen is a neutral atom that possesses an unshared pair of electrons. For this reason, *amines and ammonia are both basic and nucleophilic.* If they function in either capacity, they acquire a positive charge (Fig. 10.1). Recall also that the ammonium ion is acidic (i.e., the first equation in Fig. 10.1 is reversible). In the same way, the product of the second reaction is also acidic. These facts will become prominent when we look at the reactions of amines.

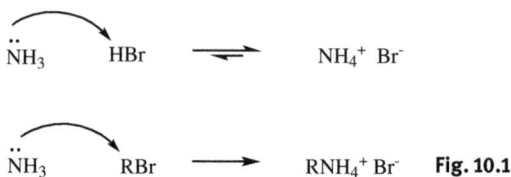

$\overset{..}{N}H_3$ HBr $\rightleftharpoons$ NH_4^+ Br⁻

$\overset{..}{N}H_3$ RBr $\longrightarrow$ RNH_4^+ Br⁻   **Fig. 10.1**

In general, replacement of one, two, or three hydrogens on ammonia by *alkyl groups increases the basicity.* However, *aryl (i.e., aromatic) groups strongly decrease the basicity.* These facts can be seen from the typical examples shown in Fig. 10.2. What makes aniline such a weak base relative to a primary alkyl amine? The same type of reasoning can be applied to this question as we used in explaining the acidity of phenol (Chapter 9).

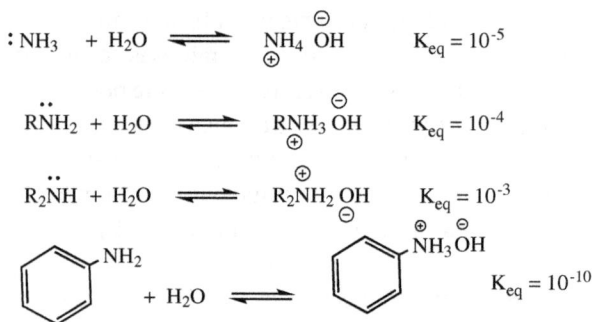

$: NH_3$ $+ H_2O$ $\rightleftharpoons$ $\overset{\oplus}{N}H_4 \overset{\ominus}{O}H$ $K_{eq} = 10^{-5}$

$\overset{..}{R}NH_2 + H_2O$ $\rightleftharpoons$ $\overset{\oplus}{R}NH_3 \overset{\ominus}{O}H$ $K_{eq} = 10^{-4}$

$\overset{..}{R_2}NH + H_2O$ $\rightleftharpoons$ $R_2\overset{\oplus}{N}H_2 \overset{\ominus}{O}H$ $K_{eq} = 10^{-3}$

[aromatic ring]—NH_2 $+ H_2O$ $\rightleftharpoons$ [aromatic ring]—$\overset{\oplus}{N}H_3 \overset{\ominus}{O}H$ $K_{eq} = 10^{-10}$

Fig. 10.2

Consider the resonance possibilities for both the neutral and ionized (protonated) form of aniline (Fig. 10.3). As long as the pair of electrons on nitrogen is available, it can participate in resonance and thus add stability to the molecule aniline. However, if the electron pair is used to form a new N–H bond, it can no longer contribute to the resonance hybrid and the stability is decreased relative to the unprotonated form. Aniline shows a marked reluctance to function as a base.

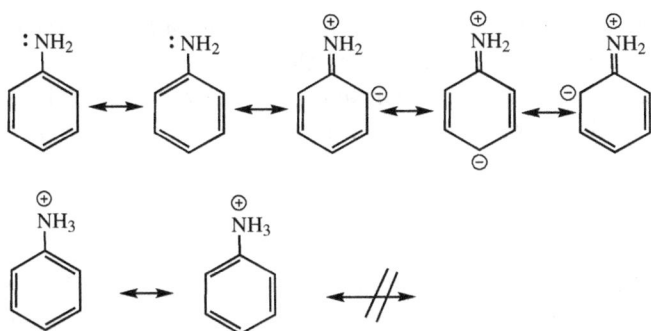

Fig. 10.3

10.2.2 Preparation of Amines

Several preparations of amines have already been seen (Fig. 10.4). Also, because ammonia and amines are nucleophilic, their use in Sn2 reactions is feasible. (Fig. 10.5) The products of such reactions are *ammonium salts*.

It is instructive at this point to look at a series of interrelated compounds derived from ammonia (Fig. 10.6). The first *horizontal* row shows the series of amines formed by sequential replacement of the hydrogens of ammonia by alkyl groups. The second

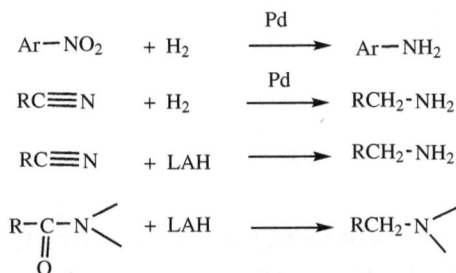

Fig. 10.4

Fig. 10.5

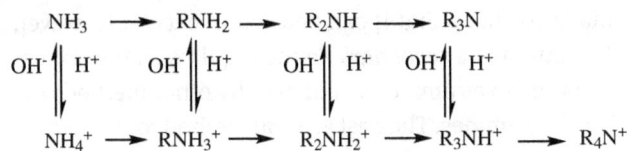

Fig. 10.6

horizontal row shows the series of ammonium salts formed by the same sequential replacement of the hydrogens of the ammonium ion by alkyl groups. When an alkyl halide reacts with ammonia, the product is a primary ammonium salt. However, since ammonia is basic, a second molecule of NH_3 can convert this into a primary amine.

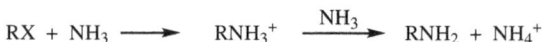

$$RX + NH_3 \longrightarrow RNH_3^+ \xrightarrow{NH_3} RNH_2 + NH_4^+$$

The ammonium salt is neither basic nor nucleophilic (why?). However, the primary amine is both basic and nucleophilic and so can react with a second molecule of alkyl halide. The result is the replacement of more than one of the hydrogen atoms by alkyl groups. This sequence can be repeated over and over until, finally, the quaternary ammonium salt is formed (Fig. 10.7). Therefore, if one mole of alkyl halide is treated with one to four moles of ammonia, a mixture of primary, secondary, tertiary, and a quaternary ammonium salt will be formed, a synthetically undesirable situation. How can this be avoided?

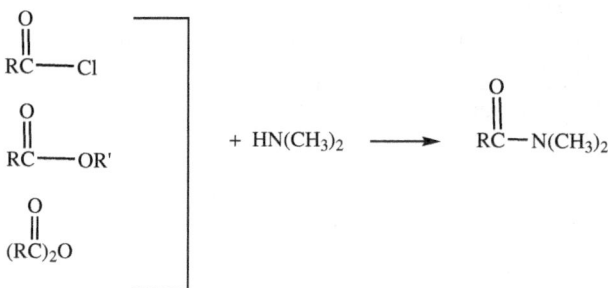

Fig. 10.7

The same difficulty was faced in the F/C alkylation of benzene. A solution to that problem used the fact that, when two bimolecular reactions are in competition, increasing the concentration of one of the reactants will favor the reaction in which it is involved (Sect. 9.9). Since the rate of the first alkylation of ammonia is dependent on the concentration of ammonia, but the rate of alkylation to give *secondary* amine is not, using a large excess of ammonia will cause mainly primary amine to be formed. (You should write the rate equations to prove this point to yourself.)

The result of all this is that the formation of *primary* amines by alkylation of ammonia is feasible since ammonia is so cheap that it can be used in excess. However, to use the same technique for formation of a *secondary* amine requires a large excess of primary amine. These are not as cheap as ammonia and usually other methods are used to prepare secondary and tertiary amines. The best general method for the preparation of secondary and tertiary amines is by reduction of amides using lithium aluminum hydride (LAH). This will be discussed more fully in Sect. 10.3.3.

10.2.3 Reactions of Amines (Other Than alkylation)

The lone pair of electrons on the nitrogen atom of amines controls their reactions. In addition to nucleophilic substitution reactions (Sect. 10.2.2) which leads to higher amines, they can function as nucleophiles toward the carbonyl group. As we have seen, acyl halides and esters can be converted to amides by primary or secondary amines (Fig. 10.7). If a primary or secondary amine reacts with an aldehyde or ketone, the following reaction, which is usually acid-catalyzed, occurs (Fig. 10.8):

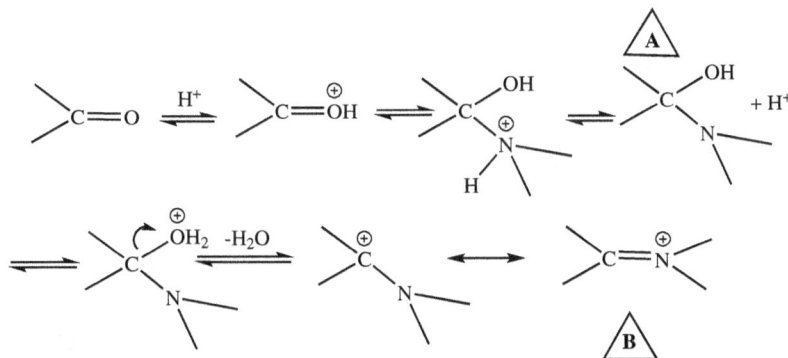

Fig. 10.8

Compare the sequence in Fig. 10.8 with the acid-catalyzed reaction of carbonyls with alcohols. You will see that A in Fig. 10.8 is analogous to a hemiacetal or hemiketal. Also, the intermediate B in Fig. 10.8 is analogous to the intermediate formed during the formation of acetals and ketals (Fig. 7.13). In that case, this carbocation was attacked by a second molecule of alcohol. In the nitrogen case, different reactions occur depending on whether or not the nitrogen atom in B has a hydrogen bonded to it (Fig. 10.9). As noted, the product containing the C=N double bond is called an *imine*, while that with an amine attached directly to an alkene is called an *enamine* (Can you see the derivation of this name?).

Amines R–NH$_2$ in which R is another nitrogen atom (i.e., molecules with the structure N–NH$_2$) are called hydrazines. Hydrazine itself (H$_2$NNH$_2$) was one of the original rocket fuels in WWII. Hydrazines are more nucleophilic than simple amines. *Can you explain why?*

Several important classes of imines are formed when hydrazine and some of its derivatives are reacted with carbonyl compounds. Some of these reagents are called, respectively, hydrazine, 2,4-dinitrophenylhydrazine, semicarbazide, and hydroxylamine. They react very readily with aldehydes and ketones to form imines as shown in Fig. 10.10. Because these products have a clear structural relationship to the starting aldehyde and ketone, and because they are generally solid materials with well-defined melting points, *they are often used to identify liquid aldehydes and ketones.* A liquid

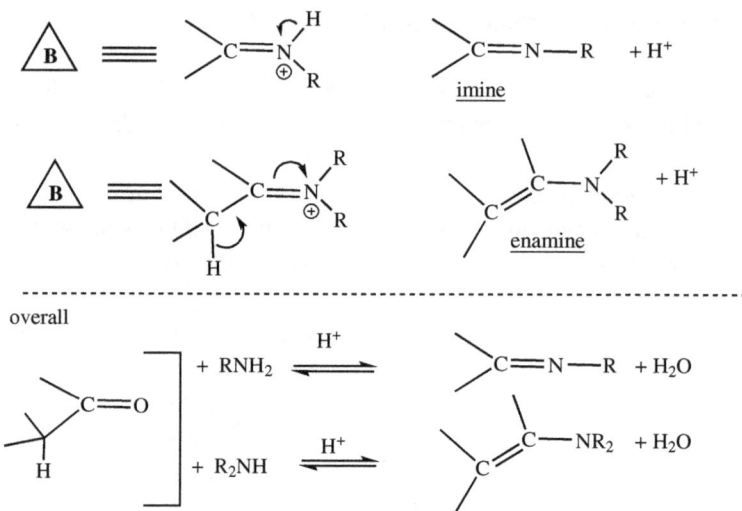

Fig. 10.9

aldehyde of unknown structure can be converted to any one of these "derivatives" and its melting point determined. (This can be determined much more accurately than the boiling point of the aldehyde. Do you know why?). Comparison of this piece of data with those in a table of melting points will frequently identify the unknown aldehyde.

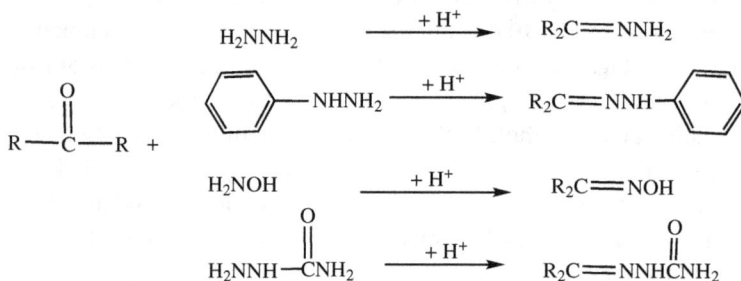

Fig. 10.10

10.2.4 Heterocycles

Heterocycles are cyclic materials that contain a noncarbon as one of the ring atoms. Nitrogen is a very common atom in such situations. Some very common heterocyclic amines are shown in Fig. 10.11. Students who pursue organic chemistry or biochemistry will meet this type of molecule many times in the future.

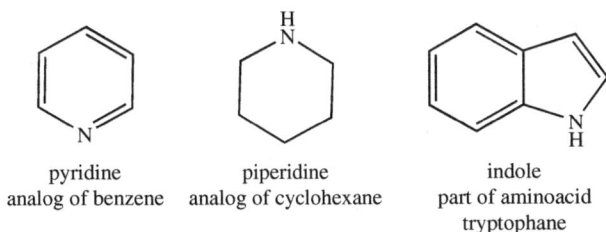

pyridine   piperidine   indole
analog of benzene   analog of cyclohexane   part of aminoacid
                                                 tryptophane

Fig. 10.11

10.2.5 Separations of Organic Mixtures

It is instructive at this point to ask the question, "How might I separate a mixture of an alcohol, amine, and carboxylic acid if they all have about the same boiling point and solubilities?". Many times, this kind of situation is faced when an organic reaction is carried out.

Notice that each of the three types of molecules has different acid-base properties. We might rephrase the question to ask, "How can we separate a basic, a neutral, and an acidic compound?".

Consider the scheme shown in Fig. 10.12. Treatment of the mixture with aqueous hydrochloric acid does not affect the acid or alcohol, but changes the basic amine into its salt which, being ionic, is water soluble. Extraction of this acidic mixture with (for example) ether gives an aqueous solution of the salt of the amine and an ether solution containing the alcohol and carboxylic acid. The free amine can be liberated from its salt and recovered by treating the water solution with sodium hydroxide (which, being a strong base reconverts the amine salt to the amine) and extracting with ether (in which the amine will be soluble).

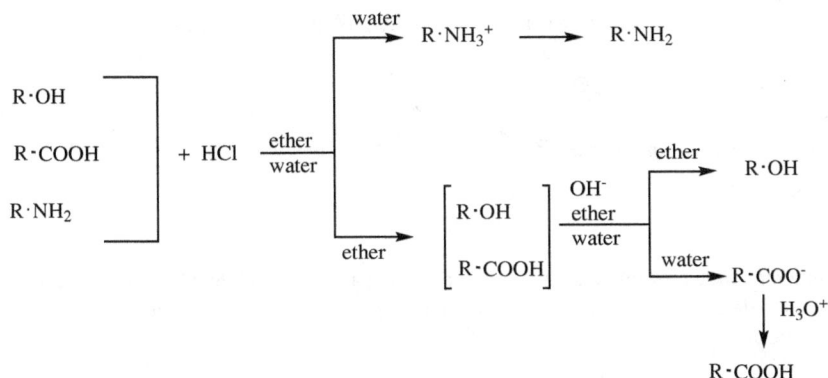

Fig. 10.12

Treating the ethereal solution of alcohol and acid with a base converts the acid to a salt that is now water soluble. Separation of this by water extraction as before, followed by adding a strong acid to the carboxylate salt completes the separation.

It is important that the *concepts behind* this type of procedure, which is based on the chemical (acid-base) properties involved and the solubility properties of ionic and neutral compounds, be understood.

10.3 Amides

We have met the class of compounds called amides before. Before proceeding you should review this material.

10.3.1 Amides – Nomenclature, Physical, and Chemical Properties

Nomenclature, as usual, is treated in your nomenclature text. Most amides are solids at room temperature. The exceptions are the amides of methanoic (formic) acid: formamide, N-methyl-formamide, and N,N-dimethylformamide (DMF). The latter is a very useful solvent.

The chemical properties may seem somewhat surprising at first glance. Unlike amines, amides are *neither basic nor nucleophilic!* Clearly, amides contain a trivalent nitrogen atom which has a lone pair of electrons, but their lack of basicity and nucleophilicity indicates that these electrons are reluctant to form new bonds to hydrogen or carbon. Why should this be so when amines show no such reluctance? Again, resonance provides the answer. The resonance forms of amides (shown below) are a stabilizing factor and would not be possible if the electrons were used to form other bonds. Recall that amides are made from amines and acyl halides, esters or anhydrides and, as we will see in the next section, amides can be hydrolyzed to amines.

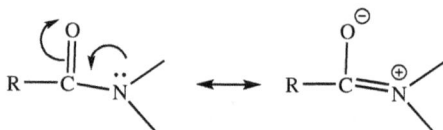

If we want to temporarily prevent an amine from functioning as a nucleophile, etc., it can be converted into an amide and then one can restore the basicity later by a hydrolysis reaction. This is another example of a protecting group (Sect. 7.12.4).

10.3.2 Preparation of Amides

We have previously outlined two preparations of amides (Sect. 8.9). The mechanisms of these transformations have also been outlined (Chapter 8). These are the only two methods of importance and the equations are written again below (Fig. 10.13).

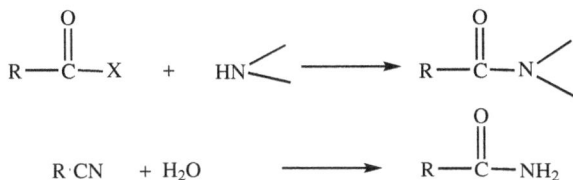

Fig. 10.13

10.3.3 Reactions of Amides

Important reactions of amides include hydrolysis, dehydration, and reduction. Acid-catalyzed hydrolysis follows the general pathway outlined in Chapter 8 (Fig. 10.14). Notice that, whereas the amide nitrogen is not basic, once the resonance is removed by the addition of water as in A, the nitrogen atom again becomes basic. Therefore, this step would be expected to be unfavorable. However, as soon as it does happen, the product A can be protonated, thus transforming what was a poor leaving group (R_2N^-) into a much better one (R_2NH). Notice that, unlike the case of ester hydrolysis, this mechanism is irreversible due to the acid-base properties of the carboxylic acid and amine that are formed.

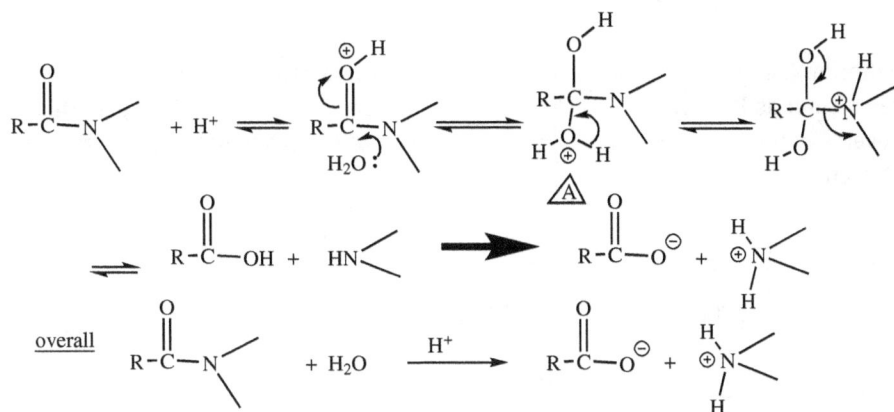

Fig. 10.14

Base-catalyzed hydrolysis follows the usual pathway (Fig. 10.15). Here the leaving group must be R_2N^-, a very poor leaving group and for the hydrolysis of amides, *acid-catalyzed* conditions are usually employed. (Compare this with the situation for ester hydrolysis that was discussed in Chapter 8).

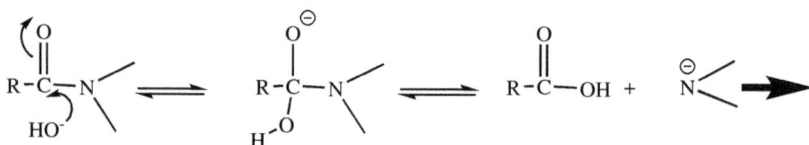

Fig. 10.15

The reduction of amides with lithium aluminum hydride has already been mentioned (Sect. 10.2.2) (Fig. 10.16). Unlike the reaction of esters which leads to two alcohols, LAH reduction of amides leads to amines by complete reduction of the carbonyl group. The formation of secondary amines is frequently achieved by treating the primary amine with an acyl halide and reducing the resulting amide with LAH (Fig. 10.17). Tertiary amines are prepared in the same way from secondary amines.

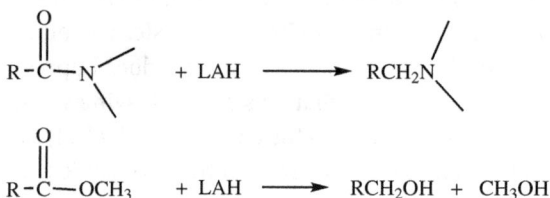

Fig. 10.16

Dehydration (i.e., loss of the elements of water) occurs when amides are treated with reagents such as P_4O_{10}, $SOCl_2$, PCl_3, etc. The products are nitriles. Note that this is the reverse of the last reaction shown in Sect. 10.3.2, although a different mechanism is involved.

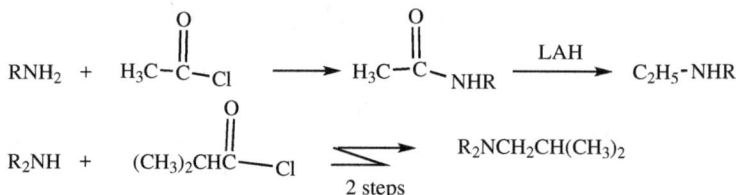

Fig. 10.17

10.4 Problems

10-1. Classify each of the following as primary, secondary, or tertiary amines.

$CH_3CH_2NHCH_3$ $(CH_3)_2CHNH_2$ $CH_3CH_2-N-CH_3$ $CH_3CH_2-N-[CH(CH_3)_2]_2$
$|$
CH_2CH_3

10-2. Arrange the following series of compounds in order of decreasing basicity. Consider resonance, steric, and inductive effects.

(a)

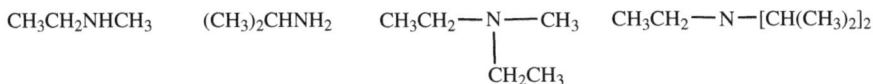

(b) CH_3NHCH_3

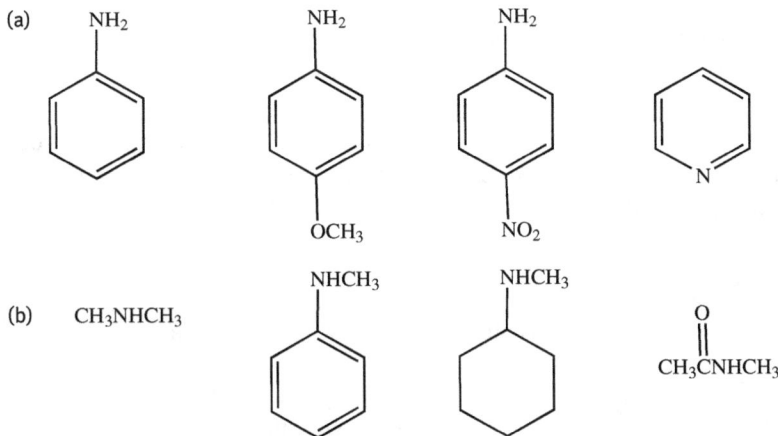

10-3. Amides are usually prepared from amines and either acyl halides or anhydrides. Why are carboxylic acids not used?

10-4. The reaction of trimethyl amine and acetyl chloride gives a compound that is very reactive. If an alcohol is added to a solution of this product, esters are formed in high yields. Show the structure of the product, account for its high reactivity, and draw a mechanism for the ester formation.

10-5. Pyrrole (I) and pyridine (II) show aromatic properties. Using Hückel's Rule, show why this is so.

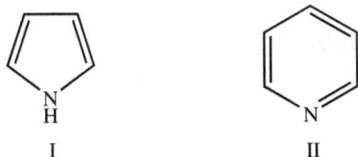

I   II

10-6. Show by equation how each of the following could be prepared from ethanamide (CH_3CONH_2).

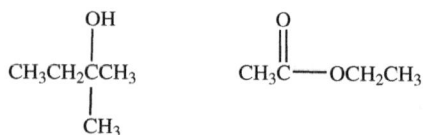

$$CH_3CH_2NH_2 \qquad CH_3CH_2\overset{\displaystyle OH}{\overset{|}{C}}HCH_2CH_2CH(CH_3)_2 \qquad CH_3COOH$$

$$CH_3CH_2\overset{\displaystyle OH}{\underset{\underset{\displaystyle CH_3}{|}}{\overset{|}{C}}}CH_3 \qquad CH_3\overset{\displaystyle O}{\overset{\|}{C}}\!\!-\!\!OCH_2CH_3$$

10-7. Semicarbazide (Fig. 10.10) has three nitrogen atoms. When it reacts with a ketone, only one of these reacts. Explain why the other two nitrogens are unreactive.

10-8. The reaction of dimethyl amine with benzyl bromide (I) gives a mixture of products as shown. Suggest how you could take advantage of the *physical* and *chemical* characteristics of these to effect their separation.

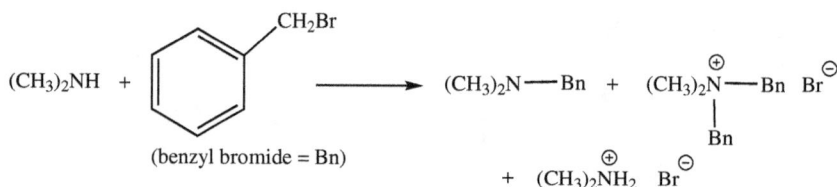

(benzyl bromide = Bn)

$(CH_3)_2NH \;+\;$ [benzene ring with CH_2Br] $\longrightarrow$ $(CH_3)_2N\!\!-\!\!Bn \;+\; (CH_3)_2\overset{\oplus}{N}\!\!-\!\!Bn \;\; Br^{\ominus}$

$\overset{\displaystyle |}{\underset{\displaystyle Bn}{}}$

$+\; (CH_3)_2\overset{\oplus}{N}H_2 \;\; Br^{\ominus}$

10-9. Replace the "?" in the following equations with the correct reagent or product. Indicate any necessary catalysts and relevant stereochemistry.

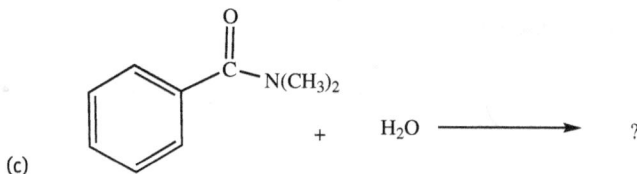

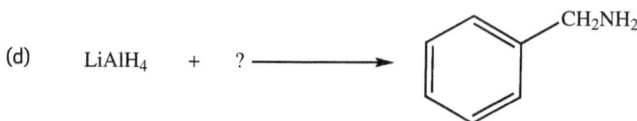

(a) CH_3CH_2COOH + NH_3 $\longrightarrow$?

(b) $CH_3CH_2CH_2Br$ + $(CH_3)_3N$ $\longrightarrow$?

(c)   [benzene ring with $\overset{\displaystyle O}{\overset{\|}{C}}\!\!-\!\!N(CH_3)_2$]   + H_2O $\longrightarrow$?

(d) $LiAlH_4$ +   ? $\longrightarrow$ [benzene ring with CH_2NH_2]

(e)

(f)

+   CH₃NH₂   ⟶   ?

(g)

CH₃CCH₂CH₃   +   ⟶   ?

(h)

+   CH₃NH₂   ⟶   ?

(i)   (CH₃)₂CHCNH₂   +   SOCl₂   ⟶   ?

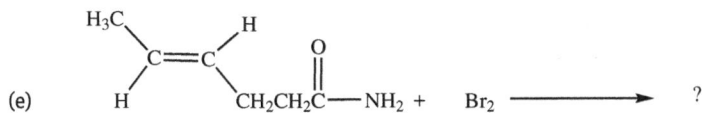

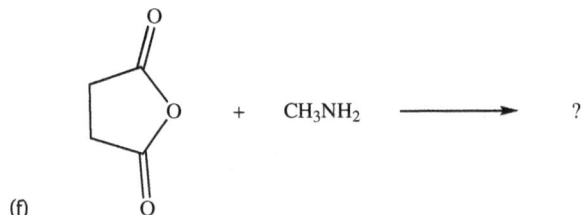

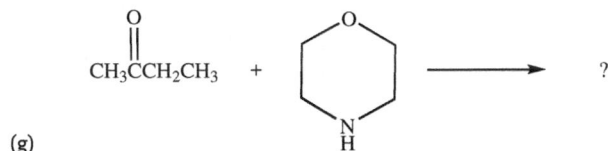

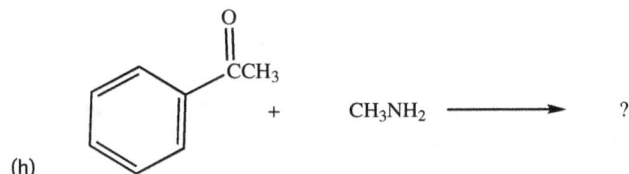

10-10. Show by equation how you could prepare:
(a) methyl ethyl amine (3 ways)
(b) pentanamide (2 ways)
(c) 1-methylpropyl diethylamine (from 2-butanol)
(d)

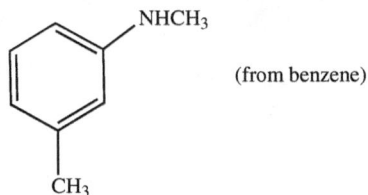

(from benzene)

10-11. When compounds I and II are reacted with one equivalent of acetyl chloride, the products shown are obtained. Explain why the alcohol reacts in one case, but the amine reacts in the other.

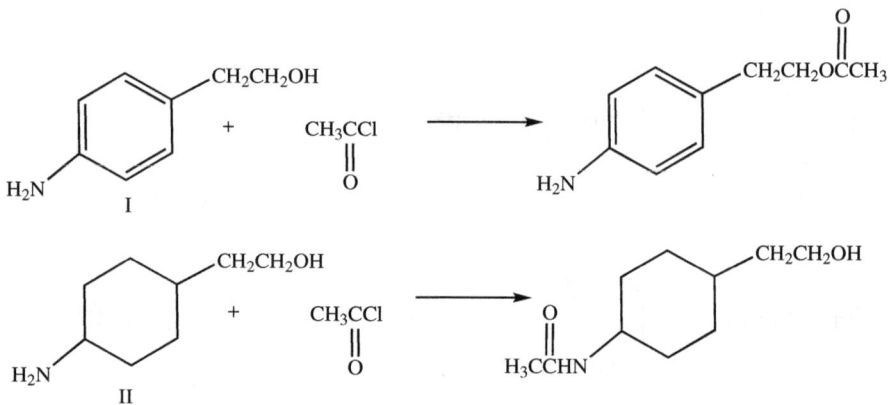

10-12. As indicated in Problem 9.3, aniline is difficult to handle in S_E2 reactions. However, conversion of aniline to acetanilide (I) avoids these problems and acetanilide can be used in S_E2 reactions quite easily. Explain why the presence of the acetyl group avoids the two problems mentioned in Problem 9.3.

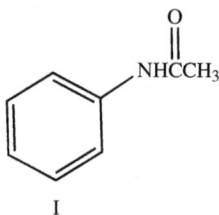

11 Introduction to Spectroscopy

11.1 Introduction

Until now, when you were told that the product of a reaction had a certain structure, you simply had to accept this as fact. However, the question of "How was that structure determined?" must have occurred to many of you. The answer to that question depends to a large extent on *when* the structure determination took place. Prior to 1900, it was frequently the culmination of many years of work. Materials of unknown structure were subjected to reactions which led to other compounds whose structures had been determined and from these data, the original structure was deduced – frequently incorrectly! Later, with the advent of more modern techniques, all of which rely on the interaction of matter and energy in some form, it became possible to relate the absorption or emission of energy to structural features. The term spectroscopy was originally used to describe the interaction of light energy with material. Today, the term has been expanded to describe the interaction of energy in any form with material. Physicists make absolute correlations between the energy absorption and structure based on complex calculations, but chemists, at least in the case of organic compounds, prefer to do this correlation on an empirical basis.

In this chapter, we will look at the basics of two techniques that are used every day by practicing chemists to determine the structure of organic compounds. The intent is not to present complete details of each technique, but rather to introduce you to the *general* aspects and show you how they can be used. If you pursue more advanced courses in chemistry, you will learn much more detail.

In addition, the details of the instrumentation used in each kind of spectroscopy will be omitted and left to future courses for elaboration.

11.2 First Things First

The most important data piece, without which structure determination is extremely difficult, is one which you have all learned about in (probably many) previous courses: i.e., the *molecular formula*.

Without this information the number of possible structures is limitless. In addition, with a knowledge of the molecular formula, the IHD (see Chapter 2) can be determined which is another very useful piece of information. Methods for determining the molecular formula include combustion analysis, where the molecule is burned and the amount of water and carbon dioxide produced is measured and mass spectrometry. The former concept has been previously introduced (Sect. 5.2.1), but the use of mass spectroscopy will have to wait for a future, more advance course.

https://doi.org/10.1515/9783110778311-011

In *any* structure determination, the knowledge of the molecular formula is the *first* piece of information that is required.

11.3 Infrared (IR) Spectroscopy

11.3.1 Introduction and General Principles

As noted above the interaction of light with material was the initial form of spectroscopy. Depending on the wavelength of light, different kinds of information can be obtained. When light in the infrared range interacts with organic compounds, information about the presence or absence of many *functional groups* is obtained. Correlation between the wavelength of light that is absorbed and the presence of a particular functional group is the result of empirical investigations: that is the spectra are obtained on many compounds of known structure and similarities in the wavelength of absorbed light with compounds that have the same functional group are noted. Tables are then drawn up and when an absorption at a particular wavelength is noted in the spectrum of a compound of unknown structure, its associated functional group can be inferred.

For this course, memorization of the table is *not* expected. When such information is required, it will be provided.

The IR spectrum range extends from $4000\,\mathrm{cm}^{-1}$ to about $600\,\mathrm{cm}^{-1}$. (The unit cm^{-1} is called a wave number and is defined as $1/\lambda$ where λ is the wavelength of light. Some texts use light *frequency* instead but this will not be used here). The part of the spectrum between $4000\,\mathrm{cm}^{-1}$ and about $1450\,\mathrm{cm}^{-1}$ is referred to as the *Characteristic Group Frequency Range* and it is here that the most useful information can be obtained for organic molecules. Functional groups such as alcohols, ketones, aldehydes, esters, acids, some alkenes and alkynes, and nitriles give absorptions in this area. The positions of these absorptions are quite well defined. If the spectrum of a molecule that contains an alcohol is obtained, it *will* show an absorption at the same place that all molecules that contain the same functional group show one.

The area of the spectrum between $1450\,\mathrm{cm}^{-1}$ and $600\,\mathrm{cm}^{-1}$ is referred to as the *Fingerprint Region*. This is an apt name since every molecule has a unique set of absorptions in this area. These are very difficult to interpret in terms of structure. However, if you have a compound for which you have a possible structure and you also have an authentic sample of the same material, the coincidence of all absorptions in the fingerprint region of the two spectra is proof that the materials are identical. This constitutes a "fingerprint" of the molecule.

IR spectroscopy gives little useful information about the parts of organic molecules that are not functional groups.

11.3.2 IR Correlations

(a) Alcohols

The region between $3500\,\text{cm}^{-1}$ and $2900\,\text{cm}^{-1}$ is where absorptions due to OH groups are found. OH groups can be alcohols or a part of carboxylic acid and the positions and appearance of these two groups are quite distinct. The OH group of alcohols ranges between $3500\,\text{cm}^{-1}$ and $3200\,\text{cm}^{-1}$. The exact position depends on the type of alcohol (primary, secondary, tertiary) and also on the concentration of the sample. It is usually reasonably broad. Conversely, the OH of a carboxylic acid usually extends from about $3300\,\text{cm}^{-1}$ to $2600\,\text{cm}^{-1}$ or even lower. It is tremendously broad and strong.

It is unusual for an organic compound not to have C–H bonds present. These all occur between $3100\,\text{cm}^{-1}$ and $2850\,\text{cm}^{-1}$. Therefore, an absorption in this range is present in essentially all organic molecules and the OH of a carboxylic acid will overlap with the CH bond absorptions.

Again, note that it does not matter whether the molecule is methanol, ethanol, propanol, or some alcohol with other functional groups present, there will be an absorption in this range. Conversely, the *absence* of such an absorption demands the absence of an OH group.

Table 11.1 gives the positions of characteristic group absorptions. Note that the numbers are generally $\pm 5\,\text{cm}^{-1}$. In addition to the positions given, there are some other, very useful correlations that can be made, particularly with regard to the *carbonyl* region.

(b) Carbonyl Absorptions

The carbonyl group (C=O) is a part of several different functional groups (aldehyde, ketone, acid, ester, amide, anhydride, etc.). All carbonyls absorb strongly in the IR range in the region between $1850\,\text{cm}^{-1}$ and $1650\,\text{cm}^{-1}$, but the place within this region where each of these functional groups absorb is quite well defined (see Table 11.1). In addition, certain structural features influence these positions in a well-defined way.

The presence of *conjugation* on any carbonyl reliably shifts the absorption 30–$35\,\text{cm}^{-1}$ lower than the "expected" location. An ester, which would normally show an absorption about $1740\,\text{cm}^{-1}$ (Table 11.1), will absorb at $1710\,\text{cm}^{-1}$ if conjugated.

$$
\underset{\substack{\text{not conjugated}\\1740\,\text{cm}^{-1}}}{CH_3CH_2CH_2CH_2\overset{\overset{\displaystyle O}{\|}}{C}OCH_3}
\qquad
\underset{\substack{\text{not conjugated}\\1740\,\text{cm}^{-1}}}{CH_3CH{=}CHCH_2\overset{\overset{\displaystyle O}{\|}}{C}OCH_3}
\qquad
\underset{\substack{\text{conjugated}\\1710\,\text{cm}^{-1}}}{CH_3CH_2CH{=}CH{-}\overset{\overset{\displaystyle O}{\|}}{C}{-}OCH_3}
$$

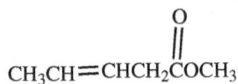

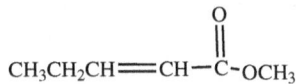

The position of absorption of a carbonyl group can be influenced by another strong effect. When it is contained in a ring that has six atoms or more, the atoms can arrange themselves to accommodate the natural tetrahedral angles of sp^3-hybridized carbon. However, if there are less than six atoms in the ring, the angles cannot be accommodated and *strain* is introduced into the molecule. This strain is reflected in the position of the carbonyl absorption. Whereas conjugation lowers the absorption, strain *raises the position by approximately* 35 cm^{-1} *for each atom of ring size decrease below six*. The two effects (conjugation and ring strain) can both be present in the same molecule.

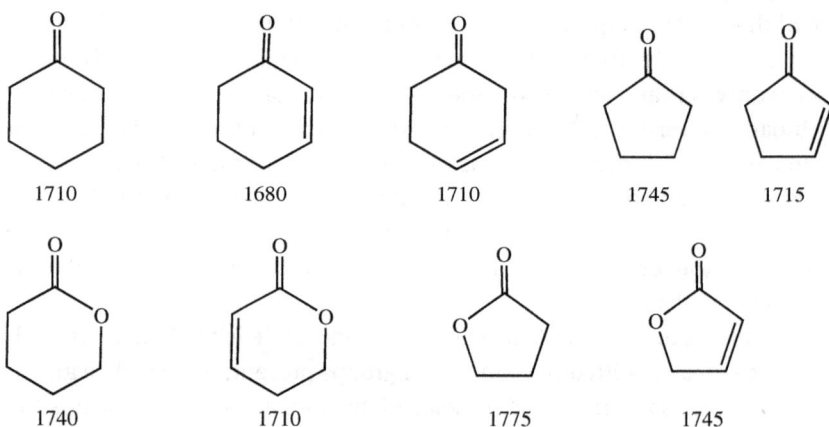

| 1710 | 1680 | 1710 | 1745 | 1715 |

| 1740 | 1710 | 1775 | 1745 |

11.4 The Appearance of an IR Spectrum

Shown below is a typical IR spectrum.

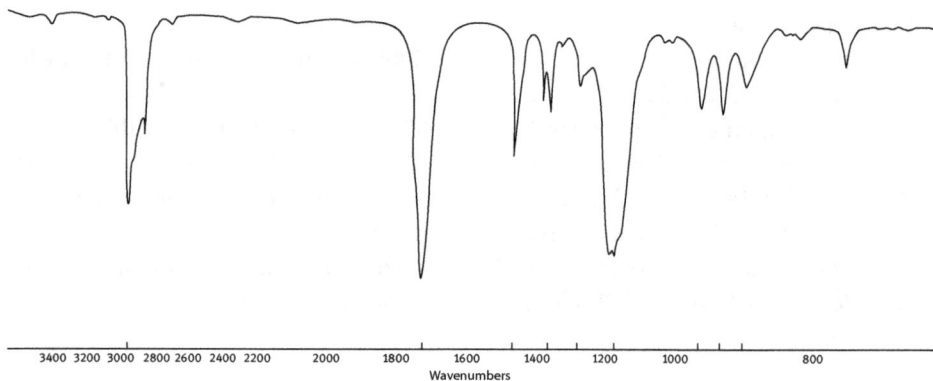

3400 3200 3000 2800 2600 2400 2200   2000   1800   1600   1400   1200   1000   800

Wavenumbers

*used by permission from NIST

The molecule is isobutyl methanoate (formate). As can be seen, in the region in which we are interested ($4000\,cm^{-1}$ to $1450\,cm^{-1}$) there are only two absorptions – one at about $3000\,cm^{-1}$ [C–H] and another at $1735\,cm^{-1}$ [ester].

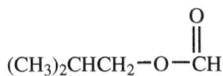

$$(CH_3)_2CHCH_2-O-\overset{\overset{\displaystyle O}{\|}}{C}H$$

The next spectrum is of levulinic acid [4-ketopentanoic acid].

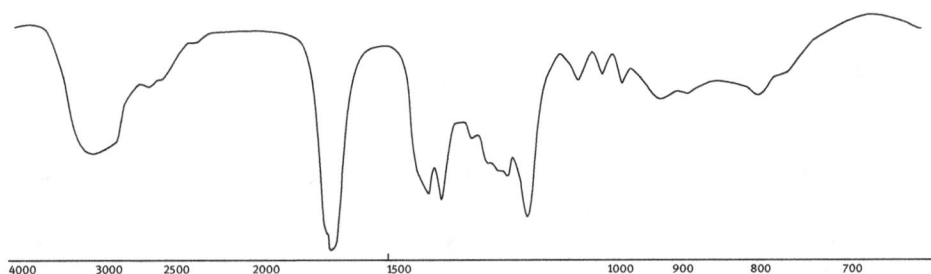

$$CH_3\overset{\overset{\displaystyle O}{\|}}{C}CH_2CH_2\overset{\overset{\displaystyle O}{\|}}{C}-OH$$

*used by permission of NIST

As you can see, there is a *huge* OH absorption stretching from $3600\,cm^{-1}$ to $2400\,cm^{-1}$, which is indicative of a carboxylic acid. This is confirmed by the carbonyl absorption at $1705\,cm^{-1}$.

Closer inspection will show a second carbonyl absorption as a "shoulder" at about $1720\,cm^{-1}$, which is due to the ketone in the molecule.

11.5 How to Use IR data

In solving the structure of an unknown molecule, the steps to be followed are well defined. The first steps involve determining (if necessary) the molecular formula and the Index of Hydrogen Deficiency. Subsequently, the IR data will reveal what functional groups are present. However, it is the function of the next spectroscopic technique to "stitch" the functional group(s) together into the complete molecular framework.

11.6 Nuclear Magnetic Resonance

Nuclear Magnetic Resonance Spectroscopy (NMR) is a relatively modern tool which is tremendously powerful in supplying information that allows the final assembly of a molecular structure. The fundamental theory of NMR is not particularly complex and an understanding of the basics will be required for our use. It is fundamentally different from other forms of spectroscopy in that the *nucleus* and not the electronic shells of the atom is of central interest.

The nuclei of some atoms have magnetic properties that allows them to be viewed as bar magnets. Everyone knows that a magnet, when placed in a magnetic field, will align itself "with the field" – that is the "north pole" of the magnet will turn to the "south pole" of the external field. Any other alignment is of higher energy – which can be confirmed by the fact that it requires "work" (in the physics sense) to alter this alignment. If an atom with such a nucleus is placed in a magnetic field, it will align itself as expected. If energy (in the form of radio-frequency radiation) of exactly the correct frequency is introduced into the system, the bar magnet will absorb energy and align itself "against the field." The absorption of the energy can be detected and correlated with molecular structure.

The most common nuclei with this "bar magnet" property are ^{1}H, ^{13}C, and ^{31}P. The first two will be of interest to us but it must be noted that the latter is the basis for MRI instruments (MRI = NMR but if patients were told they were going into a *nuclear* machine, it might cause panic!).

If a bar magnet is put in the magnetic field of another magnet (an applied magnetic field) it will naturally align itself "with the field." If one tries to turn the magnet to any other orientation, the force required will depend on the strength of the applied field; the stronger the field the more force is required and therefore the more work must be done to reorient the magnet. In other words, the work required is proportional to the strength of the applied field. Similarly, when a nucleus is placed in a magnetic field, the energy (i.e., the frequency) required to cause the nucleus to reorient itself against the field will depend on the strength of the applied magnetic field. This frequency is called the *resonance frequency*.

The frequency required for resonance is proportional to the strength of the applied magnetic field.

If all nuclei were absorbed at the same frequency, not much information would be obtained from the spectrum. Even if ^{13}C and ^{1}H nuclei absorbed at different frequencies (and they do) the only information this technique would supply would be that carbon and hydrogen were present – not very useful stuff! Fortunately, other factors intervene to change the resonance energy by very slight amounts and it is these differences that supply the information that is useful is structure determination.

For the next few sections, we will refer specifically to the spectra of *hydrogen* atoms (often referred to as proton spectra). However, you should note that the principles also apply to other atoms with the same magnetic properties and we will be applying the same principles to carbon atoms later in this chapter.

11.7 The Scale

The typical frequencies that are used in today's instruments range from 300 MHz (300 million cycles per second) to 500 MHz – very large numbers. The frequency *differences* that will give the information in which we are interested are generally in the 1 to 10 Hz range – relatively very small numbers. Since it is very difficult to measure small differences between large numbers accurately, a standard is always added to the samples to be measured. This standard is $(CH_3)_4Si$, tetramethylsilane or *TMS*. Its protons resonate at higher field than the protons in almost all other organic compounds. It is arbitrarily assigned the value zero and everything is measured relative to its absorption. The difference is called the *chemical shift*.

A scale is defined by the expression

$$\delta = \frac{\text{freq (sample)} - \text{freq (TMS)}}{\text{operating freq. of spectrometer}}$$

The units of δ are parts per million (ppm).

NMR spectra can be obtained by holding the magnetic field constant and varying the frequency or by holding the frequency constant and varying the magnetic field. When discussing the spectra, it is usual to consider the field as being varied.

11.8 The Chemical Shift

What causes the small differences in resonance frequency that give us our structural information? You will remember that the things we call bonds between atoms are actually electrons that are circulating around nuclei in space called orbitals. You should also recall from Physics courses that electrons moving in a magnetic field establish their own field which is in the *opposite direction to the applied field*. Therefore, the field experienced by the nucleus is the vector sum of the two fields – the applied field and that caused by the bonding electrons. Since the field produced by the electrons *reduces* the effect of the applied field on the nucleus, it is said to "shield" the nucleus. *The strength of the shielding is proportional to the electron density around the atom.* Consider a proton attached to the oxygen atom of an alcohol. Since oxygen is more electronegative than hydrogen, the bonding electrons will spend more time near the oxygen – the bond is polarized. Therefore, the shielding effect of the electrons on the hydrogen atom will be small (low electron density = small shielding). As usually plotted on an NMR spectrum, shielding *increases* toward the right end of the spectrum and *decreases* toward the left. The universal terminology used is that

Shifts to the left are *downfield* and shifts to the right are *upfield*.

In general, atoms that are attached to electronegative atoms will be shifted downfield relative to atoms that are not. There are many other effects that change chemical shifts. Such things as hybridization are extremely important and we will see many of these in upcoming sections.

General areas for ^{1}H peaks

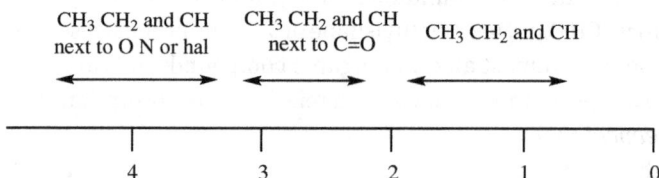

CH$_3$ CH$_2$ and CH next to O N or hal	CH$_3$ CH$_2$ and CH next to C=O	CH$_3$ CH$_2$ and CH

4   3   2   1   0

General areas for ^{1}H peaks
Part B

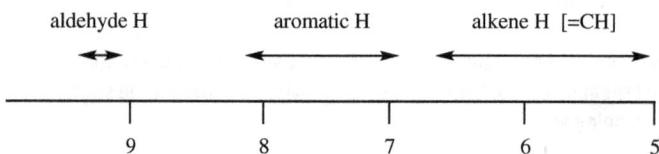

aldehyde H	aromatic H	alkene H [=CH]

9   8   7   6   5

At this point let's look at a typical ^{1}H NMR spectrum. This is just to give you a sense of what they look like. Don't worry yet about the detailed appearance of each signal.

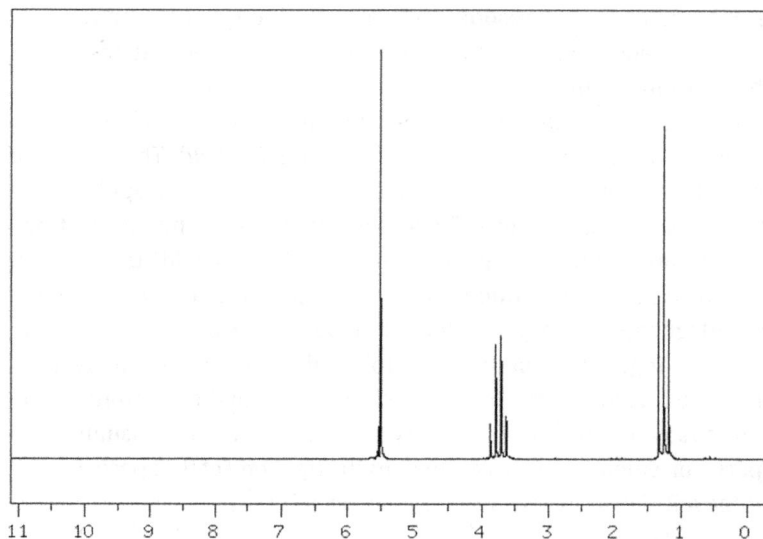

11   10   9   8   7   6   5   4   3   2   1   0

The molecule is $Cl-CH_2-O-CH_2-CH_3$. As you can see there are three different kinds of hydrogens and three signals. The signal at $\delta = 5.5$ is *downfield* (to the left) from the signal at $\delta = 3.7$

At this point what could we learn from an NMR spectrum? Since each kind of hydrogen atom in a molecule will have a different electronic environment (the differences may be *very* small!) there should be one absorption for each kind of hydrogen. For example, one would expect two signals for propane, two signals for butane (note the symmetry in the molecule!), five signals for 2-butanol, and two signals for 3-pentanone. Make sure you can see these relationships.

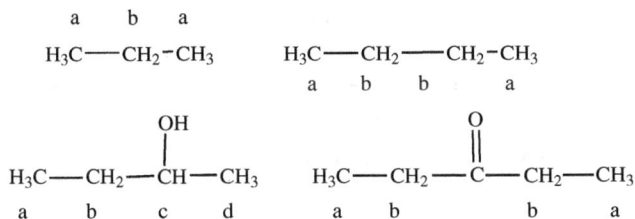

This would be useful information, but the NMR spectrum actually gives much more.

11.9 The Integral

A second piece of useful information comes from the fact that the *area under an absorption is proportional to the number of hydrogen atoms causing it*. Therefore, using the example of the four compounds shown above, the spectrum of propane would show two absorptions in the ratio of 3:1 (or 6:2), that of butane would show two absorptions in the ratio of 3:2 (or 6:4), that of 2-butanol would show five absorptions in the ratio of 3:3:2:1:1, and the spectrum of 3-pentanone would show two absorptions in the ratio of 3:2. The NMR spectrometer has the capability of measuring the peak areas electronically. They are often shown as both numbers below the peaks and curves superimposed on the peaks themselves. Measuring the rise in the integral curve with a ruler can also be used to determine the peak ratios.

Note that the integrals give only the *ratios* of areas. A knowledge of the actual molecular formula that give the *absolute* number of hydrogens is necessary to assign absolute values to the peak areas.

With this information we can now determine, not only how many different kinds of hydrogen atoms are present, but also how many of each kind.

What would the NMR of a simple molecule look like at this point? A schematic cartoon of methyl ethyl ether ($CH_3CH_2-O-CH_3$) is shown below.

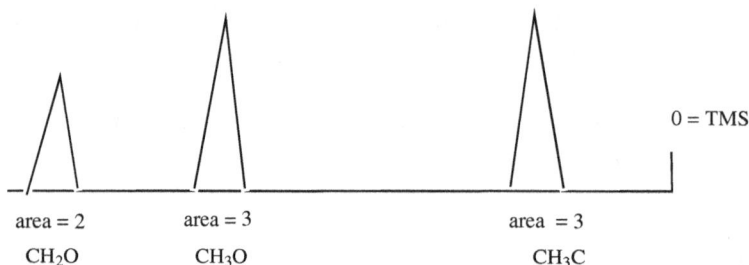

0 = TMS

area = 2   area = 3   area = 3

CH_2O CH_3O CH_3C

Let's see how this occurs.

The *methyl* group of CH_3CH_2 is farthest away from the electronegative atom (oxygen) and so its electrons will be only slightly affected. Therefore, the shielding will be maximized and the signal will be toward the right end of the spectrum. Conversely, both the methyl group and the methylene group attached to oxygen are much more affected and will be much more deshielded and occur more to the left (downfield) in the spectrum. The methylene group is somewhat more affected than the methyl group as is shown by the chemical shifts.

11.10 ^{13}C NMR Spectra

Because ^{13}C NMR spectra are fundamentally simpler than proton spectra, we will stop here and see what the ^{13}C NMR spectrum looks like and what information can be obtained from it. Recall from previous courses that the natural abundance of ^{13}C is only about 1% of the total abundance of carbon – almost all carbon atoms are ^{12}C and because this isotope does not have the required magnetic properties to take part in the NMR process, it is invisible to the NMR instrument. Therefore, when a molecule is subjected to the NMR conditions, the signals that are seen are due to only 1% of the available nuclei and so are very weak. In addition, for reasons that are beyond the scope of this discussion, the concept of integration just introduced *does not apply to carbon spectra*.

On the first level, the principal information that can be derived from ^{13}C spectra is *the number of different kinds of carbon atoms* present in the molecule. Recall the question you were asked repeatedly when you were learning about isomerism: "How many different sets of hydrogens are in this molecule." The question now is: "How many different kinds of carbon are in this molecule"? For each different kind you will see one absorption. Therefore, if the number of lines is *less* than the total number of carbons as given by the molecular formula, there must be some symmetry in the molecule. The two spectra shown below are the ^{13}C nmr spectra of 2-pentanol and 3-pentanol. As you can see, there are five absorptions in the former and only three in the latter. The reason for this should be obvious from the symmetry in the molecules.

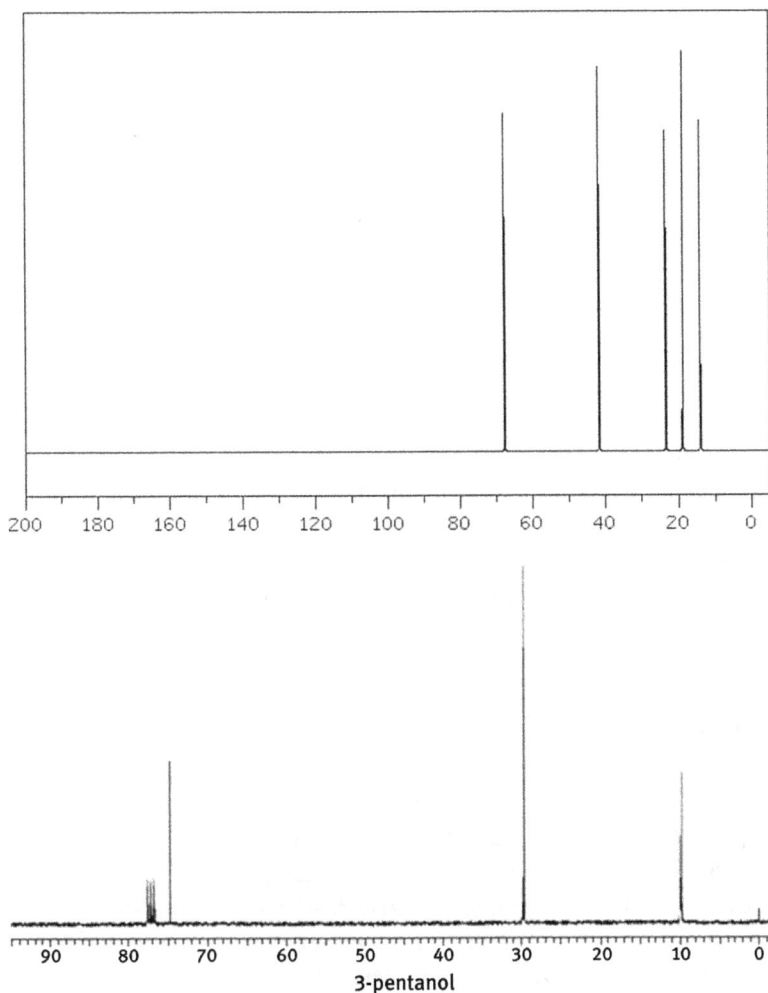

3-pentanol

Important note: In essentially all ^{13}C spectra, you will see the peak for TMS at $\delta = 0$ *and* a series of three peaks at $\delta = 77$. These latter peaks are due to the solvent commonly used to dissolve the samples (CDCl$_3$) and should be ignored when interpreting the spectrum.

In addition to providing information about the number of unique carbon atoms in a molecule, the chemical shift of each peak gives information about the nature of the atom. The general regions where typical types of carbons absorb are shown in the chart in Fig. 11.1. Particularly useful is the carbonyl region. Recall from the discussion on IR spectra that ketones will absorb in the same general region as carboxylic acids and conjugated esters. Fortunately, these groups are easily distinguished by the position of their ^{13}C absorptions. In addition, sp^2- and sp-hybridized carbons have quite different chemical shifts compared to their sp^3-hybridized counterparts.

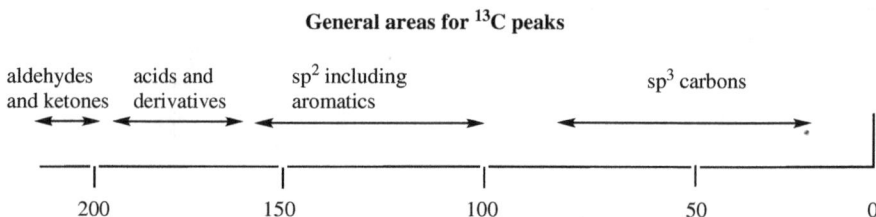

General areas for ^{13}C peaks

aldehydes and ketones	acids and derivatives	sp² including aromatics		sp³ carbons	

```
←→        ←——→        ←——————————→        ←——————————→
```

```
 ┬                      ┬              ┬              ┬              
200                   150            100             50              0
```

Fig. 11.1

To summarize: ^{13}C NMR spectra give information about the *number of unique carbon atoms* in a molecule and *what type of carbons* they are.

Before proceeding to the next section, note that the absorptions in ^{13}C spectra are single lines. As we will see shortly, this is not true for the ^{1}H spectra. While this is a seriously complicating factor for ^{1}H spectra, it also affords the opportunity to obtain much more information from a spectrum.

Table 11.2 at the end of this chapter gives a numerical tabulation of the expected positions of ^{13}C absorptions.

11.11 Spin–Spin Coupling

Go back to the initial discussions about bar magnets and consider the methyl group in the molecular fragment CH_3-CH. The protons of the CH_3 group are in an applied magnetic field that is reduced (shielded) by the electrons bonding it to the CH group. *But the proton on the CH group is also a bar magnet* and its field will influence the net field experienced by the protons on the CH_3 group. As noted previously, the lowest energy alignment for all protons is "with the field." This would be the case at absolute zero, but at room temperature, thermal energy is large and at that temperature, there are nearly equal numbers of protons in each alignment. Therefore, the protons of the CH_3 group will have a neighbor, half of which *add* to the applied field and half of which *subtract* from it. Since there will be two different fields experienced by the CH_3 protons, there will be two different frequencies that will cause resonance and the signal will be split into two lines. If the molecular fragment was CH_3CH_2, now the methyl group has *two* neighbors, each of which can be aligned with or against the field and so there are *four* different possibilities. However, since the two hydrogens on the CH_2 group are identical, the spacing between the signals will be the same and two of the lines will coincide and only three absorptions will occur. This is shown schematically in the figure below. The protons in the CH_3 are said to be *coupled* to those on the CH_2.

Ha

coupling between Ha and Hb

coupling between Ha and Hc

Hb

H₃C—C—X
a
Hc

peak height ratios = 1:2:1

Ha

coupling between Ha and Hb

coupling between Ha and Hc

coupling between Ha and Hd

Hb

H—C—CH₂
d   a
Hc

peak height ratios = 1:3:3:1

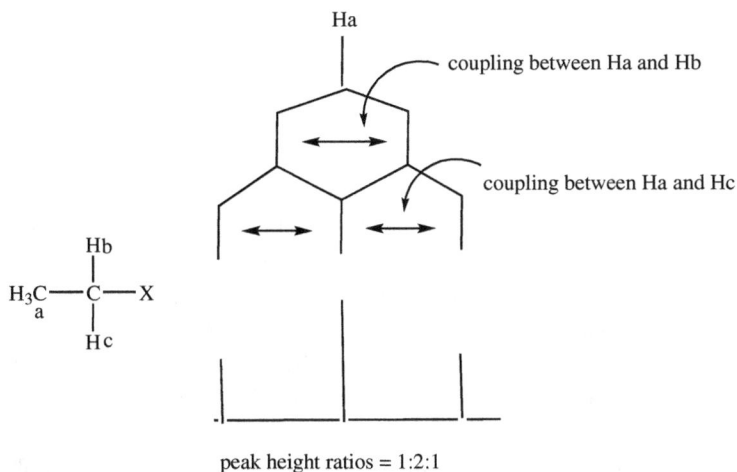

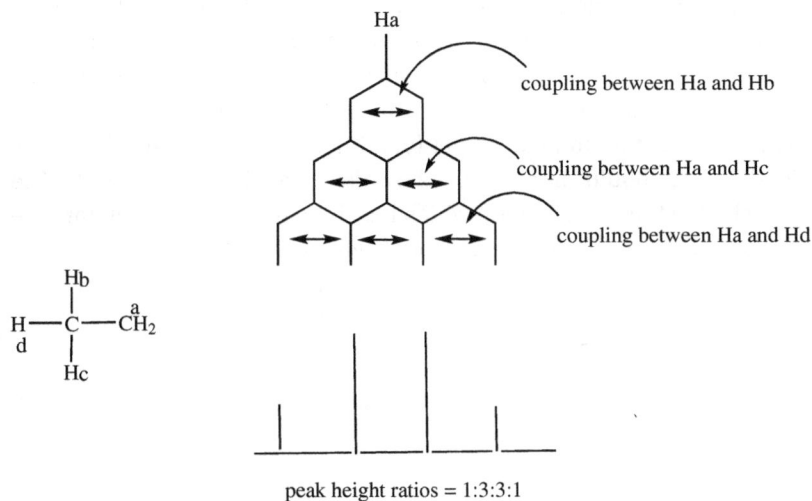

If one looks at the CH_2 of this fragment, it becomes obvious that, if it is coupling to the CH_3 as we have just learned, then the CH_3 must be coupling to the CH_2 in exactly the same way. Applying the same principles to that coupling, we see that the CH_2 must be split into four absorptions. This is shown schematically in the figure.

In general, if a proton has N nearest neighbors, its absorption will be split into N+1 lines.

The intensity ratios of the lines will be those of the coefficients of the binomial expansion.

Caveat: The N+1 rule is called *First-Order Coupling* and applies only in cases where the size of the coupling is small compared to the chemical shift difference between the coupling protons. If this is not the case, the coupling patterns are much more complex.

> It is extremely important to realize that the *integral* gives the *number of protons causing the absorption* but the *number of lines* is determined by the number of hydrogens *on the attached carbons!!*

At this point it is worth explaining why the ^{13}C spectra do not show this same coupling phenomenon. The natural abundance of hydrogen isotope #1 is nearly 100% whereas that of ^{13}C is about 1% and only those isotopes have the required magnetic properties of a bar magnet. Therefore, only those isotopes will cause the coupling phenomenon. Since the likelihood of *two* ^{13}C isotopes being side-by-side and able to couple is so small, coupling between two carbons can be ignored. Also the method under which ^{13}C spectra are obtained erases the coupling between carbons and their attached hydrogen atoms (termed decoupling) and therefore the absorptions appear as single lines.

So what would we expect the spectra of some simple organic compounds to look like? The ^{1}H spectrum of 2-butanone is shown below. It consists of three signals (which would be expected because there are three different sets of protons – two different methyl groups and one methylene group) which will be in the ratio of 3:3:2. One methyl group – the one that is directly attached to the carbonyl carbon has *no* protons on the attached carbon and so it should be a singlet (one line). The other methyl group is attached to the CH$_2$ and should therefore be shown as three lines ($n + 1 = 3$). The CH$_2$ group is attached to the methyl group which has three protons and therefore the signal should be split into four lines. This is exactly what is seen.

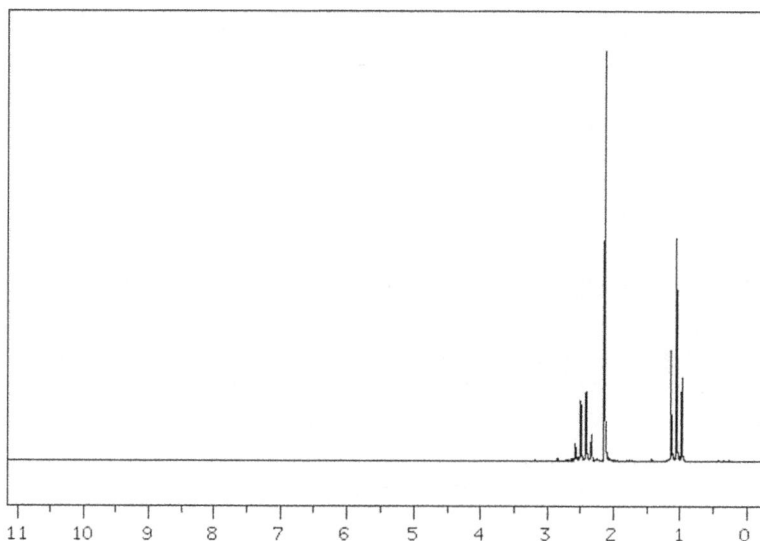

It may not be obvious from the small drawing above, but the *separation of the lines* in the signal at $\delta = 1.1$ is exactly the same as that between the lines of the signal at $\delta=2.4$. This separation is referred to as the *coupling constant* and is given the symbol J.

The separations are the same because whatever effect is being exerted on the methyl group by the methylene must be the same as that exerted on the methylene by the methyl group.

Note the chemical shifts of the three signals. The singlet methyl group is farther downfield (to the left) than the triplet methyl group because it is closer to the electronegative carbonyl group and therefore the electron density around it is reduced. The CH_2 group is farther downfield than the CH_3 group even though they are both attached directly to the C=O. This is a general result – the fewer hydrogens that are on the carbon being affected, the farther downfield the signal will appear.

An important exception: For reasons that do not concern us now, it must be noted that *protons on atoms other than carbon* (usually oxygen or nitrogen) *do not couple with adjacent protons* as you would expect. These signals are frequently quite broad.

11.12 Shoolery's Rules

In Table 11.4 at the end of this chapter there are numbers listed that can be used to calculate the *approximate* chemical shift of a *disubstituted* methylene $(X–CH_2–Y)$ where X and Y are functional groups using the formula

$$\delta = 0.23 + \sigma_X + \sigma_Y$$

This relationship is referred to as Shoolery's Rules. As an example, the σ value for OR on a CH_2 is 2.36 and that for Cl is 2.53. On this basis, the δ value for the atom attached to both oxygen and chlorine should be 5.12. In fact, the value for chloromethyl ethyl ether is 5.5 so you can see that the values are quite approximate.

11.13 Putting It All Together

Let's look at some other simple examples. The ^1H spectrum of 3-pentanone consists of only two signals – one triplet and one quartet. These are in the ratio of 3:2, but from the molecular formula which shows ten hydrogens, they must really be in the ratio of 6:4. The two ethyl groups are identical.

To show you how useful the NMR data can be in solving structures, consider the following situation. From the molecular formula and the IR spectrum, you deduce that one of the two structures I or II must be the correct one. Note that these two molecules, ethyl ethanoate and methyl propanoate are isomers and would show the same functional group in the IR spectrum.

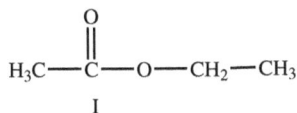

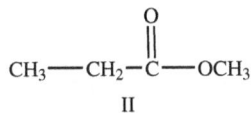

$$\underset{\text{I}}{H_3C-\overset{\overset{\displaystyle O}{\|}}{C}-O-CH_2-CH_3} \qquad \underset{\text{II}}{CH_3-CH_2-\overset{\overset{\displaystyle O}{\|}}{C}-OCH_3}$$

The spectrum of the molecule is shown below.

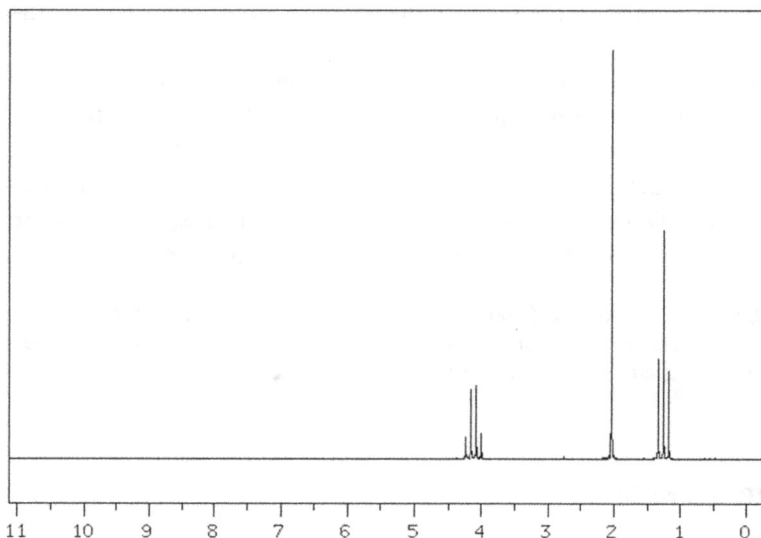

There is clearly one methyl group and one ethyl group – which is consistent with either structure. However, the singlet methyl group occurs at about $\delta = 2.1$ and this is the chemical shift that is consistent with a methyl group attached to a carbonyl. If the singlet methyl had been attached directly to the oxygen (as in structure II), it would resonate at about $\delta = 3.9$. Conversely, the quartet CH_2 is at $\delta = 4.2$ which is consistent with it being attached to an oxygen atom. Both pieces of information demand that the structure be I and not II. The spectrum of II is shown below.

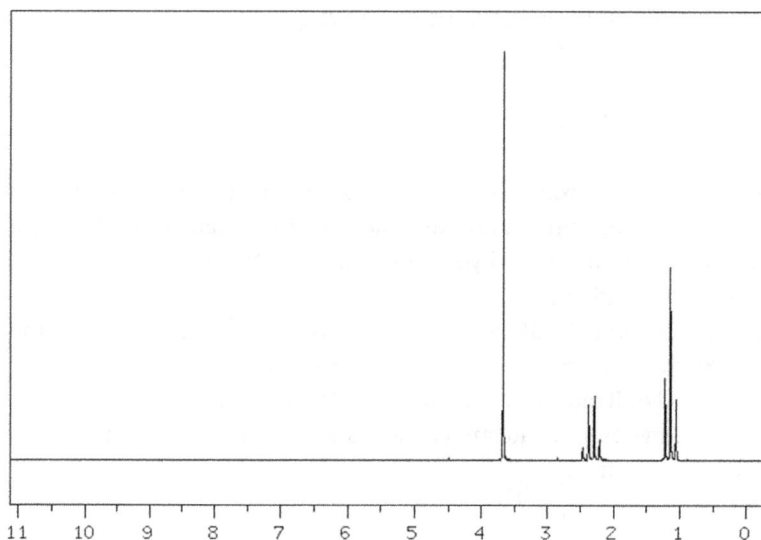

Table 11.3 gives a list of approximate chemical shifts for proton spectra.

It should be noted that the NMR technique is not limited to organic chemistry. Many inorganic compounds contain nuclei that have the required magnetic properties for the NMR experiment. Such elements as P, Sn, B, and many others follow all the same principles as those just presented. As you proceed to learn more chemistry, NMR will become more and more useful in both structure determination and other situations (e.g., kinetic experiments).

How are the various spectroscopic techniques best used to solve the structure of an unknown compound? The best way is to follow the following simple steps.

1. Look at the IR spectrum and determine what functional groups might be present. Remember that these must fit within the parameters of the *molecular formula*! (If there is no nitrogen in the formula, the carbonyl absorption at $1690\ cm^{-1}$ cannot be an amide!!)

2. Look at the ^{13}C NMR spectrum to obtain the following information:
    -   Is the number of peaks equal to the number of carbons in the molecular formula? If not, there must be some symmetry in the molecule.
    -   If a carbonyl absorption is present in the IR, can the specific functional group be established from the chemical shift of the C=O in the ^{13}C spectrum?
    -   Are there peaks in the region of $\delta = 100–160\ ppm$? If so the molecule is either aromatic or at least contains a double bond.

3. Consider the ^1H NMR to see what contiguous fragments can be identified ($CH_3\ CH_2$), etc., and especially what singlet signals can be seen. Pay attention to the chemical shifts when it comes to assembling the structure from the fragments you have generated!

Some problems that combine the IR and NMR techniques to solve structure problems are given at the end of this chapter.

Tab. 11.1: Diagnostic infrared absorptions.

$u(cm^{-1})$		Comments
3000–3400	O–H stretching	Alcohols – unassociated OH's – two bands around 3600 (sharp) H-bonded – broad absorption at 3400 Acids – very broad centered at about 3000 cm^{-1}
3400–3200	N–H stretching	Amines – unassociated NH's – two bands around 3400 (sharp) H-bonded – broad absorption at 3200 (weaker than OH)
3300	C–H stretching of alkyne	
3100–2850	C–H stretching	sp^2-hybridized >3000; sp^3-hybridized <3000
2900–2700	C–H stretch of aldehyde	
2250–2225	C≡N of nitrile	2250 if not conjugated, 2225 conjugated. May be weak
2250–2100	C≡C of alkyne	*Usually* weak unless conjugated to C=O
1800	C=O of acid chloride	
1735–1740	C=O of Ester	*Conjugation lowers these*
1710	C=O of aldehyde or ketone	C=O *bands by* 30 cm^{-1}.
1700	C=O of acid	*Being in a 5-membered ring*
1660	C=O of amide	*raises these bands by* 35 cm^{-1}.
1650–1600	C=C stretch	May be weak. Intensity increases with conjugation, especially to C=O

Tab. 11.2: C^{13} NMR absorption regions.
The following values are approximate and can be *drastically* affected by nearby substitution and especially the presence of electronegative atoms and pi-electron systems (double bonds and aromatic systems). The values are for the starred (*) carbon atoms.

Carbon type	Chemical shift (δ)
$CH_3-, -CH_2-$	15–25
CH	15–45
CH_2O-	25–60
$C=CH_2$ *	105–120
$C=CH-$ *	115–140
Aromatic	120–150
C=O (ester, acid)	160–175
C=O (aldehyde and ketone)	180–220

Tab. 11.3: ^{1}H nuclear magnetic resonance chemical shifts.
Chemical shifts in ppm downfield from tetramethylsilane (TMS) (defined as = 0.00 ppm).

General regions:

0–1	Cyclopropyl hydrogens and methyl groups not shifted by electronegative atoms
1–2	Methyl groups – to O or N atoms, attached to C=C or attached to aromatic rings; methylene groups
2–3	Methyl and methylene groups next to carbonyls or attached directly to nitrogen of amines
3–4	Methyl and methylene groups attached to oxygen or halogens (Br, Cl). C=CH$_2$ groups
4.5–6.5	Hydrogens on sp^2-hybridized carbons of alkenes (not aromatics)
6.8–8.5	Aromatic protons
9–10	Aldehyde protons

Chemical shifts in ppm downfield from tetramethylsilane (defined as = 0.000 ppm). Values are approx. ±0.2 ppm.

Protons on a carbon adjacent to a functional group

Functional group	CH$_3$	CH$_2$	CH
Saturated system	0.8	1.2	1.6
M– – –C=C	1.6	2.0	–
M– – –C≡C	1.7	2.2	2.8
M– – –Phenyl	2.2	2.6	2.8
M– – –Cl	3.0	3.4	4.0
M– – –Br	2.7	3.4	4.1
M– – –I	2.2	3.1	4.2
M– – –OH	3.2	3.4	3.8
M– – –OR0	3.2	3.4	3.6
M– – –O–Phenyl	3.9	4.1	4.5
M– – –OC(=O)R	3.6	4.1	4.5
M– – –OC(=O)Ph	3.8	4.2	5.0
M– – –CH=O (aldehyde)	2.2	2.4	2.5
M– – –C(R)=O (ketone)	2.1	2.3	2.6
M– – –COOH (acid)	2.1	2.3	2.5
M– – –COOR (ester)	2.0	2.2	2.5
M–NR$_2$	2.4	2.6	2.9
M– – –NHC(=O)R	2.9	3.3	3.9

Tab. 11.3: (continued)

Protons on a carbon once removed from a functional group			
Functional group	CH$_3$	CH$_2$	CH
M–C–CH$_2$	0.8	1.2	1.6
M– – –C–C=C	1.0	1.55	1.8
M– – –C–C≡C	1.2	1.5	1.8
M– – –C–Ph	1.2	1.6	1.8
M– – –C–Cl	1.5	1.8	2.0
M– – –C–Br	1.8	1.9	1.9
M– – –C–I	1.8	1.8	2.1
M– – –C–OH(orOR)	1.2	1.5	1.8
M– – –C–OPh	1.3	1.6	2.0
M– – –C–OC(=O)R	1.3	1.6	1.8
M– – –C–CH=O	1.1	1.6	2.0
M– – –C–C(R)=O	1.1	1.6	2.0
M – C–CO$_2$R	1.1	1.7	1.9
M–C–NR$_2$	1.0	1.5	1.7
M– – –C–NH–C(=O)R	1.1	1.5	1.9

Protons on sp^2- and sp-hybridized carbons			
R$_2$C=CH$_2$	4.7–5.3	C=CH–C=O	6.0
R$_2$C=CHR	5.1	C=CH–Cl	6.5
RCH=CHR	5.3	C=CHBr	6.5
Cyclohexene	5.6	CH=CH–C=O	6.9
ArCH=C–C=O	7.7	RCH=O	9.1
R–C≡C–H	2.3–3.3	R–OH (alcohol)	0.5–5.5
Aromatic hydrogens	6.0–9.0 (mostly 6.7–8.2)	R–NHR (amine)	0.5–5.0
R–C(=O)OH	12–14	R–NH–C(=O)R (amide)	5–8

Tab. 11.4: Shoolery's Rules
(For prediction of chemical shifts of mono- and disubstituted carbons).

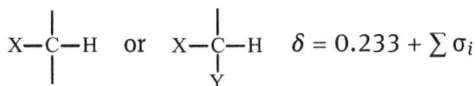

$$X-\overset{|}{\underset{|}{C}}-H \quad or \quad X-\overset{|}{\underset{Y}{C}}-H \quad \delta = 0.233 + \sum \sigma_i$$

Functional group (X,Y)	σ_i
–Cl	2.53
–Br	2.33
–I	1.82
–OH	2.56
–OR	2.36
–OAr	3.23
–OC(=O)R	3.13
–SR	1.64
–NR$_2$	1.57
–CH$_3$	0.47
–C(=O)R (ketone)	1.70
–C(=O)OR (ester)	1.55
–CN	1.70

11.14 Problems

11-1. For each of the following six infrared spectra, decide which of the two given structures best fit the spectroscopic data. (*All infrared spectra are used by permission of NIST)

(a) Molecular formula $C_5H_{10}O_2$

IR

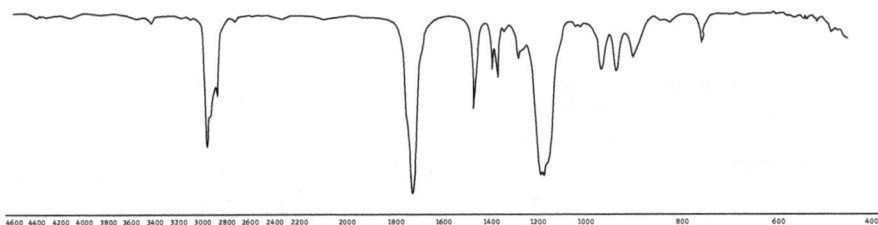

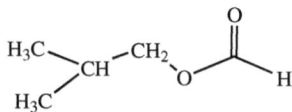

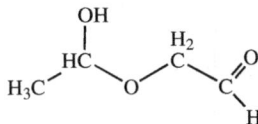

(b) Molecular formula $C_6H_{10}O_4$
IR

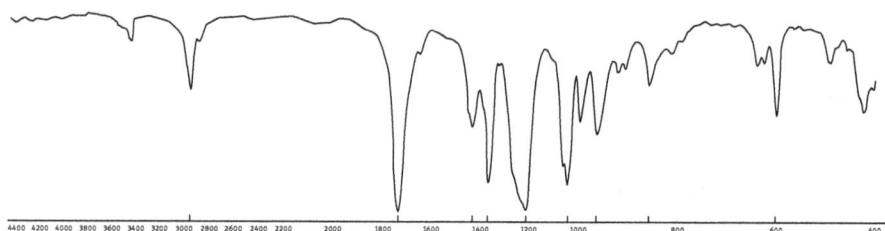

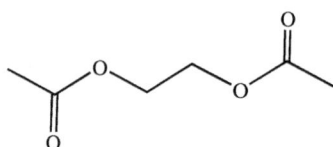
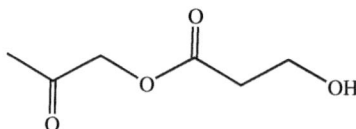

(c) Molecular formula $C_5H_8O_2$
IR

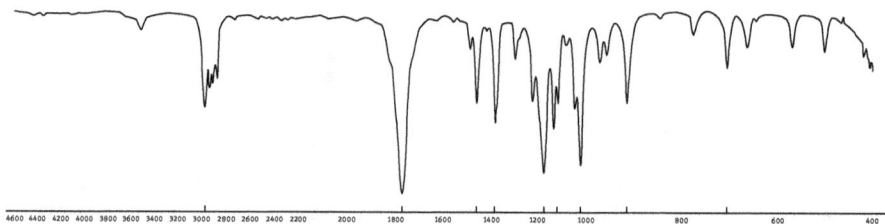

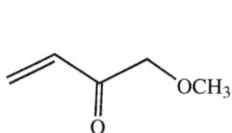
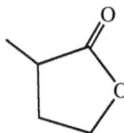

(d) Molecular formula C_5H_9N
IR

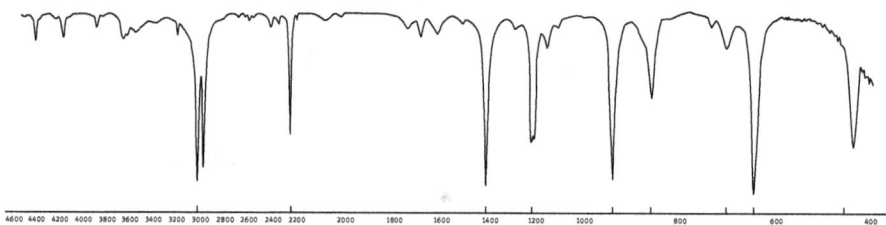

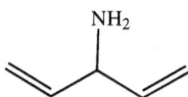

(e) Molecular formula $C_3H_4Cl_2O$
IR

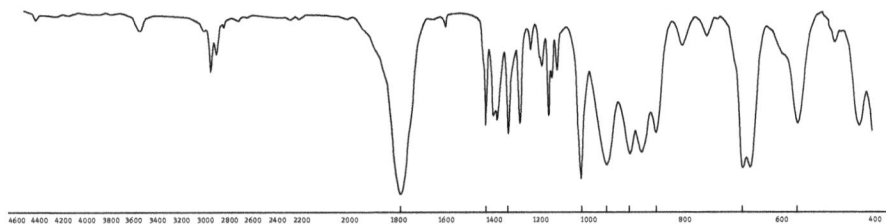

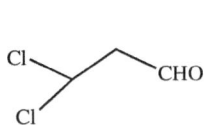

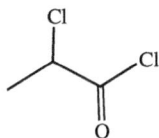

(f) Molecular formula $C_3H_7NO_2$
IR

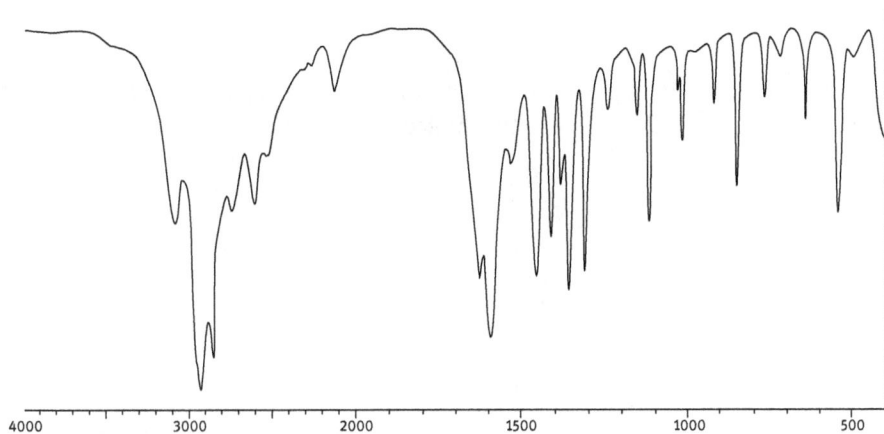

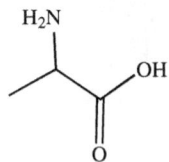

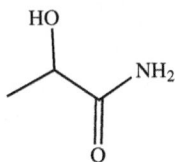

11-2. The ^{1}H NMR spectrum of the molecule $CH_3CH=CHCO_2CH_3$ is shown below. Can you determine which of the hydrogens bonded to the sp^2 carbons causes the signal at $\delta = 7.1$ and which causes the signal at $\delta = 5.8$? Explain your reasoning.

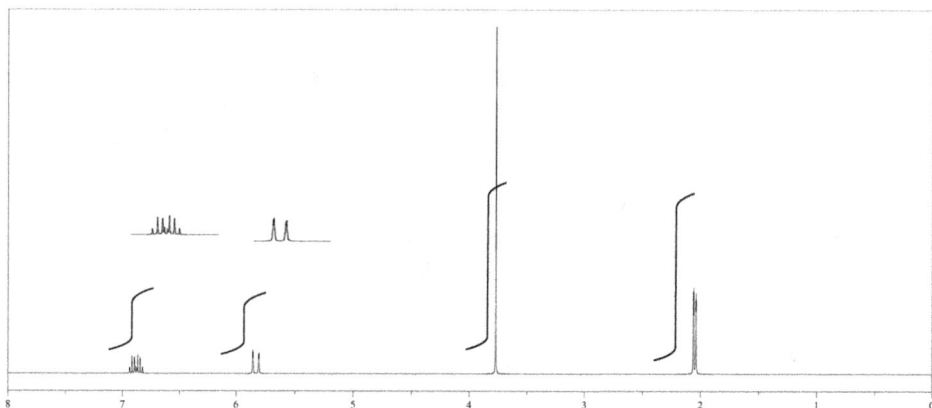

11-3. The proton NMR of compound I is shown below. Assign each signal in the spectrum to the appropriate protons in the molecule.

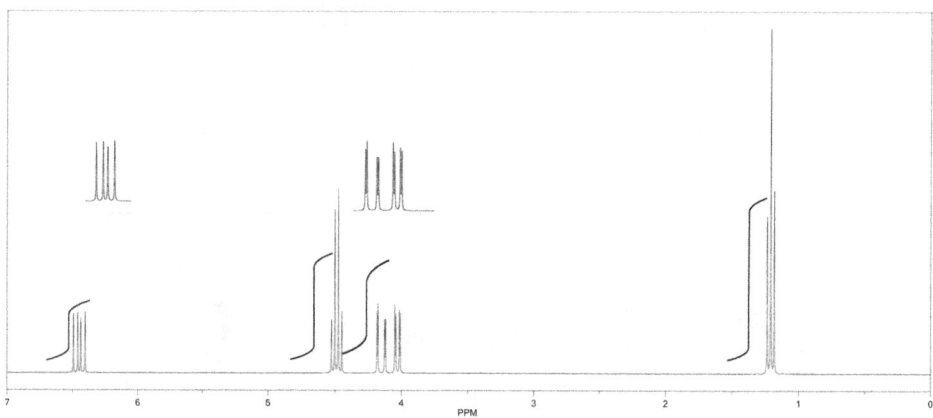

Compound I = $CH_3CH_2OCH=CH_2$

11-4. Shown below are the ^{13}C NMR spectra of the three possible isomers of dimethyl benzene (xylene). Assign one isomer to each spectrum and show how you arrived at your decisions.

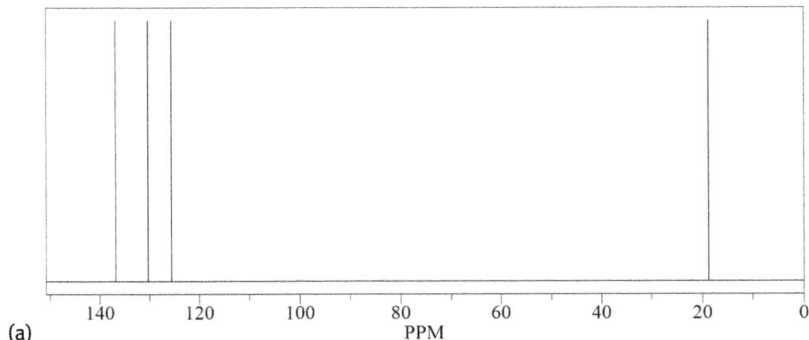

(a)

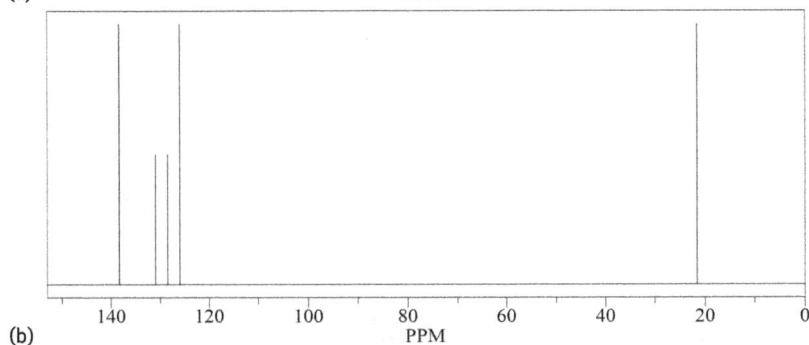

(b)

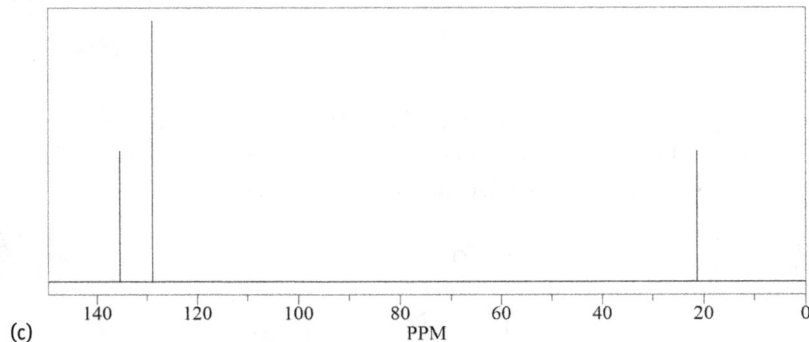

(c)

11-5. For the structure shown, *sketch* the ^{1}H NMR showing the approximate chemical shifts (δ) and the expected multiplicities of each signal. The protons on the marked carbon have $\delta = 3.7$

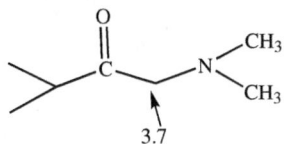

3.7

11-6. A compound with unknown structure was discovered to have a molecular formula of $C_6H_{12}O_2$. It showed an infrared absorption near $1740\ cm^{-1}$ and had the following NMR spectra. Derive a structure that fits these spectroscopic data.

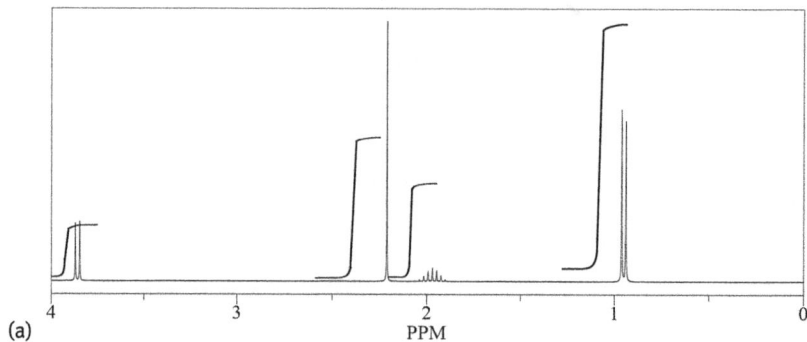

(a)

PPM

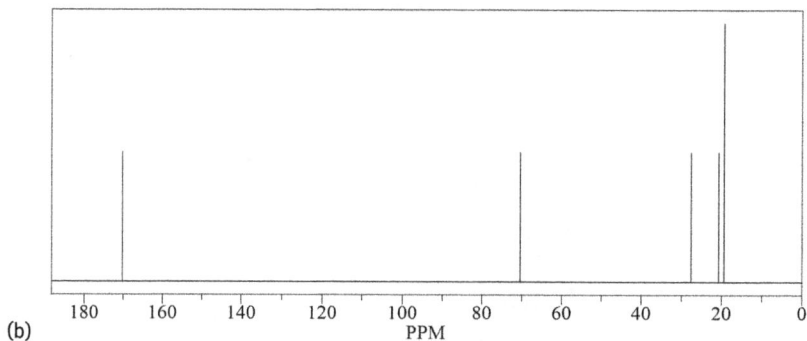

(b)

PPM

11-7. Indicate what spectroscopic technique (IR, ^{1}H NMR, or ^{13}C NMR) you would use to distinguish between the following pairs of compounds. *Briefly* indicate what difference(s) you would look for and what each compound would show.

(a)

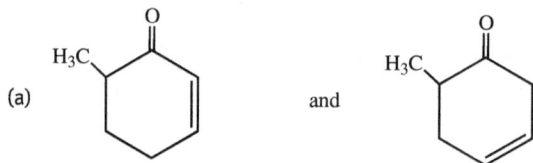

and

(b)

and

(c)

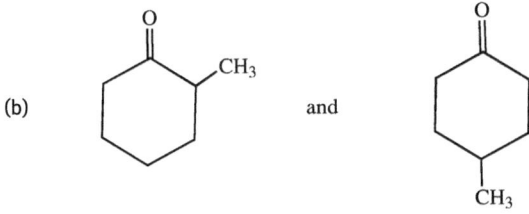

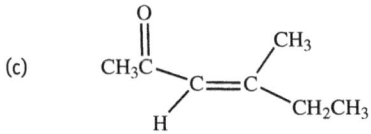

11-8. The formula is C_5H_9ClO and the IR shows a prominent band at $1720\,\text{cm}^{-1}$.

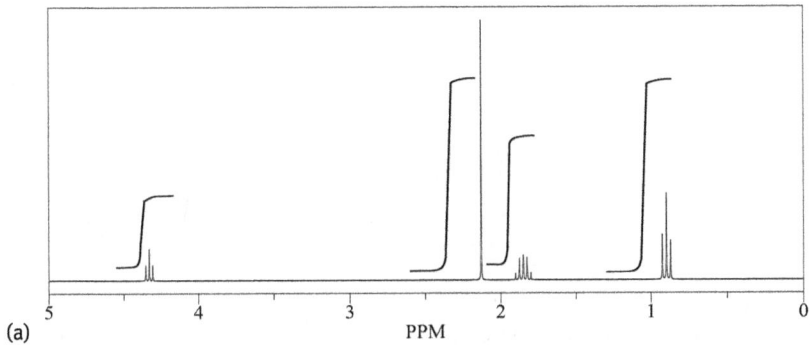

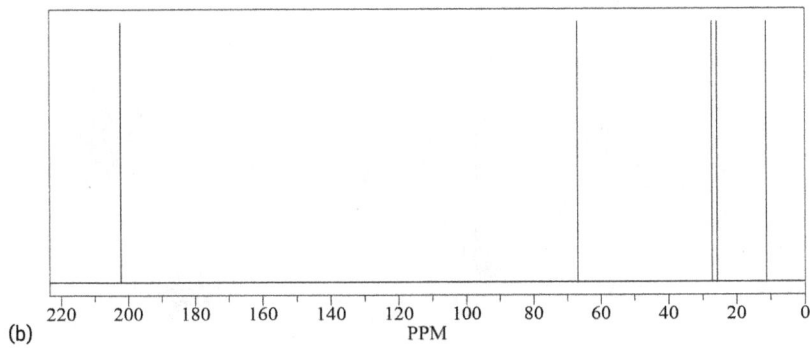

(a)

PPM

(b)

PPM

11-9. The molecular formula is $C_4H_8O_2$ and the IR shows strong bands at both 3400 and $1715\ cm^{-1}$.

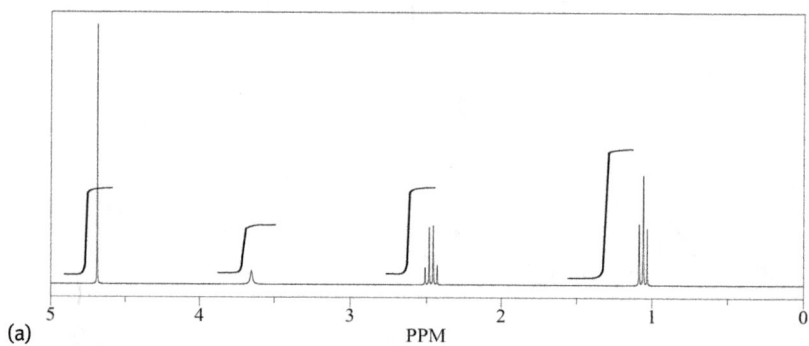

(a)

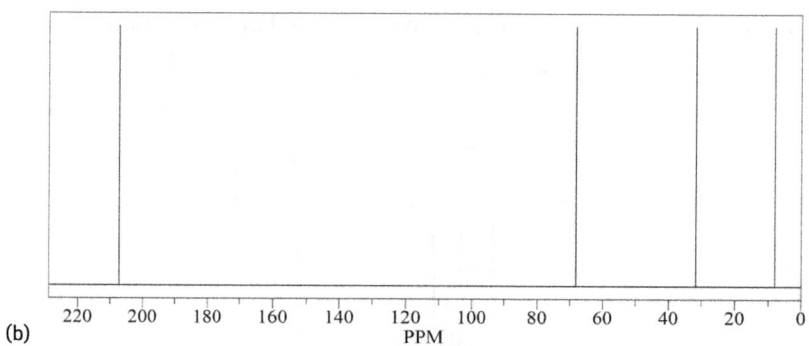

(b)

11-10. The formula is $C_6H_{12}O_3$ and the IR shows a band at $1720\ cm^{-1}$.

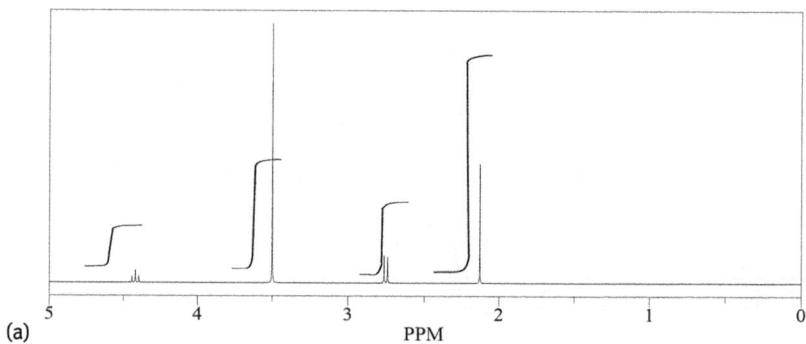

(a)

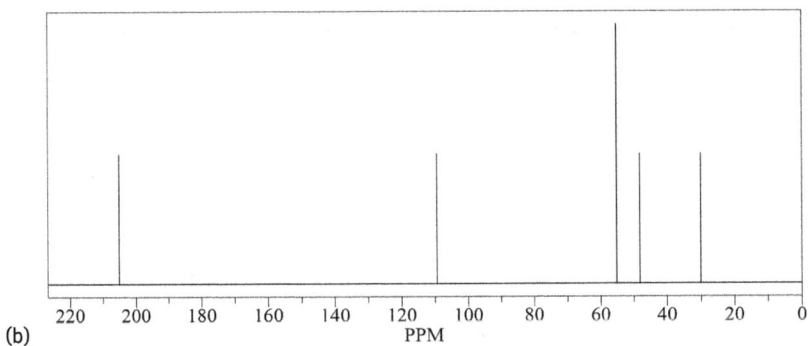

11-11. The formula is $C_4H_7BrO_2$ and the IR shows a band at 1745 cm^{-1}.

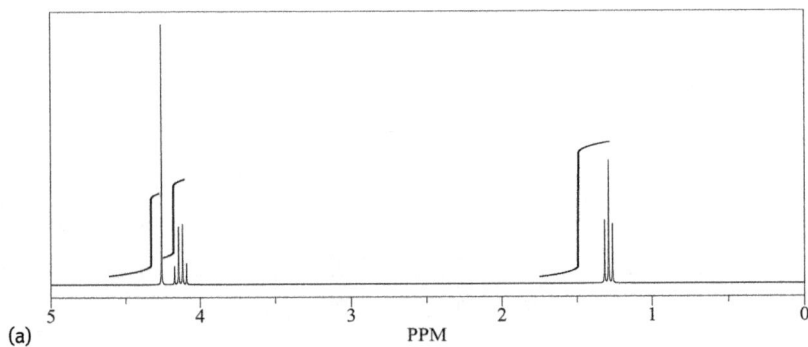

(a)

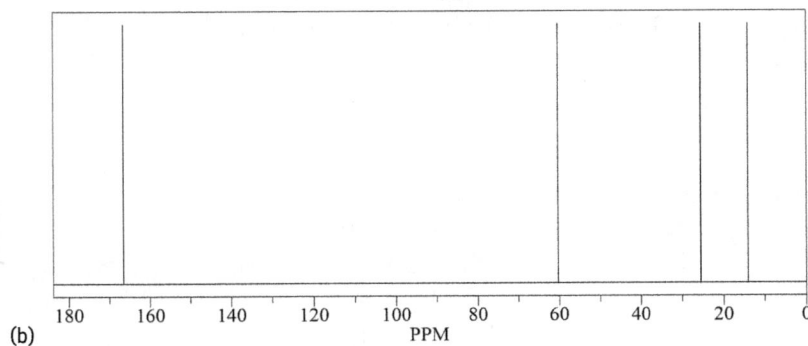

(b)

11-12. The formula is $C_9H_{19}NO$. There is an IR absorption at 1715 cm^{-1}, but nothing between 4000 and 3100 cm^{-1}.

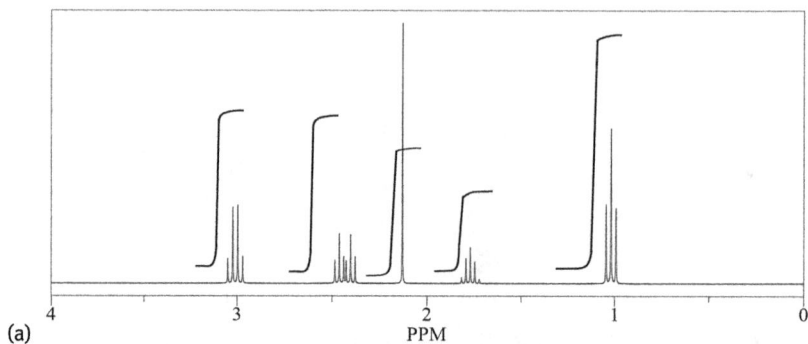

(a)

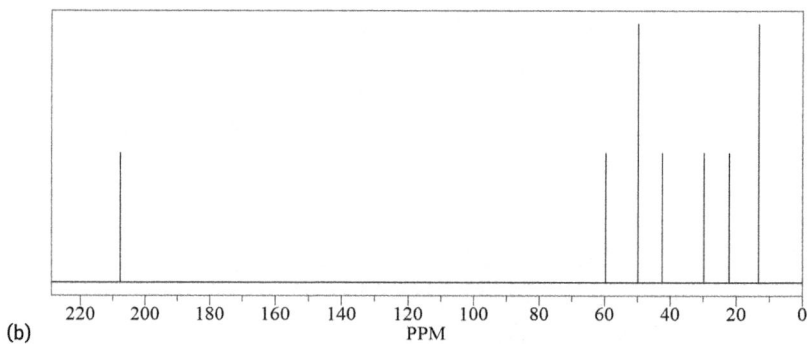

(b)

11-13. Formula is $C_{10}H_{12}O_2$. There is an IR absorption at 1740 cm^{-1}.

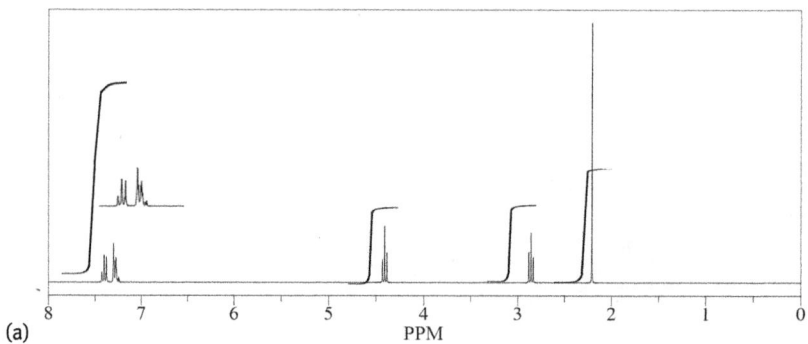

(a)

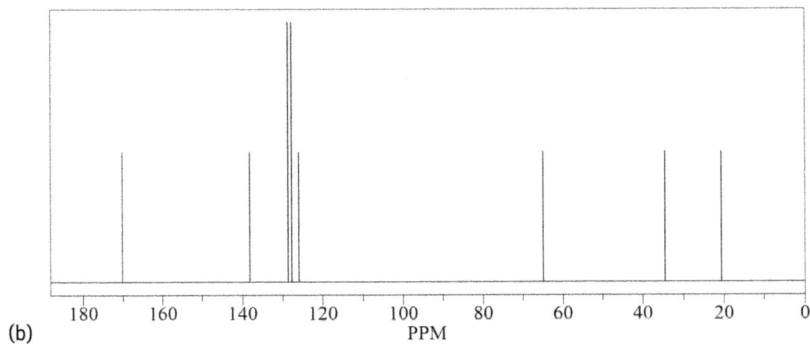

(b)

11-14. Formula = $C_9H_{10}O_3$; IR 3330, 1740 cm^{-1}.

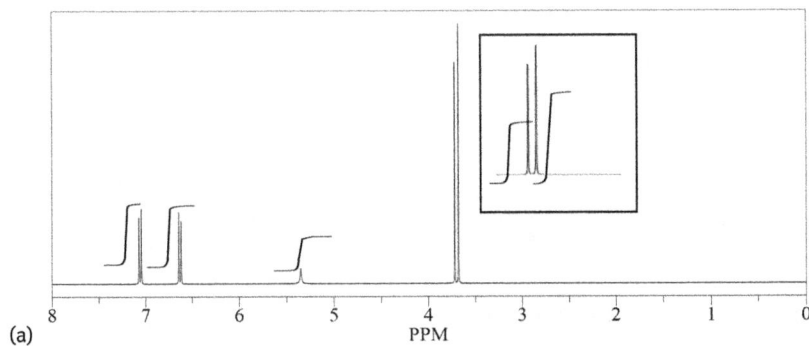

(a)

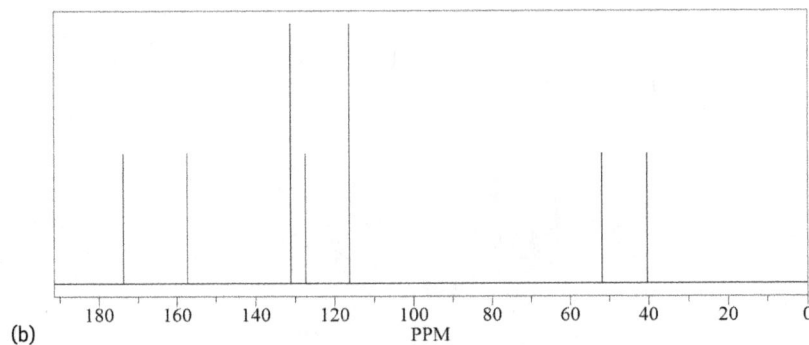

(b)

11-15. Formula = $C_{10}H_{12}O_2$; IR 1735 cm^{-1}.

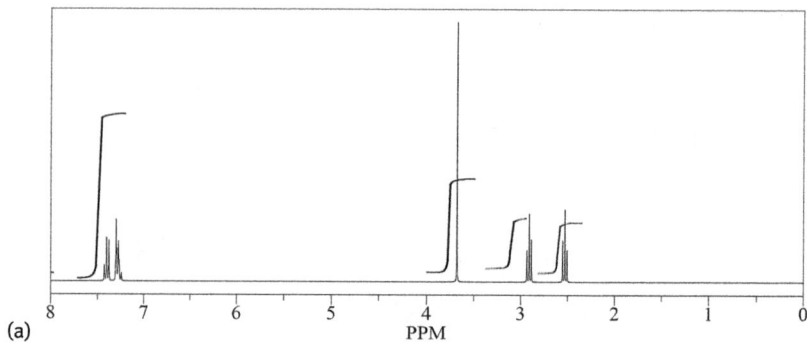

(a)

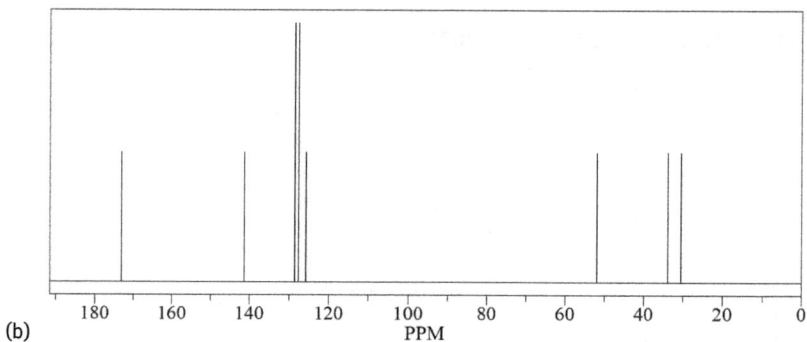

(b)

11-16. Formula = $C_{11}H_{14}O_2$; IR 1740 cm^{-1}.

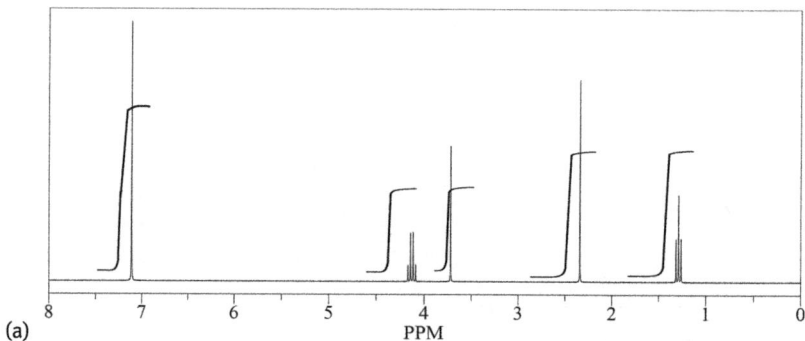

(a)

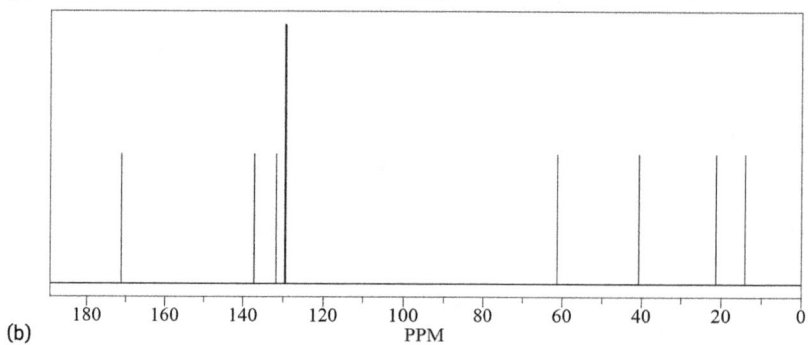

(b)

12 Elimination Reactions

12.1 Introduction

If you recall, the chapter on alkenes (Chapter 5) discussed many of their character-istic reactions, but included very little information about how they are made. Since alkenes are one of the most fundamentally important and widely encountered func-tional groups, their preparation is an important consideration.

One of the most common alkene preparation methods is as follows (Fig. 12.1):

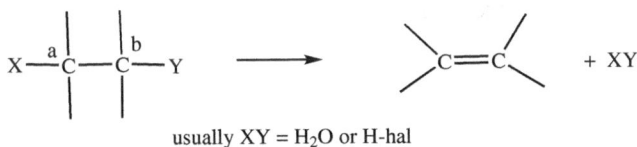

usually XY = H$_2$O or H-hal

Fig. 12.1

It doesn't *have to* be, but X–Y is most often H$_2$O or H–halogen.

This is an *elimination reaction*. More specifically, this is a *β-elimination* or *1,2-elim-ination* reaction; this is because the groups leaving are 1,2- (or *vicinal*) relative to each other. If this looks like the reverse of some of the alkene addition reactions seen in Chapter 5, this is absolutely right.

12.2 Possible Mechanisms – Background and Review

An understanding of the details of how elimination reactions occur is critical to predict when they will occur and to be able to get them to work when desired. For eliminations, there are two very important limiting mechanisms; a third one that is occasionally found will be mentioned briefly at the end of the chapter.

It may help at this point to recall the S$_N$2 (Fig. 12.2) and S$_N$1 (Fig. 12.3) *substitutions* (Chapter 6), since there are several analogous features with the eliminations here:

S$_N$2

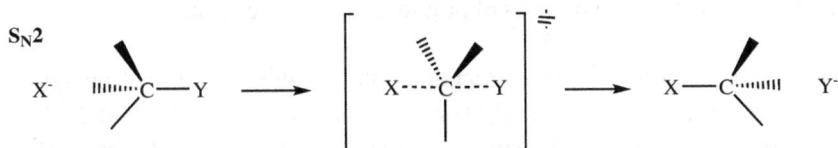

Fig. 12.2

https://doi.org/10.1515/9783110778311-012

S_N2 Features:
1. The rate of reaction is *bimolecular,* meaning both the nucleophile and the substrate enter into the rate equation – specifically...

$$\text{Rate} = k[X^-][R_3C–Y]$$

2. Inversion of configuration occurs at the carbon undergoing attack in an S_N2.

S_N1

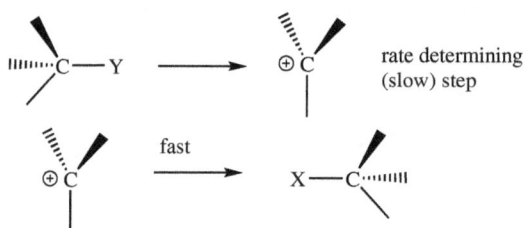

Fig. 12.3

S_N1 Features:
1. The rate of reaction is *unimolecular,* meaning *just* the substrate enters into the rate equation – specifically....

$$\text{Rate} = k[R_3C–Y]$$

2. Racemization of configuration occurs at the carbon undergoing attack in an S_N1.

By comparison, for *elimination* reactions, the two main mechanisms are called *E1* and *E2*.

12.3 E1 Mechanism – The Carbocation Route

By analogy with the S_N1, the E1 mechanism is a two-step process (Fig. 12.4):
Step 1) The substrate undergoes ionization to give carbocation; this is identical to the first step in an S_N1.
Step 2) The carbocation undergoes loss of proton $(H^+)\beta$– to a cation.

There is no added base necessary to abstract the proton in Step 2, as the solvent (even though it is quite a weak base such as H_2O or an alcohol) is basic enough to do this. As an aside, some instructors and texts prefer to show the solvent actually abstracting the H^+ in the 2nd step. This is technically correct.

STEP 1

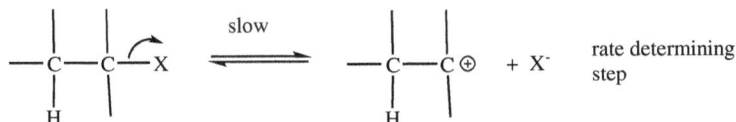

slow

rate determining
step

+ X⁻

STEP 2

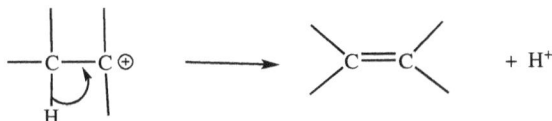

+ H⁺

Fig. 12.4

In the vast majority of cases, the rate follows first-order kinetics, with Step 1 being the slow step.

Since the first step of an E1 elimination is absolutely identical to the S_N1 substitution's first step, S_N1 substitution is always a potential competitive reaction to the E1 elimination. In fact, if one changes the leaving group (X-) of an otherwise identical reaction, and the elimination-to-substitution ratio stays the same, this is excellent evidence that E1 (and S_N1) mechanisms are taking place. The reaction below (Fig. 12.5) is a good example of this feature:

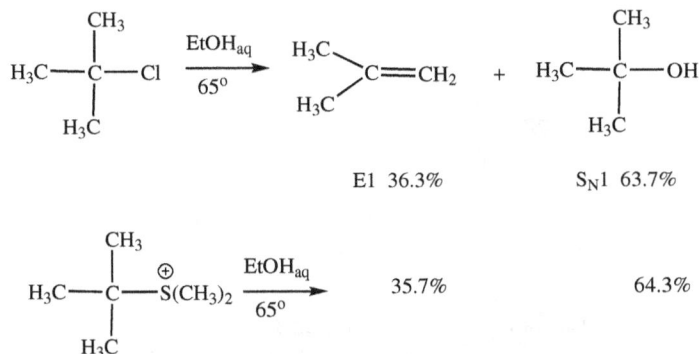

E1 36.3% S_N1 63.7%

35.7% 64.3%

Fig. 12.5

Although there are many different examples of E1 eliminations, the prototypical E1 elimination reaction is the acid-catalyzed elimination of an alcohol (Fig. 12.6). In this specific elimination reaction there's actually an extra step, where the alcohol OH (as ⁻OH is a mediocre leaving group) is protonated and therefore converted into an $^+OH_2$ group (and H_2O is an excellent leaving group). The H⁺ (or H_3O^+) is regenerated in the final step, so H⁺ is a *true* catalyst.

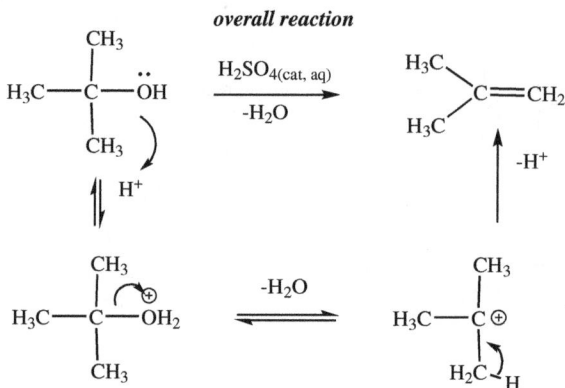

overall reaction

Fig. 12.6

12.4 The E2 Mechanism – The Concerted Route

In a fashion analogous to the S_N2 mechanism, the E2 elimination mechanism is a one-step process, with a transition state rather than an intermediate (Fig. 12.7).

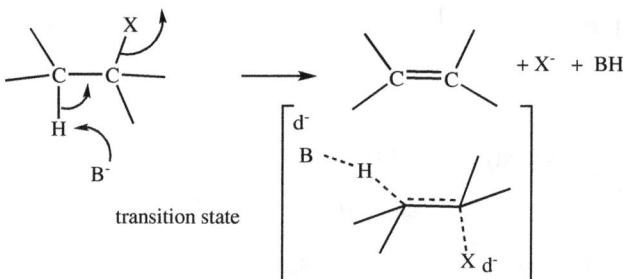

transition state

Fig. 12.7

Since all of the bond breaking and bond making going on (i.e., the proton abstraction by base, the double bond formation, and the loss of X^-) in the E2 is happening simultaneously, the term *concerted reaction* is used.

There are two reagents (the base and the substrate) that appear in the rate-determining (and only) step of the E2 elimination; as a result, second-order kinetics are observed for reactions of this type. This is also the source of the "E2" name.

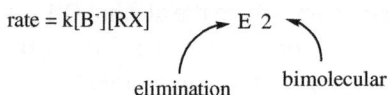

$$\text{rate} = k[B^-][RX] \qquad \text{E 2}$$

elimination   bimolecular

12.4.1 Prototypical E2 Elimination Reactions

There are probably even more examples of E2 elimination reactions than for E1's, but the most commonly encountered class of E2 eliminations are the base-induced elimination of alkyl halides. A very simple example is shown below (Fig. 12.8):

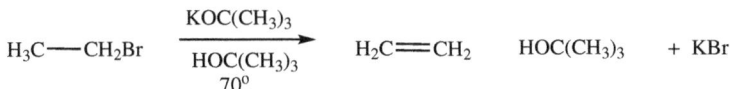

$$H_3C\!-\!CH_2Br \quad \xrightarrow[\substack{HOC(CH_3)_3 \\ 70^\circ}]{KOC(CH_3)_3} \quad H_2C\!=\!\!=\!\!CH_2 \quad HOC(CH_3)_3 \quad + \; KBr$$

Fig. 12.8

The halides are common because of their leaving group abilities, which completely parallel the leaving group ability in S_N2 substitutions....in other words,

$$I^- > Br^- > Cl^- \gg F^-$$

Many bases have been employed in E2 eliminations, including neutral ones (amines NR3 especially), but the most commonly used bases tend to be anionic ones such as:

Alkoxide bases ^-OR
Hydroxide bases ^-OH
Amide ion bases $^-NR_2$

(Note: amide ions are *not* the same as amide functional groups)

12.5 Regiochemistry and Stereochemistry of Eliminations

Unlike the two examples that have been given so far, there is often more than one site from which H^+ could be lost in an elimination reaction. For example, consider...

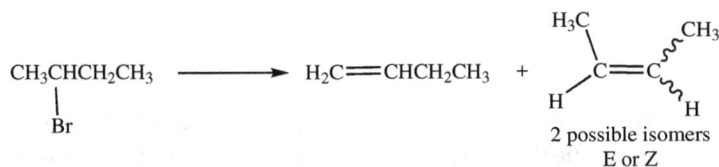

$$CH_3CHCH_2CH_3 \quad \xrightarrow{\hspace{1.5cm}} \quad H_2C\!=\!\!=\!\!CHCH_2CH_3 \quad + \quad$$

(with structure:)

$$\underset{\substack{| \\ Br}}{}$$

H3C, CH3 on a C=C with H and H
2 possible isomers
E or Z

Which one of these is obtained predominantly depends on the mechanism and how the reaction is performed. It's best to go through these by mechanism.

12.5.1 E1 Eliminations

For E1 eliminations the cationic intermediate breaks down so as to give the more stable alkene product (predominantly). This usually translates into meaning the alkene with the more substituted C=C (Fig. 12.9).

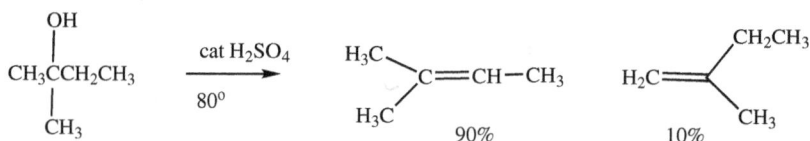

Fig. 12.9

This "rule" is called Zaitsev's Rule (or Saytzeff or Saytzev) regardless of whether the mechanism at hand is E1 or E2. This rule was originally formulated as:
–   *"The alkene formed in greatest amount is the one that corresponds to the removal of the hydrogen from the α-carbon having the fewest H substituents."*
but more commonly now phrased as:
–   *"The double bond goes mainly toward the most highly substituted carbon."*

There are exceptions that occur when the more substituted alkene is *not* the most stable. These are often situations when the less-substituted alkene is conjugated to another multiple bond – the stability gained from the conjugation then overwhelms the amount from greater substitution (Fig. 12.10).

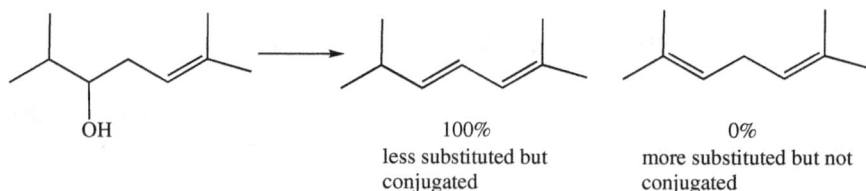

Fig. 12.10

Corollary: Since the site at which this reaction occurs has significant selectivity, the reaction therefore is referred to as one that is *regioselective*. This term is used extensively in all areas of synthetic chemistry, and is not restricted to elimination reactions.

As for the stereochemistry of the newly formed C=C, if the reaction can go to give products that have both E- and Z-isomers, such as the following one, the E- isomer is normally obtained predominantly, since the E-isomer is usually more stable than the Z- isomer (Fig. 12.11).

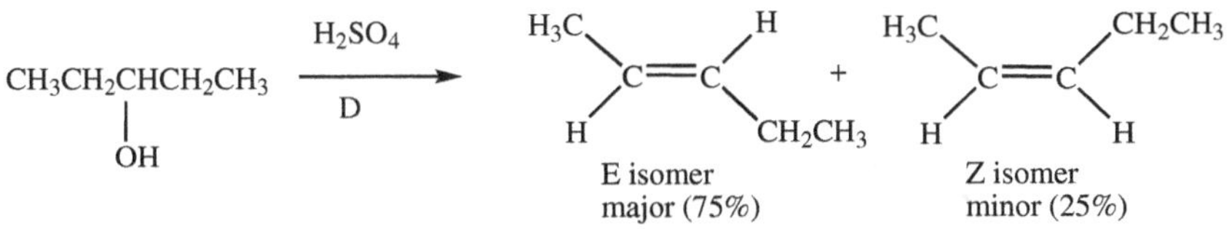
H2SO4
D
CH3CH2CHCH2CH3
OH
H3C
H
C=C
H
CH2CH3
+
H3C
CH2CH3
C=C
H
H
E isomer
major (75%)
Z isomer
minor (25%)

Fig. 12.11

Corollary: Since this reaction is one that mostly gives one stereoisomer of product, it is called a *stereoselective* process. Both this term (*stereoselective*) and the term above (*regioselective*) are used extensively in all areas of synthetic chemistry, and are not restricted to elimination reactions. You will likely see them often in future courses.

An exception to the predominant E-stereoselectivity, of course, occurs when the C=C is in the ring, since the E-isomer is impossibly strained (and much higher energy than the Z-isomer) (Fig. 12.12). One doesn't even observe an E- isomer unless the ring has more than 10 carbons.

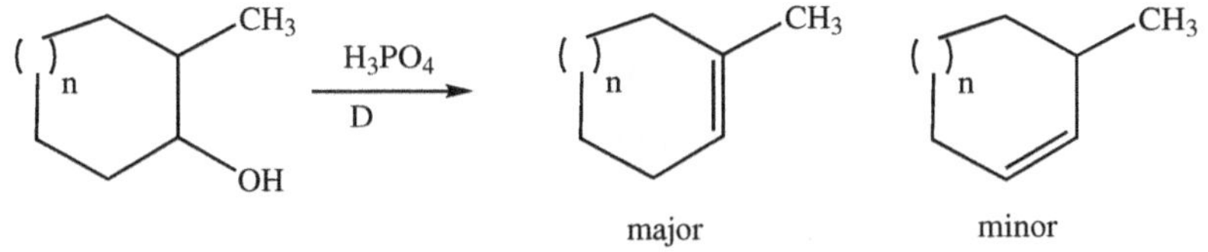
CH3
( )n
OH
H3PO4
D
CH3
( )n
CH3
( )n
major
minor

note: The ( )$_n$ means that there are a number of ring sizes. Most commonly, n=1 (6 membered ring) or n=0 (5 membered ring)

Fig. 12.12

12.5.2 E2 Eliminations

For E2 eliminations, the regioselectivity question is not quite as simple. The results depend on the leaving group, and also depend on the base used.

If the base employed is a less bulky base – and this is probably the *more* common situation – the product obtained tends more towards the Zaitsev (more substituted) product. Conversely, if a more bulky base is used, one tends to get predominantly the less-substituted alkene. Figure 12.13 shows a good example of this changeover.

In these latter cases, where the less-substituted alkene is obtained, the reaction is said to be following the Hofmann rule (*double bond going toward the less-substituted C*). The reason for this change is *steric*: the bulky base has trouble getting access to the H atom on the more substituted side, as the other groups block the way.

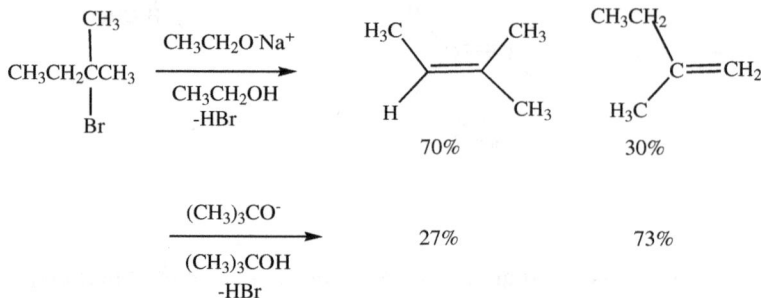

Fig. 12.13

One additional way of more reliably getting the Hofmann product is by using a specially constructed leaving group – not surprisingly, the reaction itself is called a Hofmann Elimination (Fig. 12.14).

This reaction uses quaternary ammonium salts, as follows:

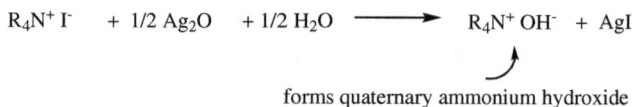

forms quaternary ammonium hydroxide

Note: For this purpose, one always uses the *trimethylammonium* halide, $R-N^+(CH_3)_3I^-$, as the starting point.

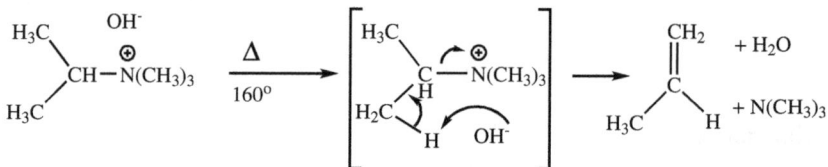

Fig. 12.14

In these cases, the proton that is removed in the elimination is from the less sterically hindered side (i.e., the Hofmann rule) (Fig. 12.15).

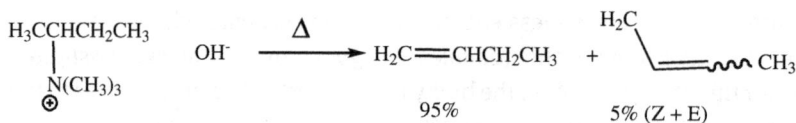

Fig. 12.15

It also should be apparent why the *trimethyl*ammonium salt, $R-N^+(CH_3)_3OH^-$, is always used here..., because if it isn't, one can get competitive abstraction from the *wrong* alkyl group (Fig. 12.16).

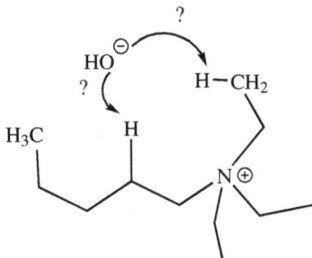

Fig. 12.16

12.6 Stereochemical Effects in E2 Eliminations

In E1 eliminations, the stereochemical disposition of the product was pretty straightforward – the most stable one is obtained. In E2 eliminations there is, by contrast, an overwhelmingly preferred orientation of the leaving H–X molecule. Since the reaction is in the process of forming a π bond, in the lowest energy transition state there must be overlap between the p-orbitals of that developing bond. In other words, the p-orbitals must be parallel to each other. This leaves two possibilities (Fig. 12.17):

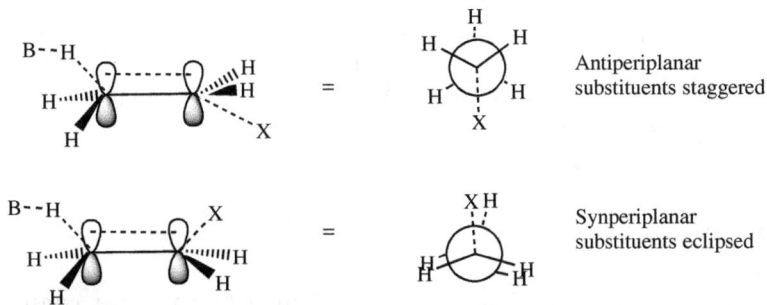

Antiperiplanar
substituents staggered

Synperiplanar
substituents eclipsed

Fig. 12.17

Of these, the antiperiplanar alignment is almost always favored, because of the energetic preference for the staggered conformation. So, for a real example, the following two diastereomers of reactions form alkenes with different specific geometric isomerism. For the *meso* diastereomer (which is also called the *anti* diastereomer) (Fig. 12.18)

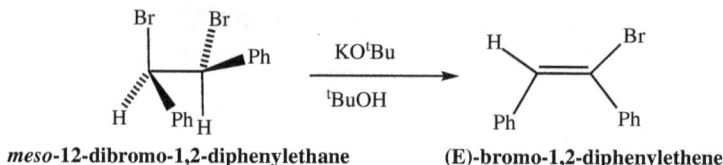

meso-12-dibromo-1,2-diphenylethane   (E)-bromo-1,2-diphenylethene

Fig. 12.18

Whereas for the *d,l*- diastereomer (which is also called the syn diastereomer) (Fig. 12.19).

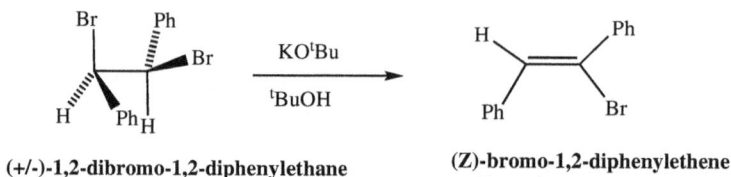

(+/-)-1,2-dibromo-1,2-diphenylethane   (Z)-bromo-1,2-diphenylethene

Fig. 12.19

Corollary: This is not only a stereoselective transformation, but this is also a stereospecific transformation.

Definition: stereospecific transformation: One in which two different stereoisomers of starting material give products that are stereoisomers of each other, i.e.,

isomer 1 ⟶ isomer A (of product)

isomer 2 ⟶ isomer B (of product)

Contrast this to a reaction that is just stereoselective, which is where the reaction of one compound gives an excess of one stereoisomer over another.

12.6.1 Consequences in Cyclic Systems

The fact that antiperiplanar orientations are preferred for E2's has other consequences. Consider, for example Fig. 12.20.

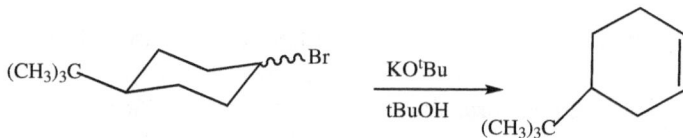

Fig. 12.20

Since the *tert*-butyl group *greatly* prefers to be equatorial, in the *cis* isomer the Br atom is axial in the ground state. It is very easy for the system to assume an antiperiplanar orientation between the leaving Br and a vicinal H, and so the elimination is rapid (Fig. 12.21).

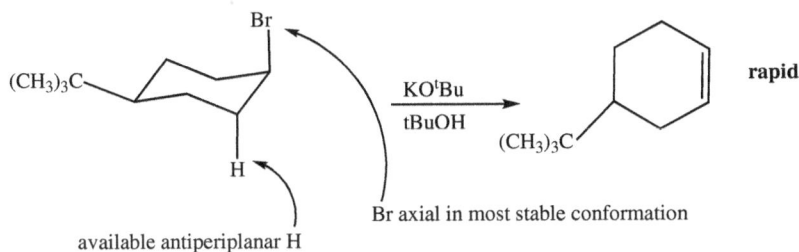

available antiperiplanar H

Fig. 12.21

In the *trans* isomer, though, the Br atom is equatorial in the ground state, and there is no vicinal H atom antiperiplanar to it (Fig. 12.22). The cyclohexane can convert to another chair in which the Br is axial, but now the *tert*-butyl is axial (very disfavored). Only a small very percentage (approximately 0.01%!) of the molecules do this at any one time, and so the reaction is relatively slow.

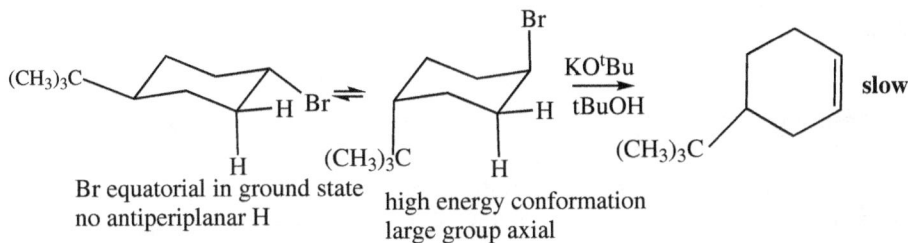

Br equatorial in ground state
no antiperiplanar H

high energy conformation
large group axial

Fig. 12.22

What has been left out is probably the most common case, acyclic cases where there is >1 H atom that could participate. So what happens here (Fig. 12.23)?

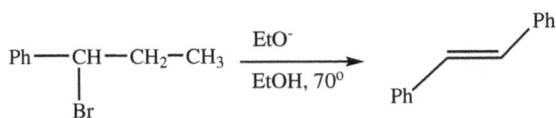

Fig. 12.23

If the conformations of the reactants that can give appropriate antiperiplanar arrangements are evaluated, one obtains the following (Fig. 12.24):

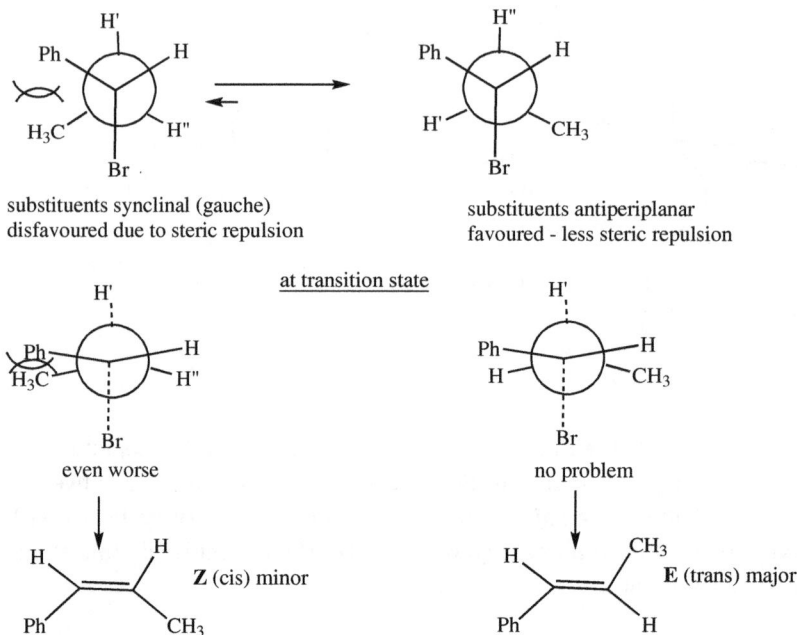

substituents synclinal (gauche)
disfavoured due to steric repulsion

substituents antiperiplanar
favoured - less steric repulsion

<u>at transition state</u>

even worse

no problem

Z (cis) minor

E (trans) major

Fig. 12.24

So in these acyclic cases, a *trans* (*E*-) C=C is preferred regardless of whether it is operating under the Zaitsev or Hofmann rule.

Aside: Although antiperiplanar E2 eliminations have been presented as the only possible type, there *are* methods of doing *synperiplanar* E2 eliminations. These are synthetically very useful, but they are beyond the scope of this course. They are usually based on the idea that the base is covalently linked to the leaving group, such as in Fig. 12.25.

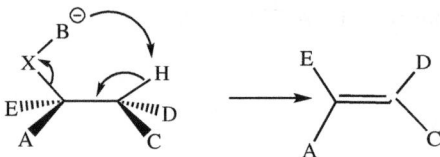

Fig. 12.25

12.7 E1 Versus E2 Eliminations

Much like the S_N1 versus S_N2 competition, it is possible to have a range of mechanisms, from pure E1 to varying degrees of intermediate character and pure E2. The leaving group, base, substrate structure, and solvent all have an effect on which mechanism is operating.

12.7.1 Leaving Group

The leaving groups that particularly favor the E1 are really, really good leaving groups. Curiously, alcohols usually fit in this category, because the reaction is normally done under acid catalysis, which means the leaving group is actually H_2O: (from the protonated alcohol) – these are "always" E1 (Fig. 12.26).

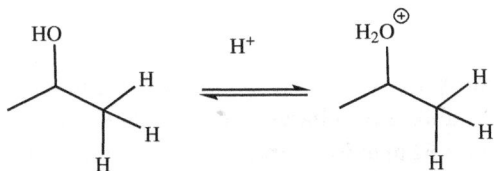

Fig. 12.26

Conversely, good, solid leaving groups such as $^+NR_3$ (from an ammonium salt), Cl^-, Br^-, and I^- usually eliminate by E2, but *can* eliminate by E1 if the other factors warrant. Since HO^- is not a very good leaving group, alcohols rarely if ever eliminate by an E2.

12.7.2 Base

If there is no base (or more properly, the base is the solvent, such as H_2O or an alcohol), this favors an E1 mechanism, really because the absence of base works *against* the E2. Conversely, moderate and strong bases, such as amines NR3 (even though they're neutral), ^-OH, ^-OR (KOtBu, NaOEt), and $^-NR_2$, tend to favor the E2 mechanism. This is probably the single most important thing to look for.

12.7.3 Substrate Structure

Consider a substrate, in which the carbon bearing the leaving group is called the α-site, and the carbon bearing the H atom that is eliminated is called the β-site (Fig. 12.27).

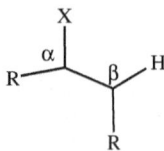

Fig. 12.27

Substitution on the α-carbon (either alkyl or aryl) favors the E1 mechanism, since greater substitution stabilizes the cationic ("+") intermediates. Alkyl substituents at the β-carbon (but *not* aryl groups) favors the E1, actually because it slows down the E2 process. An aryl group at this β-site actually favors the E2 process, because such groups make that β-H atom more acidic.

12.7.4 Solvent

Highly polar solvents, and especially hydroxylic solvents (alcohols, water), tend to favor the E1. This is because such solvents stabilize the charged intermediates ("+") in the E1 mechanism. They certainly don't harm E2 reactions, and many E2 eliminations are done in alcohol solvents.

12.8 Elimination Versus Substitution (E2 vs S_N2)

Since E2 eliminations are also in competition with S_N2 substitutions, it is also important to understand the conditions affect which of these occur. The base and the substrate structure particularly are worth consideration here.

12.8.1 Nature of Base

Remember that every base is a nucleophile, and vice versa. Since proton abstraction is occurring during an E2 elimination, strongly basic nucleophiles tend to favor elimination (E2) over substitution (S_N2). Conversely, weakly basic nucleophiles favor substitution. Sterically hindered bases/nucleophiles tend to favor elimination (E2) even more, since sterics hinder S_N2 substitutions *more* than the elimination.

For example (Fig. 12.28), consider sodium acetate (weakly basic), sodium ethoxide (more basic but sterically about the same), potassium *tert*-butoxide (slightly more basic than sodium ethoxide, but much more sterically hindered), and lithium diisopropylamide (much more basic than sodium ethoxide, and considerably more hindered too) with the same substrate.

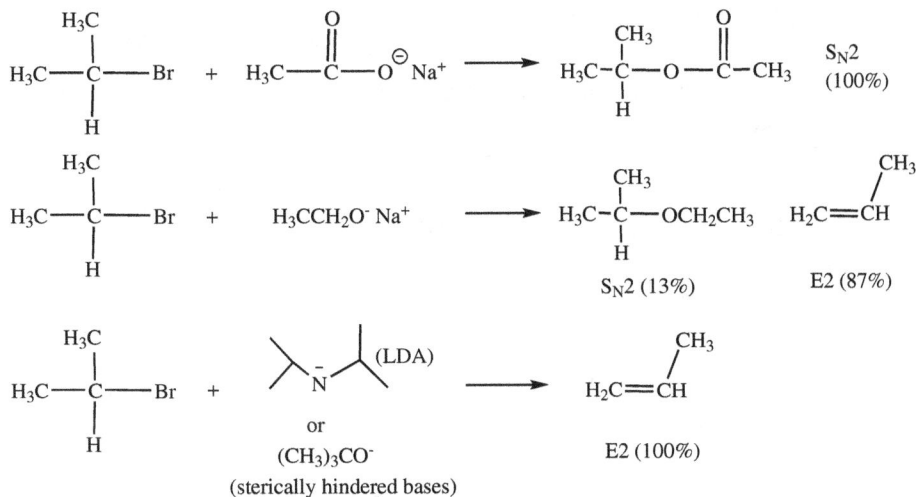

H₃C

H₃C—C——Br + H₃C—C—O⁻ Na⁺ → H₃C-C—O—C-CH₃ S_N2 (100%)

H

CH₃

O

CH₃

H

H₃C

H₃C—C——Br + H₃CCH₂O⁻ Na⁺ → H₃C-C—OCH₂CH₃   H₂C=CH

H

CH₃

H

CH₃

S_N2 (13%)   E2 (87%)

H₃C

H₃C—C——Br + (LDA) → H₂C=CH

H

or

(CH₃)₃CO⁻

(sterically hindered bases)

CH₃

E2 (100%)

Fig. 12.28

12.8.2 Nature of Substrate

The tendency of elimination (E2) versus substitution (S_N2) increases with increasing substitution of the organic substrate. So in terms of the amount of elimination versus substitution, III° (tertiary – never S_N2) ≫ II° (secondary) > I° (primary). If we consider using the same base/nucleophile (Fig. 12.29), then

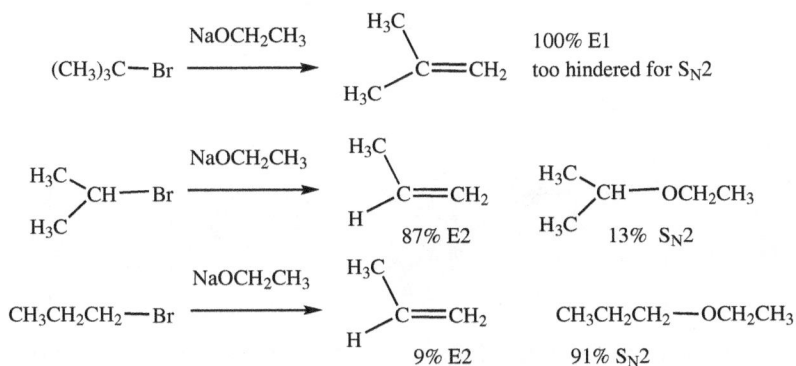

$(CH_3)_3C-Br$ $\xrightarrow{NaOCH_2CH_3}$ H₃C\C=CH₂/H₃C   100% E1 too hindered for S_N2

H₃C\CH—Br/H₃C $\xrightarrow{NaOCH_2CH_3}$ H₃C\C=CH₂/H   87% E2   H₃C\CH—OCH₂CH₃/H₃C   13% S_N2

$CH_3CH_2CH_2-Br$ $\xrightarrow{NaOCH_2CH_3}$ H₃C\C=CH₂/H   9% E2 $CH_3CH_2CH_2-OCH_2CH_3$ 91% S_N2

Fig. 12.29

12.9 The E1cB Mechanism

There is at least one other mechanism encountered on occasion. Since there are a lot of details to this one once it is looked at carefully, it will just be mentioned very briefly here, for interest's sake and since future courses have important examples of this elimination.

　　If we consider what has been seen so far, the elimination mechanisms are really a continuum, and we have really only seen one extreme (where the leaving group departs first) and the perfect middle case (where the proton and leaving group depart simultaneously). So what happens if the H atom (as a proton) is removed first (Fig. 12.30)?

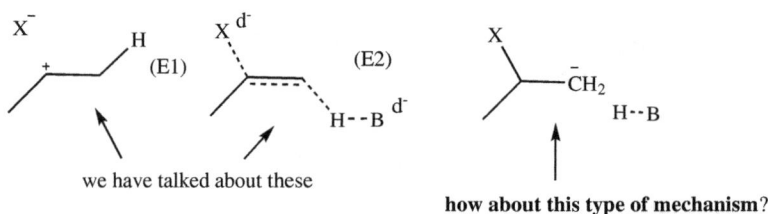

Fig. 12.30

The latter exists and is called the *E1cB mechanism*, where cB = conjugate base. It is a two-step mechanism, as follows (Fig. 12.31):

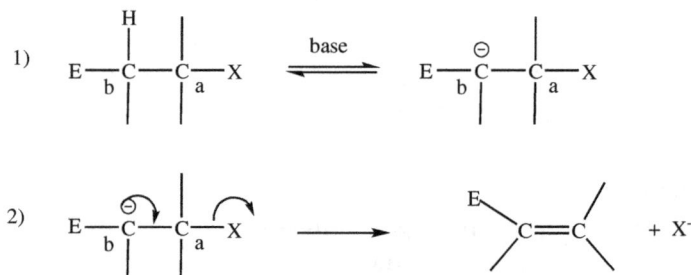

Fig. 12.31

This is favored when there are electron-withdrawing groups (especially carbonyl compounds) on the β-C (E), because it stabilizes the anionic intermediate. In subsequent courses, you'll see a reaction (called the aldol condensation) that has an E1cB elimination step.

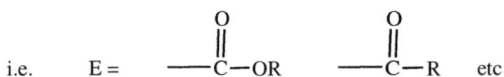

Since this anionic intermediate is stabilized and can exist for a while, the X- doesn't have to be that good a leaving group. So, not only can the halides participate in E1cB eliminations, but alcohols (X = OH) do so very commonly, too.

In truth, most mechanisms that are considered E2 are either a little bit towards E1 or a little towards E1cB. To go further into E1cB, however, is beyond the scope of this course.

12.10 Problems

12-1. Choose the reagent from the given pair that will lead to the desired product in each of the following reactions.

(a) 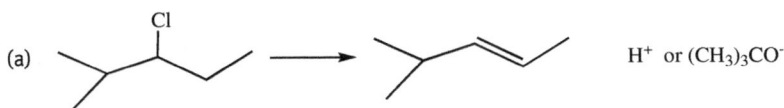 H^+ or $(CH_3)_3CO^-$

(b)  H^+ or $(CH_3)_3CO^-$

(c) 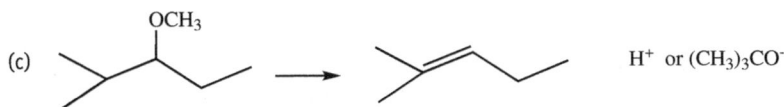 H^+ or $(CH_3)_3CO^-$

12-2. Predict the carbon NMR of the major product obtained from each of the following. Include the approximate chemical shift of each signal.

(a)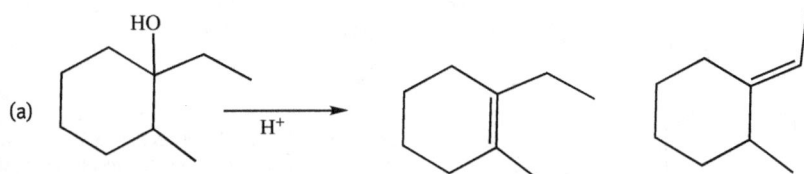

(b)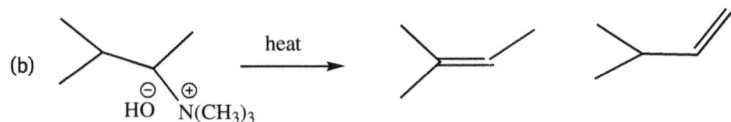

(c)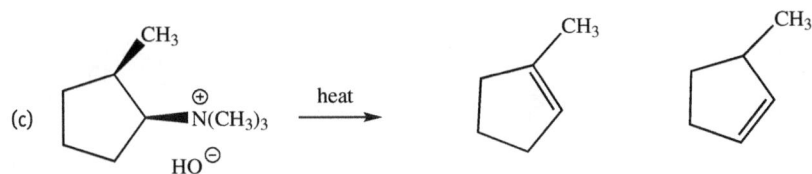

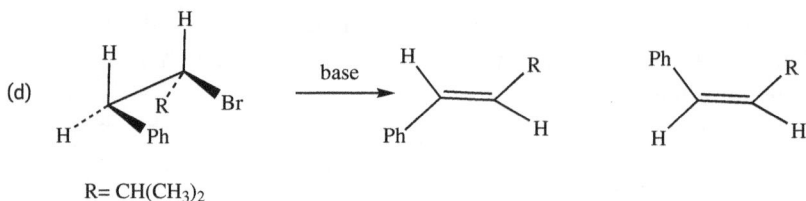

(d)

base

R= CH(CH3)2

12-3. Predict the form of the olefinic region of the proton NMR of the major product formed from each of the following. Be sure to show the multiplicities of the signal(s).

(a)

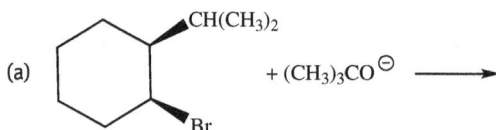

$+ (CH_3)_3CO^{\ominus}$ ——→

(b)

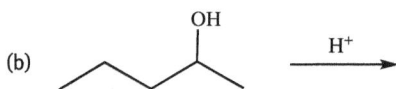

H^+ ——→

12-4. In each of the following reactions, some information about the product formed is given. Give a structure that corresponds to this information.

(a)

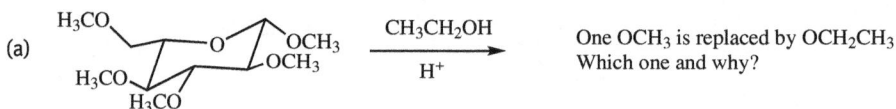

$\dfrac{CH_3CH_2OH}{H^+}$

One OCH₃ is replaced by OCH₂CH₃
Which one and why?

(b)

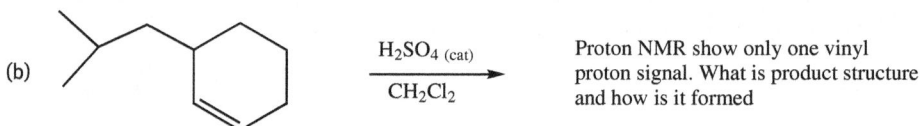

$\dfrac{H_2SO_4 \ (cat)}{CH_2Cl_2}$

Proton NMR show only one vinyl proton signal. What is product structure and how is it formed

(c)

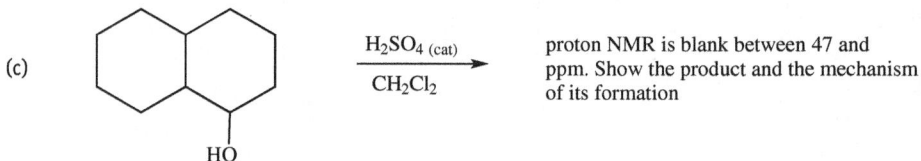

$\dfrac{H_2SO_4 \ (cat)}{CH_2Cl_2}$

proton NMR is blank between 47 and ppm. Show the product and the mechanism of its formation

12-5. A synthetic organic chemist wished to effect the following transformation.

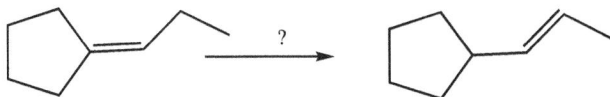

Drawing on the knowledge obtained from this course how would you proceed. (*Hint –* more than one step will be required and you should consider reactions you have learned previously!)

12-6. A chemist tried to effect the following transformation

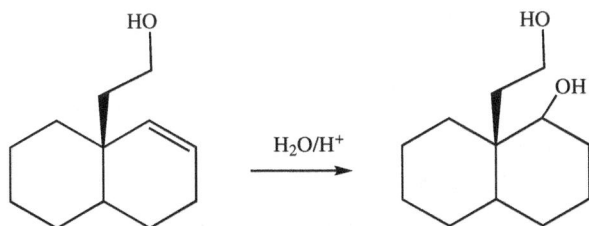

The product obtained showed no olefinic protons in the proton NMR and no OH groups in the IR. The molecular formula of the product was $C_{12}H_{20}O$. Propose a structure for this compound and show how it would be formed.

13 Introduction to the Molecules of Nature

13.1 Introduction

Almost all of the compounds produced by plants and animals are organic in nature. Some biologically important compounds do contain metal atoms (e.g., Mg^{+2} in chlorophyll and Co^{+2} in vitamin B12) but, by and large, the so-called "natural products" consist of arrays of carbon, hydrogen, nitrogen, oxygen and the occasional sulfur atoms. Plants and animals produce a bewildering array of compounds that often seem to have little to do with their requirements for life. Many have strong medicinal effects – both positive and negative – on humans. Many are generated as defense mechanisms against predators (e.g., tetrodotoxin from the Japanese puffer fish is one of the most toxic materials found in nature) and others seem to have no purpose (e.g., aflatoxin which is produced by a fungus that grows on peanuts and other grains and is highly toxic and carcinogenic). The structures shown in Fig. 13.1 will give you a taste of some of the more interesting and complex structures.

The medicinal properties of some plants have been known for centuries. Ancient civilizations used them to cure a variety of ailments without knowing the nature of the actual active ingredient. In the 19th century, efforts started to determine the structure

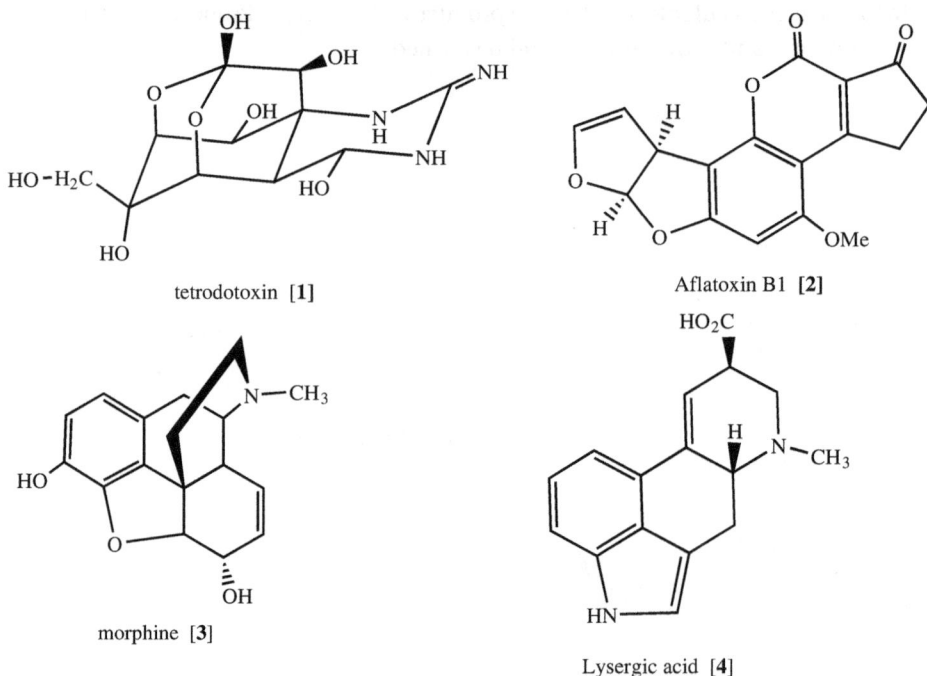

tetrodotoxin [1]

Aflatoxin B1 [2]

morphine [3]

Lysergic acid [4]

Fig. 13.1

https://doi.org/10.1515/9783110778311-013

of these active ingredients. Before this could happen, methods for isolating them from the natural source were required – a challenging task with the tools that were available at the time. Separation methods were limited essentially to crystallization and distillation. Not surprisingly, most of the early work was centered on molecules that had a basic nitrogen which allowed the preparation of a salt that was more crystalline than the actual material. These separations were very laborious and the isolation of a pure chemical from the plant was considered a major accomplishment.

Once isolated, the chemical structure of the material still had to be determined – a daunting task considering the methods that were available. Spectroscopy as we know it today was unknown except for ultraviolet spectroscopy which gave little information on non-aromatic materials. Despite these handicaps, many famous chemists were successful in determining the structure of some natural products. These ranged from rather simple materials to amazingly complex ones. Once the structure of a compound was determined, the challenge was to be able to duplicate Mother Nature's efforts by devising ways to synthesize these molecules – and many other related ones – in the laboratory.

The history of this era in organic synthesis is a fascinating one, dominated by some famous chemists, some of whom won Nobel prizes for their efforts. Names like Woodward, Eschenmoser, Jeger, Willstater and Robinson were successful in carrying out long complicated syntheses. Their work led to an explosion of knowledge about the ways to connect smaller molecules together to make larger and larger ones (synthesis), to eventually arrive at the natural product. This knowledge provided the basis for the synthesis of many non-naturally occurring materials which are of tremendous importance in today's commerce. Invariably, these synthetic routes involved low temperatures, solvents like ether and chloroform and many other materials and conditions that did not resemble nature in any way! Mother Nature essentially uses one carbon source, CO_2, and one solvent, water. Temperatures are restricted to normal ambient ones. Furthermore, most natural products are highly chiral, i.e., they contain multiple chiral centers, and the hallmark of products made in nature is that they are almost entirely ONE ENANTIOMER! It is only recently that chemists have found ways to duplicate this amazing feat, and many times a chiral molecule isolated from nature is the basis of the unique stereochemical outcome. While one has to admire the ingenuity of the chemists, the ability of nature far outstrips even their very best efforts.

In the following pages, an introduction to some relatively simple compounds is presented. The intent of this material is not to be encyclopedic, but rather to perhaps whet your appetite to further investigate this area of chemistry.

It must be noted that the early days of isolation and synthesis focused on natural products that were largely aromatic in nature. There were two reasons for this.

(1) Aromatic molecules are detectable using UV light and are therefore easier to identify in the isolation process referred to above.

(2) Aromatic molecules are FLAT – so there are no stereochemical factors to consider during their characterization and synthesis.

13.2 Amino Acids, Peptides and Proteins

We will begin with a look at amino acids and how they combine into very large molecules called proteins.

When chemists refer to amino acids, they are implicitly specifying α-amino acids of general structure I. Except for the single case of glycine where R is hydrogen, these molecules have a chiral center next to the amino group. In all but one naturally occurring example, the absolute configuration of this center is $\underline{S}$ (the lone outlier, cysteine, whose R configuration is due to the presence of a sulfur atom that changes the priorities of the substituents). Some are classified as being "essential", meaning that they must be acquired in the diet. Table 13.1 shows the structures of the side chains in the most common α-amino acids. It is common to assign a three-letter code to each amino acid. These are also shown in Table 13.1.

Peptides are short polymers of these amino acids, but much longer polymers are called proteins. These combinations of amino acids are formed by amide formation between the amino group of one and the carboxyl group of the next amino acid. These combinations are facilitated by enzymes that do this under natural conditions, unlike the methods we saw in Chapter 10. Proteins consist of hundreds of amino acids, almost all chiral, so the protein itself would have an enormous number of possible stereoisomers if it wasn't for the fact that all the amino acids are stereochemically homogeneous. Proteins appear to be linear molecules when only their molecular structure is written, but this is far from the truth. In practice, they are folded into very complex 3-dimensional structures with highly defined areas where they function as catalysts for biochemical reactions. In this context, they are referred to as enzymes – nature's catalysts.

When you get your steak from the store, you have to heat it to make it edible. One of the processes that happens during the heating is called denaturation and is destroying the folding patterns and returning the polymer to its linear form.

Some amino acids form the building blocks for other types of molecules found in nature. In particular, the ones that contain an aromatic group (tryptophan, phenylalanine, tyrosine) are the source of many types of compounds. The indole nucleus (see Fig. 13.2), which is a part of tryptophan, is especially prolific in this regard. Decarboxylation of tryptophan produces tryptamine, which is the precursor of many alkaloids including the neurotransmitter serotonin which regulates mood, memory and many other biochemical processes. The indole nucleus from tryptophan leads to many types of alkaloids, too numerous to mention here.

Smaller chains of amino acids – or fragments of proteins – are called peptides. It is interesting that while proteins are required dietary elements, peptides are the main defense mechanism of many organisms. For example, the venom of some snakes and the stinging material in many jellyfish are composed mainly of peptide material. Peptides can also have significant medical applications (insulin is a peptide).

Tab. 13.1: Structures and Codes of Amino Acids

Name	Structure of R	Code
	in H_2N	
Alanine	$-CH_3$	Ala
Arginine	$-(CH_2)_3NHC(NH_2)=NH$	Arg
Asparagine	$-CH_2CONH_2$	Asn
Aspartic acid	$-CH_2COOH$	Asp
Cysteine	$-CH_2SH$	Cys*
Glutamic acid	$-(CH_2)_2COOH$	Glu
Glutamine	$-(CH_2)_2CONH_2$	Gln
Glycine	$-H$	Gly
Histidine		His*
Isoleucine	$-CH(CH_3)CH_2CH_3$	Ile*
Leucine	$-CH_2CH(CH_3)_2$	Leu*
Lysine	$-(CH_2)_4NH_2$	Lys*
Methionine	$-CH_2CH_2SCH_3$	Met
Phenylalanine		Phe
Proline		Pro
Serine	$-CH_2OH$	Ser
Threonine	$-CH(CH_3)OH$	Thr*
Tryptophan		Trp*
Tyrosine		Try
Valine	$-CH(CH_3)_2$	Val*

asterisk (*) denotes an essential amino acid

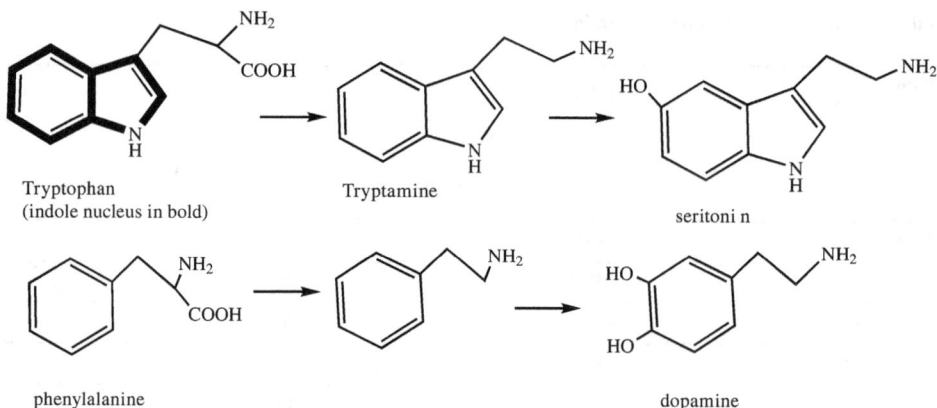

Tryptophan
(indole nucleus in bold)

Tryptamine

seritoni n

phenylalanine

dopamine

Fig. 13.2

13.3 Lipids (Fats) and Steroids

13.3.1 Lipids (aka Fats)

The term "fats" refers to esters of 1,2,3-trihydroxypropane, otherwise known as glycerol. Lipids is another term that refers to the same materials. The acids that are used to esterify the glycerol are long-chain, unbranched aliphatic acids (12–18 carbon atoms), usually with an even number of carbon atoms. The acids (fatty acids) may contain a number of double bonds (unsaturated) or none (saturated), and the physical properties of these two classes are quite different. Saturated fats are either solid (e.g., lard) or very viscous liquids; as the number of double bonds increases, the fats become increasingly liquid. Commercial preparations of highly unsaturated fats are often hydrogenated to make the product more appealing. It is well beyond the scope of this book to describe how these fats are metabolized, suffice it to say that the presence or absence of unsaturation changes the products of fat metabolism. It is perhaps surprising, based on the information previously presented about the stability of geometric isomers, that the double bonds in unsaturated fatty acids all have the Z configuration! The catalysts used in the hydrogenation process mentioned above can cause the isomerization of the remaining double bonds to the more stable E isomer, giving rise to "trans fats" which have significant negative health effects. Saturated fats are metabolized by a series of oxidations to produce acetate and then a compound called mevalonolactone, which is a direct precursor to the steroid nucleus (e.g., cholesterol), whereas the unsaturated ones are metabolized by a completely different route. People prone to high cholesterol are advised to avoid consuming saturated fats for this reason.

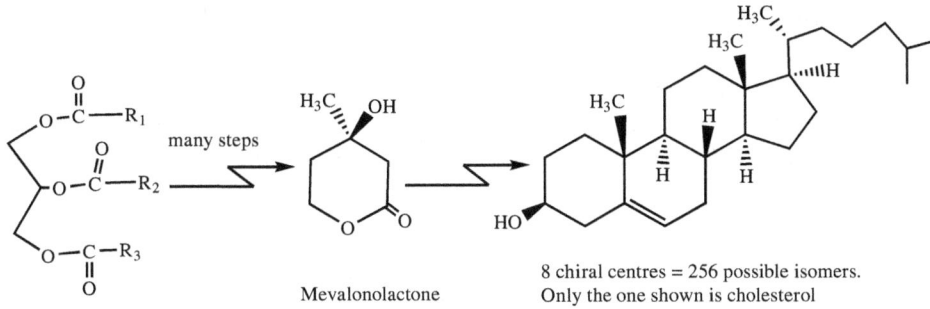

Fig. 13.3

13.3.2 Steroids

Steroids consist of 4 rings with various appendages and oxidation states. They are made by plants, animals and fungi. In addition to cholesterol, things like cortisone, prednisone and testosterone have important medical applications.

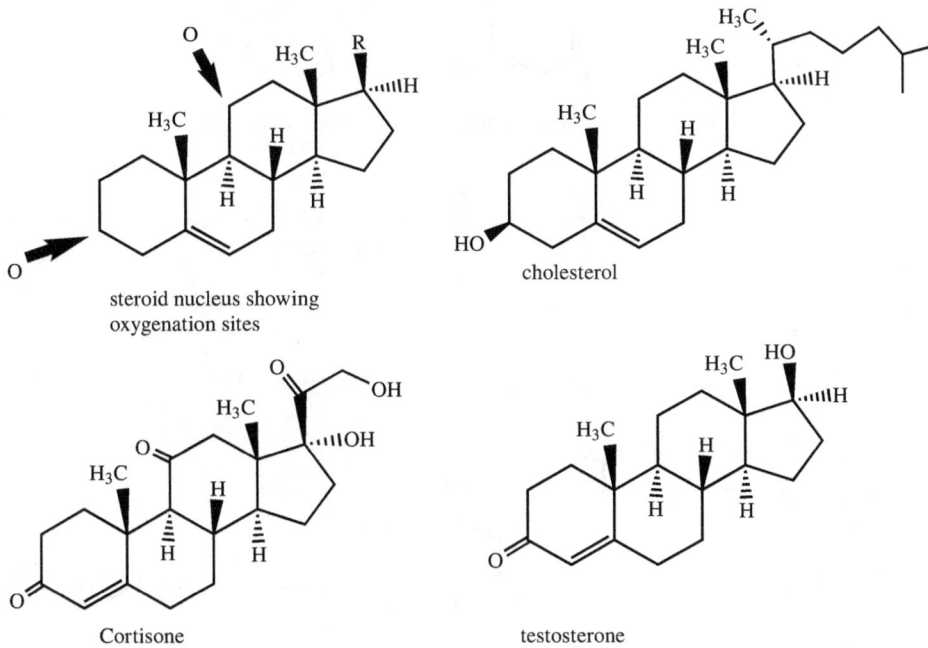

steroid nucleus showing oxygenation sites

cholesterol

Cortisone

testosterone

Fig. 13.4

13.4 Vitamins

A vitamin is an organic molecule (or set of closely related molecules), that is an essential micronutrient that an organism needs in small quantities for its metabolism to function properly. Vitamins must be obtained via the diet as the body does not make

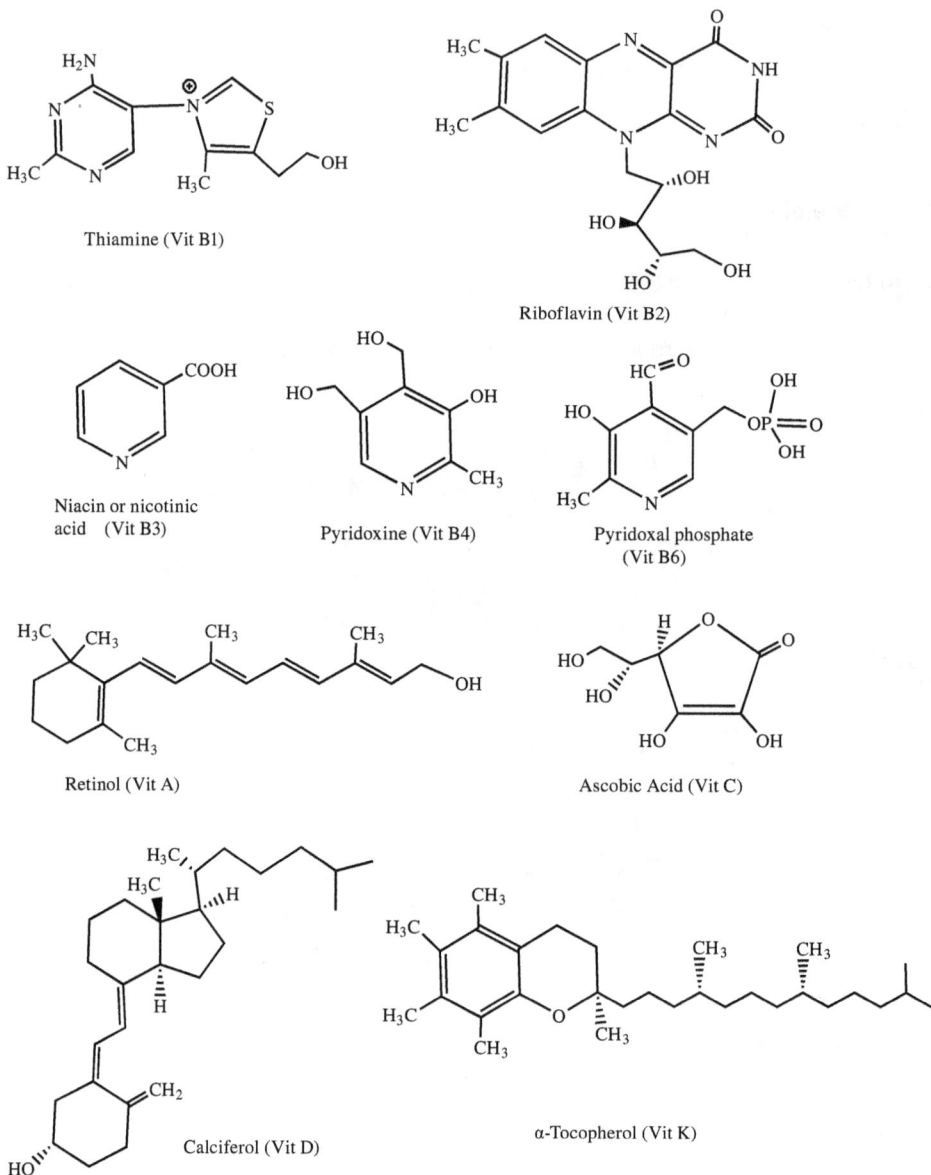

Thiamine (Vit B1)

Riboflavin (Vit B2)

Niacin or nicotinic acid (Vit B3)

Pyridoxine (Vit B4)

Pyridoxal phosphate (Vit B6)

Retinol (Vit A)

Ascorbic Acid (Vit C)

Calciferol (Vit D)

α-Tocopherol (Vit K)

Fig. 13.5

them. Some come from plants and others from animals. There are severe health consequences if any vitamin is not present in the diet. A lack of vitamin C leads to scurvy
and a lack of vitamin B1 leads to a condition called beriberi. The structure of the vitamins varies widely. Vitamin A is a long-chain molecule. There are a number of B vitamins: B1 (thiamine), B2 (riboflavin), B3 (niacin), B4 (adenine), B6 (pyridoxine) and
B12 (cobalamin). Vitamins C, D, E and K are all unique in their structures, as can be
seen in Fig. 13.5.

13.5 Alkaloids

Alkaloids is a term used to describe nitrogen-containing compounds, mainly produced
by plants. Many of these are pharmacologically active and formed an important part
of ancient traditional medicine. Alkaloids were some of the earliest materials isolated
because they are basic and can be characterized as crystalline salts. Alkaloids occur in
"families" – mixtures of closely related structures. Depending on the structure, a wide
variety of effects on humans is exhibited – analgesic (morphine, #3 Fig. 13.1), stimu-
lant (cocaine, nicotine, caffeine) and anti-malarial (quinine) are some examples. The
naturally occurring compounds are frequently modified by chemists in the laboratory
to alter or enhance their properties (Fig. 13.6).

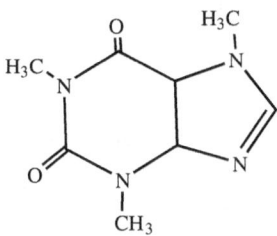

Caffeine

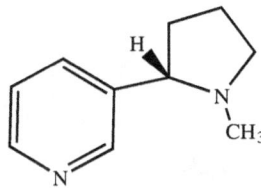

Nicotine

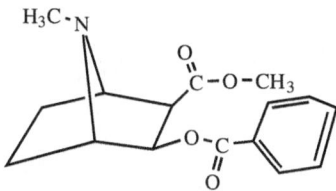

Cocaine

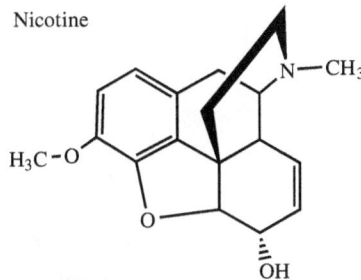

Codeine (morphine methyl ether)

Fig. 13.6

13.6 Nucleic Acids (and Nucleosides)

The molecules DNA (deoxyribonucleic acid) and RNA (ribonucleic acid) are huge polymers of nucleotides and are essential to all forms of life. The famous double-helix structure of DNA molecules discovered by Watson and Crick is probably one of the most important discoveries in the 20th century. In contrast, RNA is a single-strand helix. The helical structure of both DNA and RNA is stabilized by hydrogen bonding between base pairs.

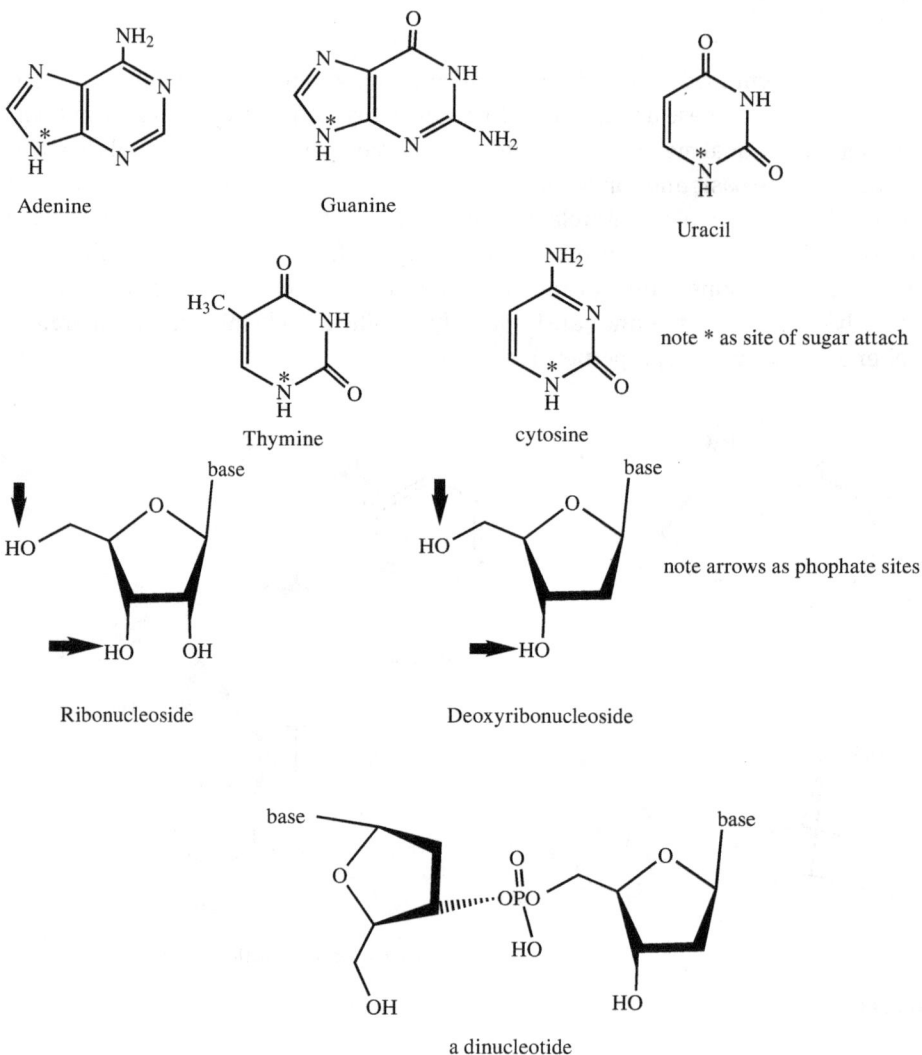

Adenine

Guanine

Uracil

note * as site of sugar attach

Thymine

cytosine

base

base

HO

HO

note arrows as phophate sites

HO   OH

HO

Ribonucleoside

Deoxyribonucleoside

base

base

OH

HO

HO

a dinucleotide

Fig. 13.7

Nucleosides are products formed by joining a nitrogenous base to a five-carbon sugar, either ribose or deoxyribose. There are five bases – adenine, guanine, thymine (only in DNA), uracil (only in RNA) and cytosine. These are linked to the sugar by bond formation between a nitrogen atom and C1 of the sugar. You should recognize this as the nitrogen equivalent of an acetal. These are then converted to nucleotides by phosphate ester formation at C5. Polymerization occurs by joining another nucleoside at the phosphorus atom through the hydroxyl group at C3 (Fig. 13.7).

13.7 Summary

The aim of this chapter was to give you insights into some of the major classes of molecules found in nature. The list is not exhaustive, but rather selective. Some will feel that important types have been left out and this is true, but to be comprehensive would require another text book. The complexity of many of the molecules is clearly many steps beyond what is typically learned in the first organic courses which is the focus of this text. Hopefully what has been included will inspire the readers to explore both more advanced organic chemistry and biochemistry. It is a fascinating world!!

Appendix: Answers to Problems From Part I

Chapter 1

1-1. Melting points are determined by the attractive forces between <u>MOLECULES</u>. In co-valent molecules, attractive forces between <u>ATOMS within a molecule</u> (i.e., bonds) are strong but intermolecular forces are relatively weak (hydrogen bonds, van der Waal's forces). Ionic compounds consist of electrically charged ions which will attract all adjacent oppositely charged ions. These interionic forces are large and the energy required (in the form of heat) to disrupt the ordered system (the crystal) is usually much larger in the ionic materials.

1-2. In order to separate molecules which are in a crystal, the attractive forces between them must be overcome. This requires energy, which is absorbed from the available thermal energy. For the same reason, molecules are more soluble in hot solvents, and crystallization of a material (the reverse of dissolution) evolves energy in the form of heat.

1-3. The fragment C-H-C implies a divalent hydrogen atom. This would require that hydrogen has a share of <u>FOUR</u> electrons which would require two orbitals. The energy difference between the 1s orbital and the next available one – the 2s orbital – is large. In addition, since the hydrogen atom has only one proton in the nucleus, forming two bonds would require the atom to become negatively charged (prove this to yourself by doing an electron count!).

1-4. This will be done for one hydrogen and one carbon atom. Each hydrogen has <u>ONE</u> bond which consists of <u>TWO</u> electrons. The nearest rare gas is helium ($Z = 2$) and therefore the hydrogen atom has a share of the same number of electrons as He. Since it has only 1/2 share of this electron pair, it has only one net negative charge and this exactly balances the nuclear charge, i.e., the atom is electrically neutral.
Each carbon forms <u>FOUR</u> bonds – one to another carbon and three to three hydrogen atoms. It therefore has 8 (4×2) electrons in its valence shell and has the electronic configuration of Ne, the nearest rare gas. It has 1/2 share of these 8 electrons which account for <u>FOUR</u> negative charges, PLUS two electrons in the 1s orbital. The total is therefore 6. This balances the nuclear charge.

1-5. The molecule CCl_4 has an sp^3-hybridized carbon and is therefore tetrahedral in shape. There are four equivalent <u>BOND</u> dipoles, one for each C-Cl bond. There are vector quantities which, when summed, give a <u>MOLECULAR</u> dipole $= 0$.

https://doi.org/10.1515/9783110778311-014

1-6. Bond #1 – σ-bond using an sp^3 orbital on one carbon and an sp^2 orbital on the other.

Bond #2 – σ-bond using an sp^2 orbital on each carbon.

Bond #3 – π-bond using a p-orbital on each carbon.

(NOTE – bonds #2 and #3 can be reversed)

Bond #4 – σ-bond using an sp^3 orbital on carbon and an sp^3 orbital on oxygen.

1-7. (a) $^{\delta+}$C–N$^{\delta-}$ (b) $^{\delta+}$C–O$^{\delta-}$ (c) $^{\delta+}$N–O$^{\delta-}$ (d) $^{\delta-}$N–H$^{\delta+}$ (e) $^{\delta+}$H–O$^{\delta-}$ (f) C–H^{same}

1-8. The table should be completed as follows:

tetrahedral	planar	linear
4	4	4
4	3	2
0	1	2
4	3	2

Chapter 2

2-1.

(a) 2,3-dimethylpentane

(b) 3-ethyl-5-methylheptane

(c) 5-ethyl-2,3-dimethyloctane

(d) 2,2-dimethylbutane

(e) 3,6-dibromo-4-chloro-2,5-dimethyloctane

(f) 1-bromo-2-pentene

(g) 5-iodo-2-methyl-1,3-hexadiene

(h) 6-chloro-4-ethyl-2-heptene

(i)   4-bromo-3-methyl-1-hexyne

2-2.

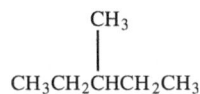

(a)

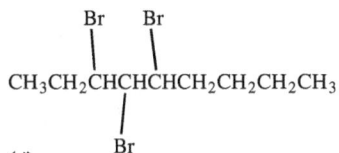

(d)

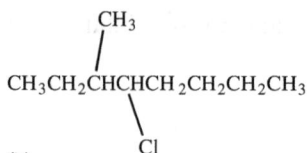

(b)

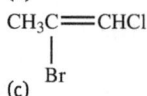

(c)

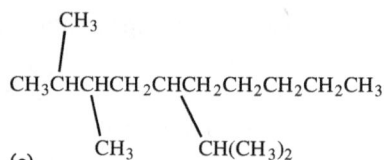

(e)

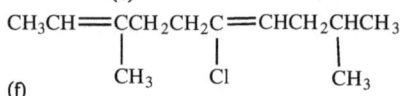

(f)

2-3.

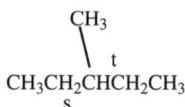

(a)

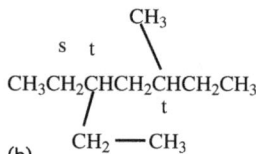

(b)

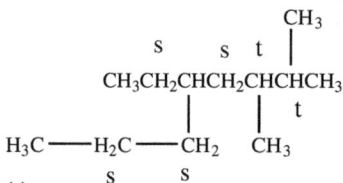

(c)

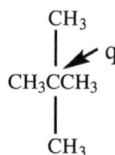

(d)

All methyl groups (CH_3) are primary

2-4. The number of <u>grams</u> of carbon in 3.2 g of CO_2 is $3.2 \times 12/44 = 0.873$ or $0.873/12 = 0.0727$ gram atoms.

The number of grams of hydrogen in 1.1 g of water is $2/18 \times 1.1 = 0.122 = 0.122$ gram atoms.

The ratio of H to C on a MOLAR basis is 1.67 or 5 : 3.

The simplest formula is therefore C_3H_5. However, hydrocarbons must have an even number of hydrogen atoms. Therefore, the molecular formula must be C_6H_{10} or some multiple of this.

2-5. (a) 0, (b) 1, (c) 7, (d) 2, (e) 2.

2-6.

CH₃CH₂CH₂CH₂CH₂CH₂Cl   1-chlorohexane

$$\underset{\displaystyle \text{Cl}}{\text{CH}_3\text{CH}_2\text{CH}_2\text{CH}_2\text{CH}_2\overset{|}{\text{CHCH}}_3}$$ 2-chlorohexane

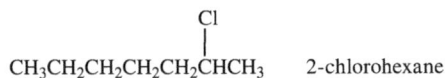

CH₃CH₂CH₂CH₂CHCH₂CH₃   3-chlorohexane
|
Cl

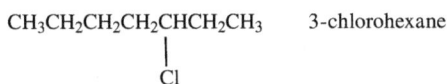

2-7. There are 17 positional isomers based on the following carbon skeletons:

C-C-C-C-C-C   3 isomers   C-C-C-C   2 isomers
                                        | |
                                        C C

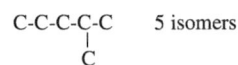

C-C-C-C-C   5 isomers   C
| |
C   C

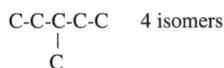

C-C-C-C-C   4 isomers   C-C-C-C   3 isomers
| |
C   C

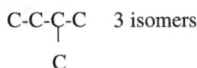

2-8.

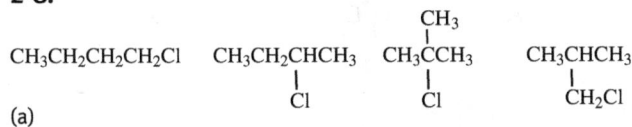

(a)   CH₃CH₂CH₂CH₂Cl   CH₃CH₂CHCH₃   CH₃CCH₃   CH₃CHCH₃
                                  Cl   Cl   CH₂Cl

CH₃CH₂CH₂CH₂CH₂-Br   1-bromopentane

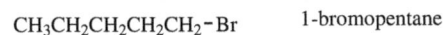

CH₃CH₂CH₂CHCH₃   2-bromopentane
          Br
(b)

CH₃CH₂CHCH₂CH₃   3-bromopentane
|
Br

CH₃CH₂CHCH₂-Br   1-bromo-2-methylbutane
       CH₃

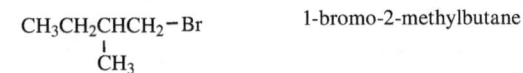

CH₃
|
CH₃CH₂CCH₃   2-bromo-2methylbutane
|
Br

Br
|
(CH₃)₂CHCHCH₃   2-bromo-3-methylbutane

(CH₃)₃CCH₂Br   1-bromo-2,2-dimethylpropane

2-9.

(a) identical

(b) positional

(c) positional

2-10. (a) sp^3 (b) sp (c) sp^3 (d) sp^2

Chapter 3

3-1. 2-methylbutane viewed down the C2-C3 bond has two methyl groups on C2 and one on C3. The rotational forms (conformations) possible are:

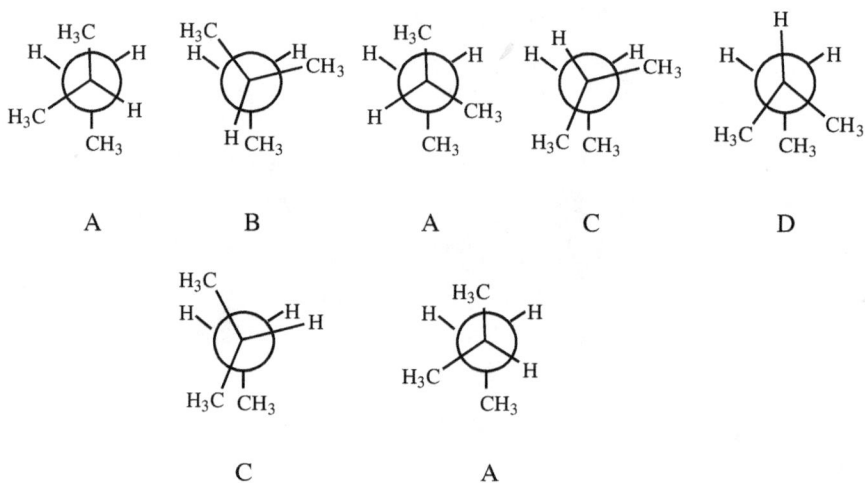

A   B   A   C   D

C   A

The energy diagram looks like:

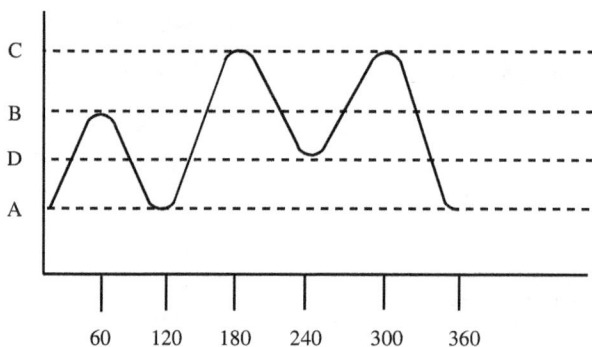

A and D are staggered and D>A in energy.
B and C are eclipsed and C>B in energy

3-2. The most stable conformation of 1,2-dichloroethane is the anti-periplanar one. In this form, the <u>VECTOR QUANTITIES</u> representing the two C-Cl BOND dipoles sum to zero. This is not true of the other, less stable conformations of this molecule. As the temperature is raised, these less stable conformations become more highly populated and the measured MOLECULAR dipole moment increases.

3-3. (a) E, (b) Z, (c) E

3-4.
(a) (E) 3-methyl-2-pentene
(b) (Z) 3-bromo-1-chloro-1-fluoro-2-isopropyl-1-butene

3-5.

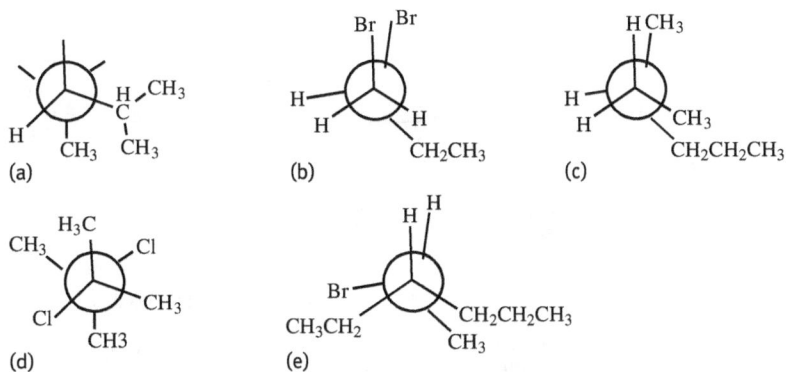

(a)   (b)   (c)

(d)   (e)

3-6. For interconversion, 180° rotation about the C2-C3 σ-bond must occur. Since, after 90° rotation, no overlap between the p-orbitals on the two carbon atoms can occur, the π-bond must be broken. Insufficient energy is usually available to break this bond and the interconversion of *cis* and *trans* isomers cannot occur.

3-7.

E   Z

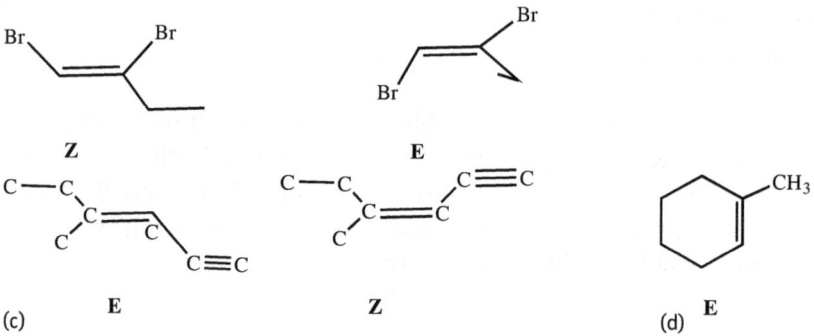

(c)   E

(d)   E

3-8.

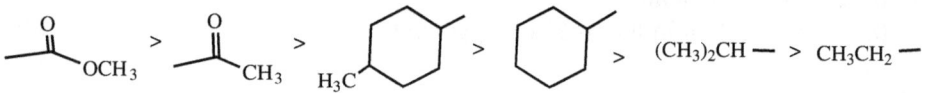

3-9.

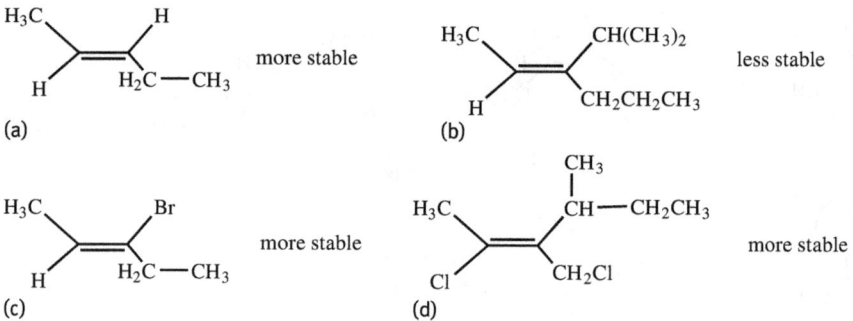

(a)   more stable

(b)   less stable

(c)   more stable

(d)   more stable

3-10.

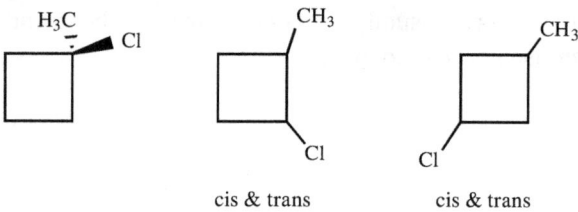

cis & trans

cis & trans

3-11. There are <u>eight</u> isomers of $C_5H_{11}Br$ based on the following skeletons:

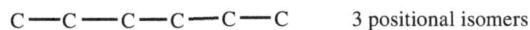

C—C—C—C—C—C   3 positional isomers

4 positional isomers

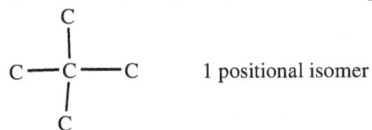

1 positional isomer

3-12.

1,2	ea	ae	cis	
	ee	aa	trans	✓
1,3	ee	aa	cis	✓
	ea	ae	trans	
1,4	ea	ae	cis	
	ee	aa	trans	✓

3-13. The most stable <u>configuration</u> of 1,3-disubstituted cyclohexanes is <u>cis</u>.

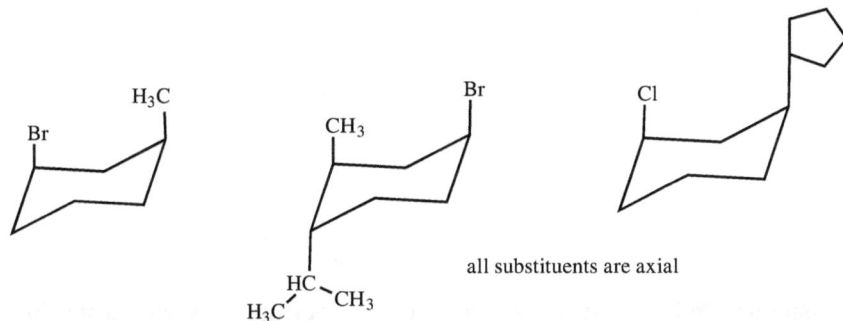

all substituents are axial

3-14.

(a) CH₃

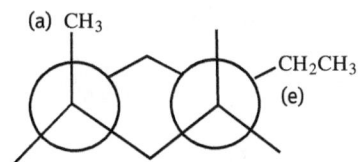

CH₂CH₃
(e)

3-15.

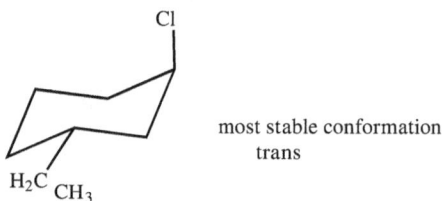

most stable conformation
trans

3-16. Yes it is the most stable conformation. It is the _trans_ isomer.

3-17. It is a very bulky group. In the axial position, it has severe skew interactions with the ring bonds. Whereas the isopropyl and smaller groups can rotate to minimize this interaction by putting a hydrogen atom in the interacting position, the tertiary butyl group cannot do this.

Chapter 4

4-1. Nucleophiles: CN^-, CH_3OH, OH^-, Br^-
Electrophiles: H^+, BF_3
Neither: H_2

4-2. All are -I

4-3. Substitution
Elimination
Reduction (addition)
Oxidation (elimination)

4-4. A transition state is the highest energy species occurring between the starting materials and product of a step in a reaction.
An intermediate is a species occurring between starting materials and product of a reaction which is a minimum on the energy diagram. It must have a transition state on both sides of it on the energy plot.

4-5.
(a) rate α [A]
(b) rate α [A][B]

4-6.

(a) Since the synperiplanar form involves the eclipsing of the two largest groups, the synclinal conformation will be more stable.

(b) The _cis_ isomer is more stable since it can be diequatorial.

(c) The _trans_ isomer (left drawing) represents the more stable isomer since the two large methyl groups are far removed from one another. All other interactions in the two systems are comparable.

4-7. Structures b, e and g cannot undergo resonance. Those that can be resonance stabilized are:

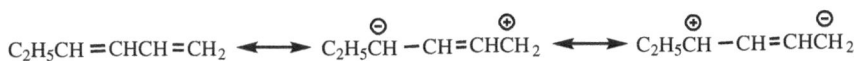

(a)

(c)   (d)

(f)

(h)

(Note that g cannot be stabilized as no electrons are available on the positively charged nitrogen atom. Also note that you cannot generate a resonance form in (d) or (f) which has the charges reversed from those shown. This would require breaking a σ-bond _and_ placing electrons on an atom which has no available orbital in which to put them.)

Chapter 5

5-1.

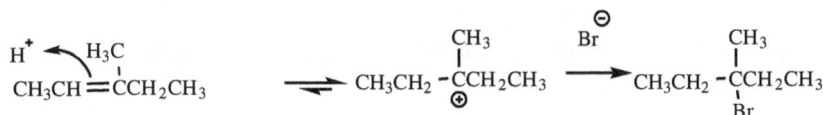

5-2. The <u>intermediates</u> are:

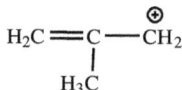

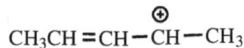

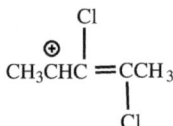

$$H_2C=CH-\overset{\oplus}{C}H_2 \qquad H_2C=\overset{\overset{\displaystyle\oplus}{|}}{\underset{\underset{\displaystyle H_3C}{|}}{C}}-CH_2 \qquad CH_3CH=CH-\overset{\oplus}{C}H-CH_3$$

$$CH_3\overset{\oplus}{C}HC=\underset{\underset{\displaystyle Cl}{|}}{\overset{\overset{\displaystyle Cl}{|}}{C}}CH_3 \qquad\qquad CH_3\overset{\overset{\displaystyle O}{||}}{C}\overset{\oplus}{C}H_2$$

The order of stability is 3>2>1>4>5.

Each of the above can be resonance stabilized. In 1,2 and 3, the relative order of stability can be determined by noting that 3 is a secondary carbocation and therefore more stable than 1 or 2 which are primary. The +I effect of the methyl group stabilizes 2 relative to 1. Both 4 and 5 can also be resonance stabilized but the electronegative element destabilizes the positive charge.

5-3. It would be expected that much more of the *cis* isomer would be obtained than in the reaction with bromine. The intermediate carbocation would be less likely to form the cyclic chloronium ion, since chlorine is more electronegative than bromine and therefore would be more reluctant to share its electrons and acquire a positive charge. Attack of the second chlorine atom (a chloride ion) can then occur from either side of the planar carbocation.

5-4. (c), (d) and (e) are conjugated.

5-5. The structure is $CH_2=C=CH-CH_2-CH=CH-C_4H_9$. C2 is therefore sp-hybridized. An orbital drawing of C2 would look like:

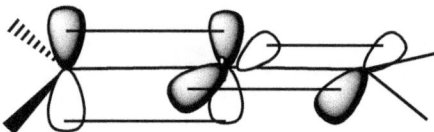

5-6.
(a) Since the positively charged nitrogen atom in I will attract electrons, the electron density in the double bond of I will be lower than that in propene. Therefore the reaction of the electrophile (H^+) with I will be slower than a reaction with propene. Since this is the slow step in the reaction, propene will react faster.

(b) The two possible intermediates between I and II are:

The second one has two positive charges on adjacent atoms, a **very** unstable arrangement. Therefore the reaction proceeds through the more stable (less energetic) intermediate to give II.

5-7. ICl (iodine monochloride) can be considered as I⁺ Cl⁻. The I⁺ is the electrophile and the following steps can occur.

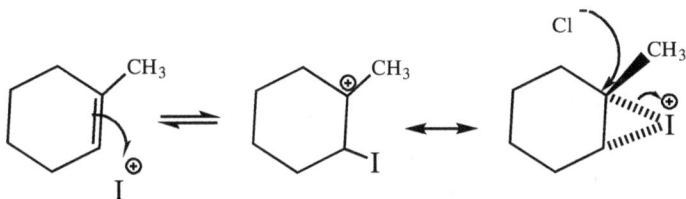

The formation of the cyclic ion will be more favorable than in the bromine case since iodine is less electronegative. This indicates the formation of the *trans* product. Markownikov's Rule predicts that the Cl will become attached to the tertiary carbon. Therefore the product will be

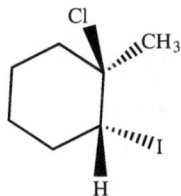

5-8. In going from I_2 to IBr to ICl, the iodine atom can form I⁺ progressively more easily as the attached atom becomes more electronegative. The increased concentration of I⁺ which results increases the reaction rate.

5-9. Bromide ion is much more nucleophilic than bisulfate ion (HSO_4). Use of HBr as a catalyst may lead to some HBr addition.

5-10. **A trick question!** If you add the pentene TO the Br_2, there will be an excess of bromine and the reaction will stay brown. You must add the Br_2 TO the organic, so that the latter is in excess.

5-11. HBr, BH₃, ICN

5-12. The two compounds are:

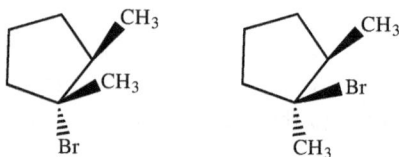

Since the intermediate has no atoms capable of forming a cyclic ion as is the case in the Br_2 addition, the two products can be formed.

5-13. (a) Z, (b) E

5-14. The structures and names are:

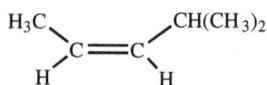   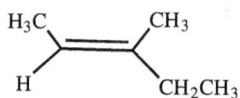

(Z) 4-methyl-2-pentene   (E) 3-methyl-2-pentene
[and its E isomer]   and its Z isomer

5-15.
(a)

i)   (CH₃)₂C=O and CH₃CH₂CH=O   ii)   CH₃C(CH₂)₄CCH₃
iii) O=CHCH₂CHCH₂CH=O   iv) CH₂=O   and   O=CHCH(CH₂)₃CH=O
                           | |
                          CH₃   HC=O

v)    =O   and   CH₃CCH₂CH₃

(b) i)   ii)   iii)

H₂C=CHCH₃   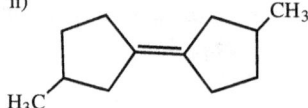   H₂C=C—C=CH₂
                                                                                                                   |
                                                                                                                  CH₃

5-16.

(a) H_2O/H^+

(b) BH_3, then H_2O_2/OH^-

(c)

$$H_2C{=\!=}CH(CH_2)_4\overset{\overset{\displaystyle Br}{|}}{\underset{\underset{\displaystyle Br}{|}}{C}}HCH(CH_3)_2$$

(d)

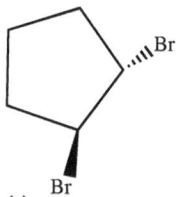

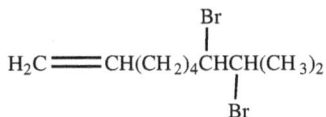

[The double bond with the more alkyl substituents (+I) is more electron-rich and there-fore reacts preferentially with an electrophile.]

(e) Br_2/CH_3OH

(f)

$$CH_3CH_2O\overset{\overset{\displaystyle Cl}{|}}{C}HCH_3$$

(g)

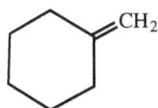

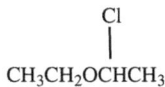

(h) Not possible due to regiochemistry! Addition of ICl would involve I^+ as the elec-trophile and therefore would give 2-chloro-3-iodo-2-methylbutane, not 3-chloro-2-iodo-2-methylbutane as shown.

5-17.

(a) H^+/H_2O

(b) BH_3, then H_2O_2/OH^-

(c) Br_2/CCl_4

(d) HBr

(e) O_3, then Zn

(f) H_2 with Ni or Pt catalyst

Chapter 6

6-1.

One projection would be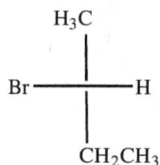

The mirror image would look like 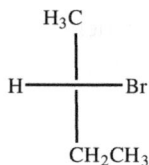 This material has the S configuration and hs a rotation of –15 degrees

The R isomer would be referred to as the (+) isomer and the d isomer because it rotates light in a clockwise direction.

There is no relationship between the absolute configuration and the sign or size of rotation. Therefore, you must do the measurement.

6-2.

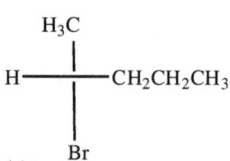

(a)

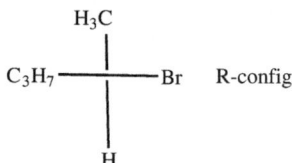 R-config

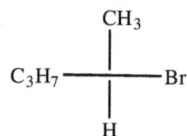

(b)

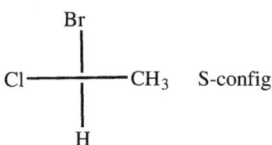 S-config

(c)

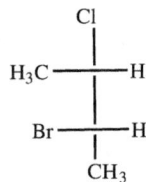

(d)

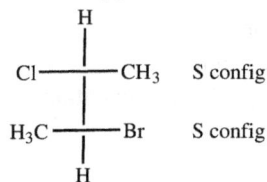 S config

S config

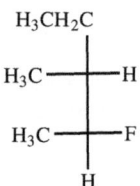

(e)

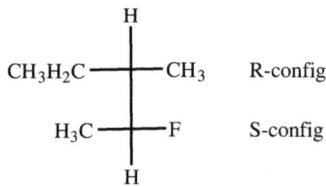 R-config

S-config

6-3.

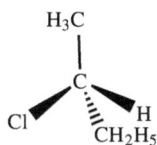

(a)

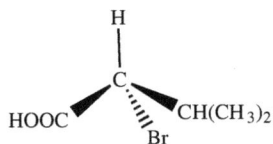

(b)

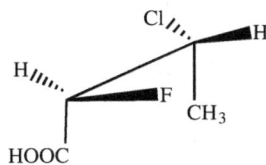

(c)

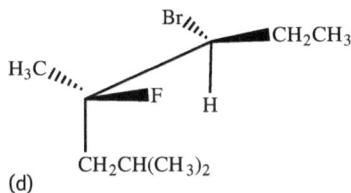

(d)

6-4.

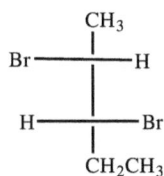

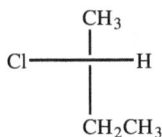

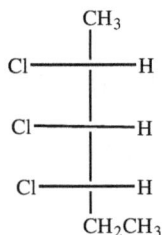

6-5. No they are not enantiomers, nor are they optically active. They are diastereomers of each other and each is superimposable on its own mirror image

6-6. Shown below are 3-D drawings of this molecule and its mirror image. It can easily be seen that the two drawings are superimposable and therefore this molecule is not optically active.

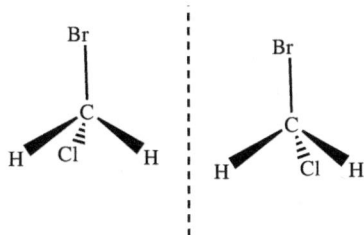

6-7.
(a) identical
(b) identical – no chiral carbon!
(c) diastereomers
(d) diastereomers
(e) diastereomers
(f) identical
(g) diasteromers

6-8.
(a) Right hand drawing of part <u>c</u>
    Left hand drawing of part <u>d</u>
(b) <u>S</u>; <u>S</u>,<u>S</u>; <u>S</u>,R

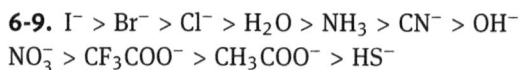

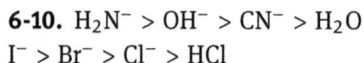

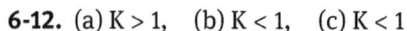

6-9. $I^- > Br^- > Cl^- > H_2O > NH_3 > CN^- > OH^-$
$NO_3^- > CF_3COO^- > CH_3COO^- > HS^-$

6-10. $H_2N^- > OH^- > CN^- > H_2O$
$I^- > Br^- > Cl^- > HCl$

6-11. (a) CH_3OH (b) CH_3CH_2OH (c) CH_3CN (d) H_2O

6-12. (a) K > 1,   (b) K < 1,   (c) K < 1

6-13.
(a) First reaction – better leaving group
(b) First reaction – tertiary carbocation
(c) Second reaction – poorer nucleophile
(d) First reaction – resonance stabilized carbocation

6-14.

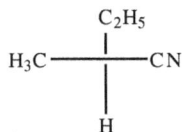

(a) S-isomer, optically active

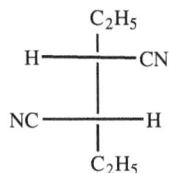

(b) S,S-isomer. Be really careful here. Change of configuration at one carbon changes config at other due to priority changes

a 1:1 mixture of 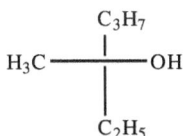 and its mirror image optically inactive

(c)

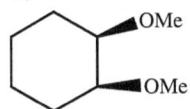

(d) optically inactive – meso form

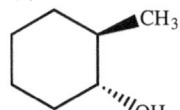

(e) + enantiomer = racemic mixture = optically inactive

6-15.

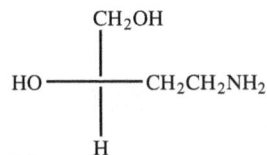

(a)

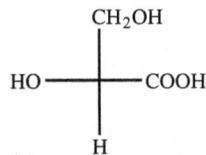

(b)

(c) The (+) isomer of II must have the R configuration. Therefore, since the oxidation of III does not affect the chiral center, (-)-III must also be R.

6-16.

(a)
OH
|
CH₃CHCH₂CH₃
S-isomer

(b)
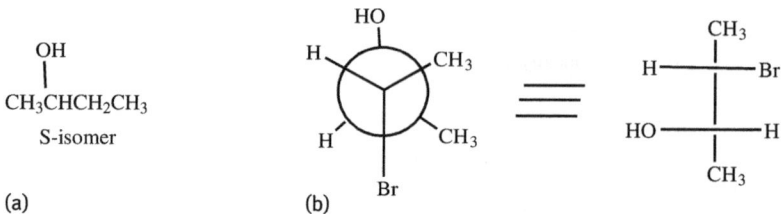

(c)
CH₃S ⊖

(d)
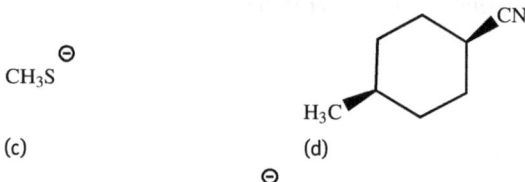

(e) BH₃ then H₂O₂/OH ⊖

(f) H₂O

(g) CH₃CH₂Br

(h)
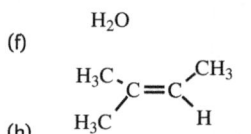

(i)
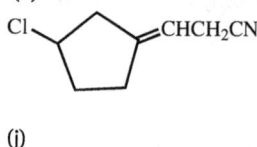

(j)
Cl—[cyclopentene]—CHCH₂CN

6-17.

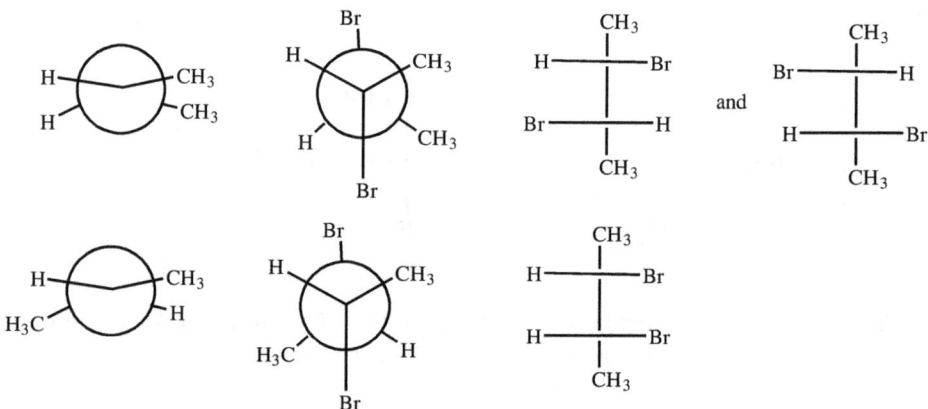

From the *trans* isomer, the product is the (2S,3R) or the (2R,3S) isomer. The products is a meso form. <u>Neither</u> of the products is optically active since one (from the *cis* isomer) is a racemic mixture and the other is a meso form.

6-18. A total of 6(!) products can be formed. These are:

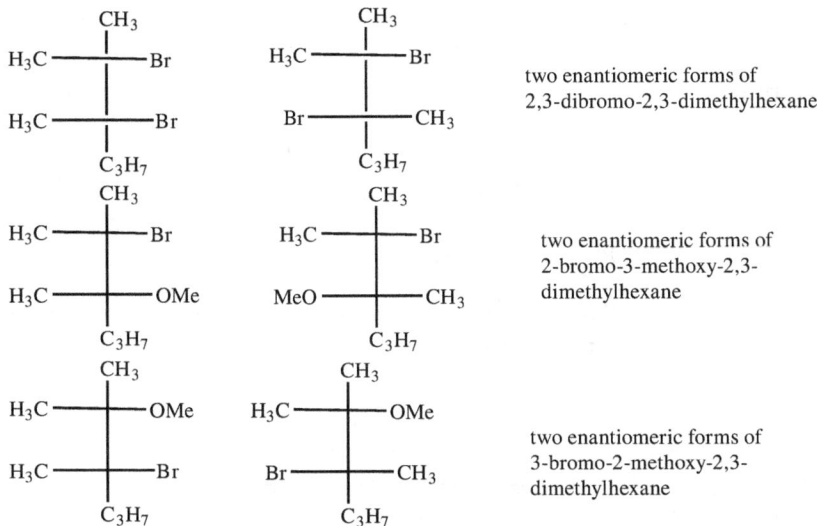

two enantiomeric forms of
2,3-dibromo-2,3-dimethylhexane

two enantiomeric forms of
2-bromo-3-methoxy-2,3-
dimethylhexane

two enantiomeric forms of
3-bromo-2-methoxy-2,3-
dimethylhexane

6-19. There are <u>ten</u> isomers of $C_5H_{11}Br$ based on the following skeletons:

For the skeleton	there are three positional isomers one of which has a pair of enantiomers	=4
For the skeleton	there are four positional isomers one of which has a pair of enantiomers	=5
For the skeleton	there is only one positional isomer and no enatiomers	=1
		———
		10

Chapter 7

7-1.

(a) Treat the sample with a dilute solution of bromine in CCl_4. If the brown colour is discharged, the sample was cyclohexene. If the solution remains brown, the unknown was cyclohexanol.

(b) Treat the sample with conc. HCl and $ZnCl_2$ (Lucas reagent). An immediate formation of a cloudiness or second layer indicates the material was 2-methyl-2-butanol. If the cloudiness or second layer is only formed after several minutes, the sample was 3-methyl-2-butanol (a secondary alcohol).

(c) Use the Lucas test. If heating is required for a second phase to appear, the sample was 2,3-dimethyl-1-hexanol. Otherwise, it is 3-methyl-2-heptanol.

(d) Use the Lucas test. In this case, the allylic alcohol (2-hexen-3-ol) can give a resonance stabilized intermediate and will react immediately even though it is secondary. 5-Hexen-2-ol will behave as a typical secondary alcohol and require 5–10 min to react.

7-2. Because water is a stronger acid than an alcohol, the equilibrium

$$ROH + OH^- \rightleftharpoons RO^- + H_2O$$

will lie to the left and little alkoxide will be formed. The product will be formed by attack of OH, i.e., an alcohol will be formed.

7-3.

(a) NaNH$_2^-$ is better. The nucleophile NH$_2^-$ is stronger than NH$_3$

(b) Ignore this question. It will be considered later in this text.

(c) Must use HBr. The proton is required for catalysis of this reaction.

(d) H$_2$O is better here. It is a weaker nucleophile BUT it is also a weaker base and the strong base would favor elimination.

(e) NaCN. HCN is a weak acid and therefore very little CN$^-$ is present in HCN. There is no nucleophile!

(f) The H$^+$ is required for this reaction. See the answer to question 7.4

7-4. The ether must be protonated to convert the methoxy group into a good leaving group.
The reaction mechanism is as follows:

* The products are methanol and 2-methyl-2-propanol and the latter contains the ^{18}O isotope because the tertiary C$^+$ is more stable than the primary one.

7-5. It would be a racemic mixture and therefore inactive. This reaction should go via an Sn1 pathway. The intermediate is planar and attack of the nucleophile (water) from either side is equally probable.

7-6.

(a) H_2O

(b)

OCH$_3$
|
CH$_3$CHCH$_2$CH$_3$

(c)

OH
|
CH$_3$CHCH$_2$CH$_3$

(d) SOCl$_2$

(e)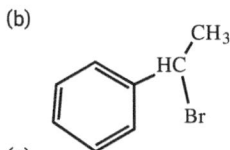

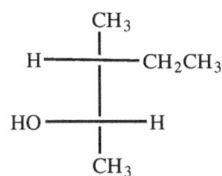

(f)

(g)

N$_3$
|
CH$_3$CH$_2$CH$_2$CHCH$_3$

(h)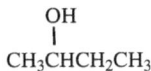

CN

(i)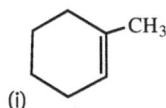

(j) CH$_3$C≡CCH$_3$

(k) The product is CH$_3$C≡CCH$_3$. The strong base NaNH$_2$ (the sodium salt of ammonia) will remove the proton attached to the sp-hybridized carbon. As noted (Section 7.6.3), this proton is slightly acidic. The product of this is CH$_3$C≡C$^-$ Na$^+$ which
behaves as a nucleophile and displaces the I$^-$ from the CH$_3$I.

7-7.

(a) Errors:

Step 1. Grignard reactions require a carbonyl group to react with. The alcohol must
first be oxidized to the ketone.

Step 2. Elimination of water will occur only with an acid catalyst, not a basic one.

Step 3. HCN is a very weak acid and there are not enough protons to cause the addition reaction. Also, if the reaction did occur, it would have proceeded by Markownikov addition and the required product is the anti-Markownikov one.

(b) Suggested solution.

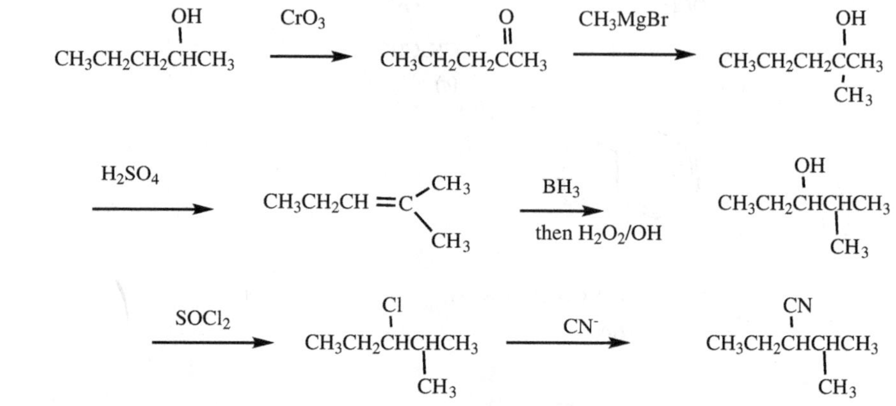

$$\underset{\underset{\displaystyle CH_3CH_2CH_2CHCH_3}{|}}{OH} \xrightarrow{CrO_3} \underset{\underset{\displaystyle CH_3CH_2CH_2CCH_3}{\|}}{O} \xrightarrow{CH_3MgBr} \underset{\underset{\displaystyle CH_3CH_2CH_2CCH_3}{\underset{\displaystyle CH_3}{|}}}{\overset{\displaystyle OH}{|}}$$

$$\xrightarrow{H_2SO_4} CH_3CH_2CH = \underset{CH_3}{\overset{CH_3}{C}} \xrightarrow[\text{then } H_2O_2/OH]{BH_3} \underset{\underset{\displaystyle CH_3}{|}}{\overset{\displaystyle OH}{\underset{\displaystyle CH_3CH_2CHCHCH_3}{|}}}$$

$$\xrightarrow{SOCl_2} \underset{\underset{\displaystyle CH_3}{|}}{\overset{\displaystyle Cl}{\underset{\displaystyle CH_3CH_2CHCHCH_3}{|}}} \xrightarrow{CN^-} \underset{\underset{\displaystyle CH_3}{|}}{\overset{\displaystyle CN}{\underset{\displaystyle CH_3CH_2CHCHCH_3}{|}}}$$

7-8.

(a) 5-methyl-3-hexanone
(b) 4-chloro-2-methylcyclohexanone
(c) 6-methyl-2-cyclohexenone
(d) 1-(trans-3-chloro-trans-4-methylcyclopentyl)ethanone
(e) (3R,6S) 6-chloro-3-hydroxy-3-methyl-4-heptanone
(f) 3,3-dichloro-4-pentenal
(g) (6S) 6-bromo-2-chlorohept-1-en-4-one

7-9.

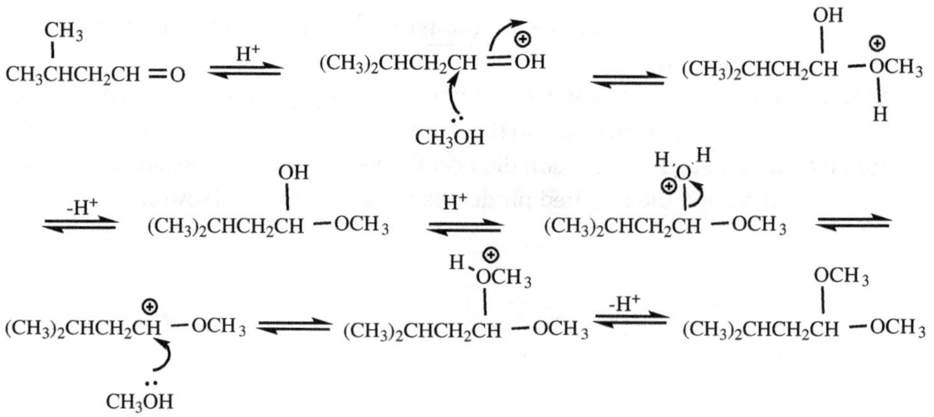

7-10.

Ni or Pd catalyst, $CH_3CH_2CHCH_3$ (with OH above) CN^-/HCN $HOCH_2CH_2OH$ H^+ catalyst

(a)   (b)   (c)

(d) $(CH_3)_2CHCH_2CHOCH_3$ (with OH below)   (e) Hg^{++}/H^+ catalyst $CH_3CH_2CCH_2CH_2CH_3$ (with O double bond above)

(f) CrO_3 (g) HO—C_6H_4—CH—CH$_3$ (h) CH_3MgBr then H_3O^+

(i) $CH_3OCH_2CH_2CH_2CCH_3$ (with O double bond)

(j) $CH_3CH_2CHCHCH_2Br$ (with OH above, Br below)   (k) $CH_3CH_2CH_2CHCH_2CH_3$ (with OH below)   (l) H_3C—(cyclohexane)—MgBr

(m) $CH_3CHCH_2CH_2CH{=}O$ (with OCH$_3$ above)   (n) O_3 then Zn

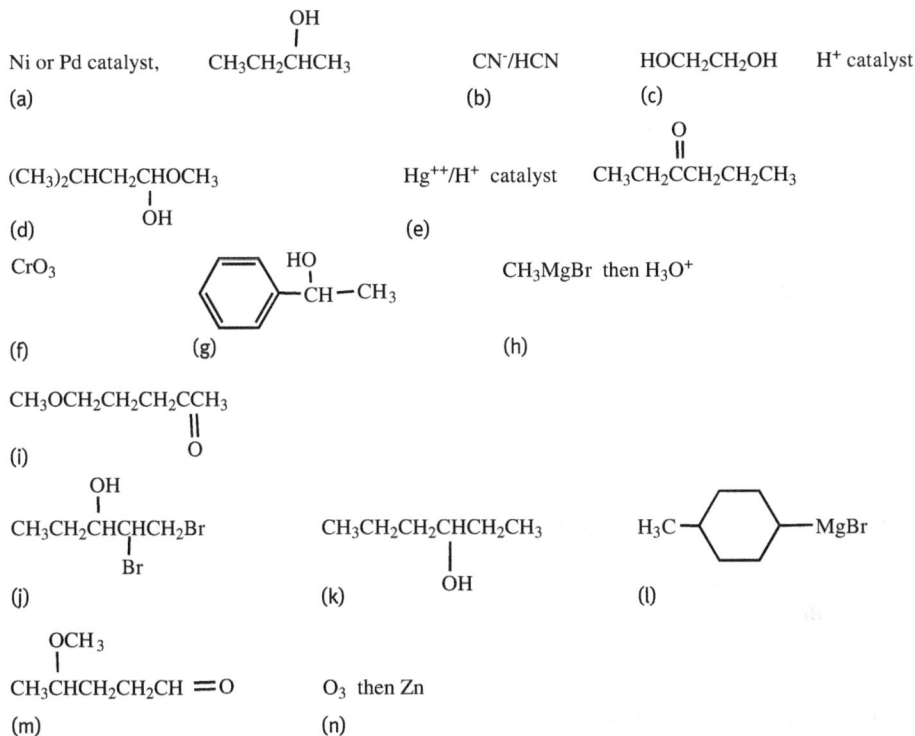

7-11. Two are possible.

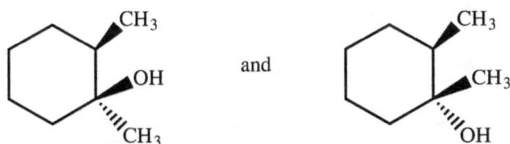

and

7-12.
(a) Add sample to sodium metal. Evolution of gas indicates the presence of an alcohol (2-butanol).
(b) Treat with Tollen's reagent. Precipitation of a silver mirror indicates that the sample was an aldehyde (cyclopentane-carboxaldehyde). No precipitation indicates 3-hexanol.
(c) Treat with I_2/NaOH. A yellow precipitate indicates 2-hexanol (a secondary methyl alcohol). No precipitation indicates 3-hexanol.
(d) Tollens' test. A precipitate means the presence of 5-hexenal -OR- treat with a solution of bromine in CCl_4. Disappearance of the brown color means the presence of 5-hexenal.

7-13. None.

7-14. The mechanism is shown in Figure 7.12. The oxygen atom of the cyclohexanone is lost in the form of water and therefore the organic product does <u>not</u> contain ^{18}O.

7-15.

(a)

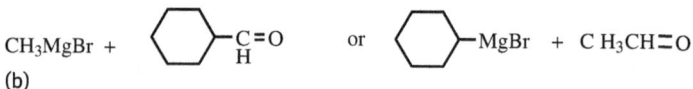

(b)

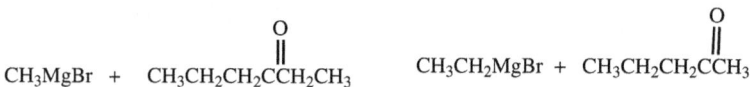

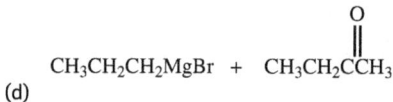

(d)

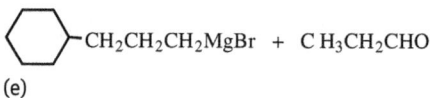

(e)

7-16. None. The reagents you used would all give <u>racemic</u> products which are therefore optically inactive.

7-17.

(a)

 + CH₃CH₂MgBr

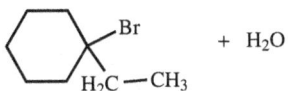 + H₂O

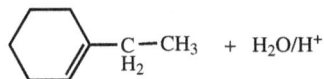 + H₂O/H⁺

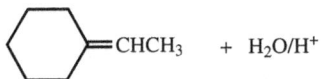 =CHCH₃ + H₂O/H⁺

(b)

OH
|
CH₃CH₂C(CH₃)₂   + HBr

CH₃CH=C(CH₃)₂   + HBr

CH₃CH₂C=CH₂   + HBr
       |
      CH₃

CH₃CH₂C(CH₃)₂   + NaBr
       |
      Cl

(c)

CH₃CH₂CH₂CH=O   + H₂/Ni

H₂C=CHCH₃   + BH₃, then H₂O₂/OH⁻

CH₃CH₂MgBr   + CH₂O

CH₃CH₂CH₂Br + OH⁻

(d)

7-18.
(a) diastereomers

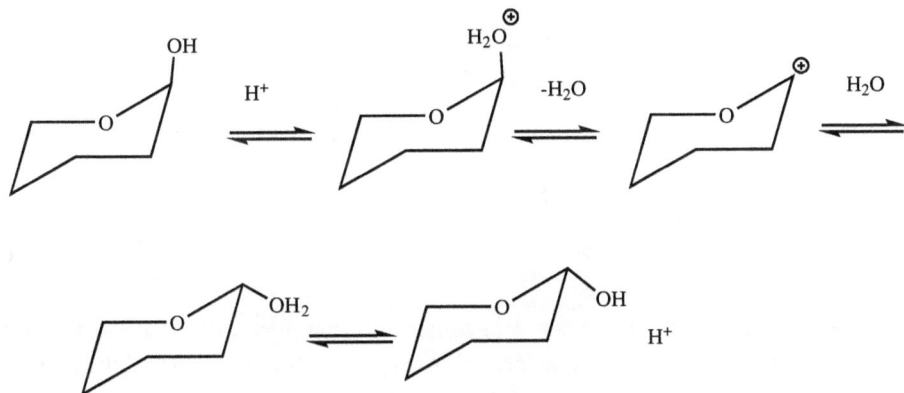

(c) Since the specific rotation falls when I is converted to II, the rotation of II must be smaller than that of I.

7-19.

(a) This OH is part of an <u>hemiacetal</u>. The others are not. Only hemiacetals react under these conditions because the intermediate carbocation can be resonance stabilized.

(b) The same mechanism as shown in 7.11(b) applies except that the attack on the carbocation is done with <u>methanol</u>.

(c) Yes.

(d) The product would still be III.

7-20. $CH_3CH_2CHF_2 + {}^- OCH_3 \quad CH_3CHCF_2 + CH_3OH$
The strongly electronegative F atoms make the C–H bond weak and allow the proton to be removed (see Chapter 5).

7-21. <u>SUGGESTED</u> answers.

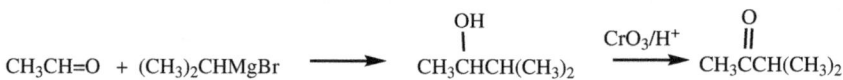

(a)

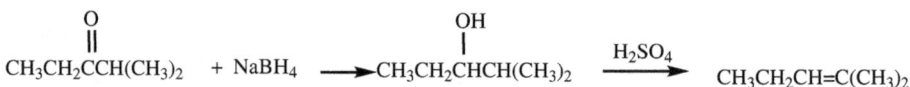

(b)

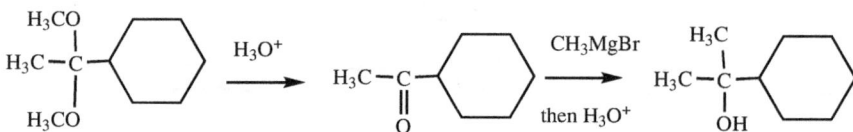

(c)

7-22.

(a) The product contains only one chiral center and therefore will exist as a pair of enantiomers. Since these will be formed in a 1 : 1 mixture (racemic mixture), the product, as isolated, will be optically INACTIVE.

(b) The product obtained has TWO chiral centers. One of them has only the R configuration (established by the configuration of the starting material). The other center is a mixture of R and S isomers. Therefore, there will be two products related as <u>diastereomers</u>, neither of which has a superimposable mirror image, and the mixture will be optically ACTIVE.

(c) The product contains NO chiral centers. Therefore, only one product will be formed, it will be superimposable on its mirror image and it will be optically INACTIVE.

(d) The product has only one chiral center which possesses the S-configuration. Therefore, only one product will be formed and it will be optically ACTIVE.

(e) Same answer as (c)

(f) Here the product has two chiral centers, but since the starting material was racemic, there will be four products related as two pairs of enantiomers. The product will be optically INACTIVE.

(g) The product has only one chiral center and since that center is racemic in the starting material, it must also be racemic in the product. Therefore, there will be two products formed in equal amounts which are related as enantiomers and the product will be optically INACTIVE.

7-23. Structures A, B and E fit the description. C would give two molecules with ozone, D would react with Na metal and F would give an aldehyde with O_3 and therefore give a positive Tollen's test.

Chapter 8

8-1.

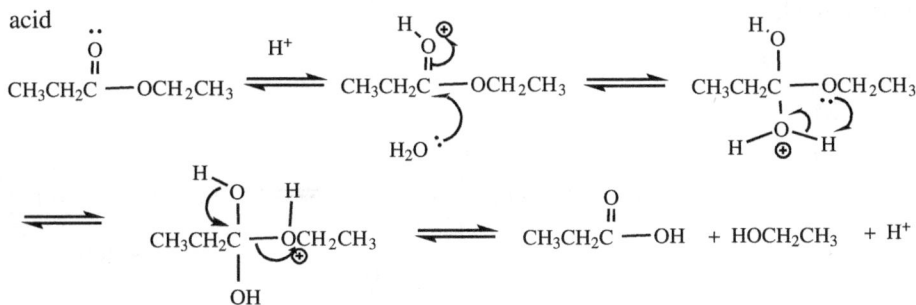

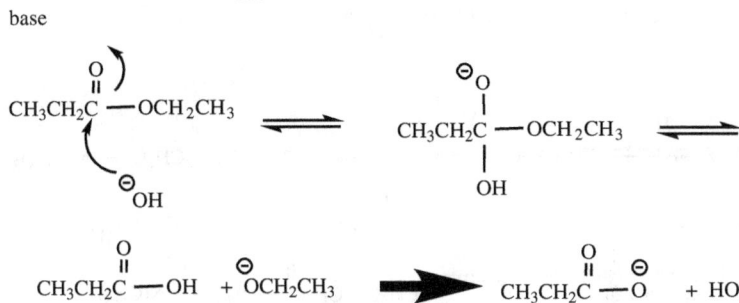

The base-catalyzed reaction is more useful since it is irreversible and therefore the yield of product is not controlled by an equilibrium constant that is frequently unfavorable.

8-2. Yes, the product does contain ^{18}O. The oxygen atom eliminated is from the <u>acid</u>. (See Fig. 8.3)

8-3.

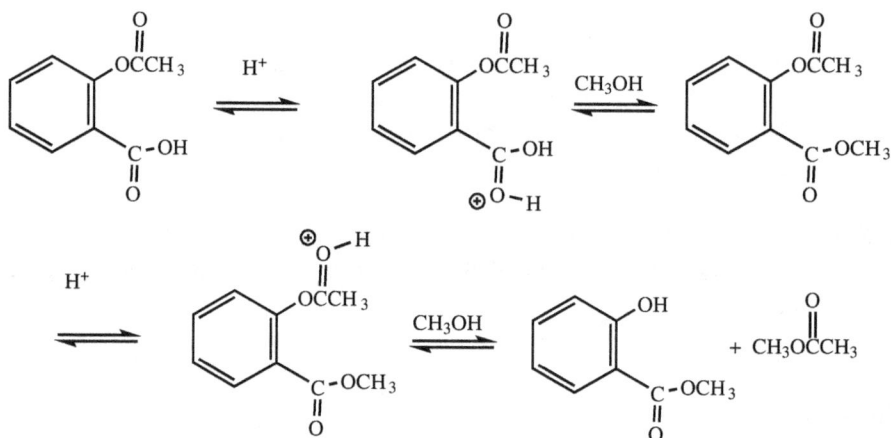

Two reactions have occurred. The acid has been esterified and the ester has been *trans*esterified.

8-4.

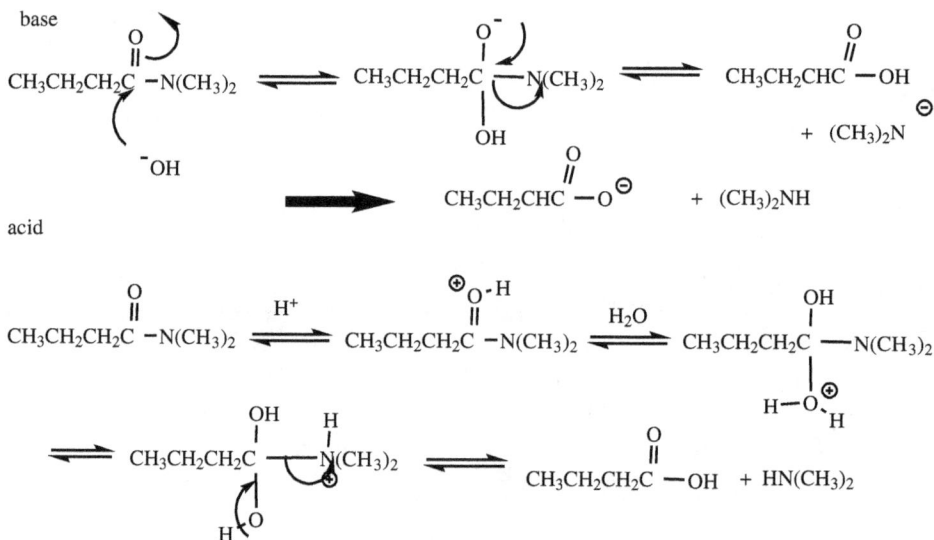

8-5.

(a) The anhydride group is a good enough leaving group that it can react with methanol as the nucleophile without added catalysts.

(b) Once the anhydride has reacted, the molecule II will not react further unless an acid catalyst is added (the molecule is now an <u>acid</u>). If such a catalyst <u>is</u> present, the carboxylic acid group can be esterified in the usual manner and III is formed.

(c) II would be formed.

8-6.

(a) $CH_3CH_2CH_2OH$ with H^+ catalyst

(b) $SOCl_2$

(c) 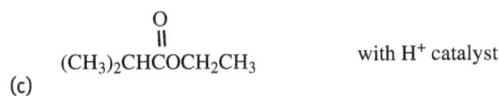   with H^+ catalyst

(d)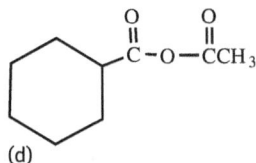

(e) H_3O^+/heat

(f) $2CH_3MgBr$, then H_3O^+

(g) I_2 + NaOH

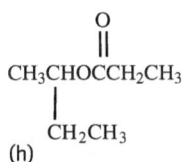

(h)

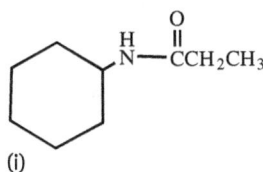

(i)

(k) $LiAlH_4$, then H_3O^+

8-7. In basic solution, carbamic acids will exist as their <u>salts</u>.

As shown above, if they were to decompose, the <u>anion</u> of the amine must be eliminated. Since this is a very poor leaving group, this equilibrium will lie heavily on the left side.

8-8. If the ester is water-soluble, it can react with the sodium hydroxide to reform the acid. This hydrolysis reaction will not occur well unless the two reactants (ester + OH^-) are together in the same solution. Since OH^- is much more soluble in water than in organic materials (because it is ionic), the hydrolysis reaction is much slower if the ester is not water-soluble.

8-9.

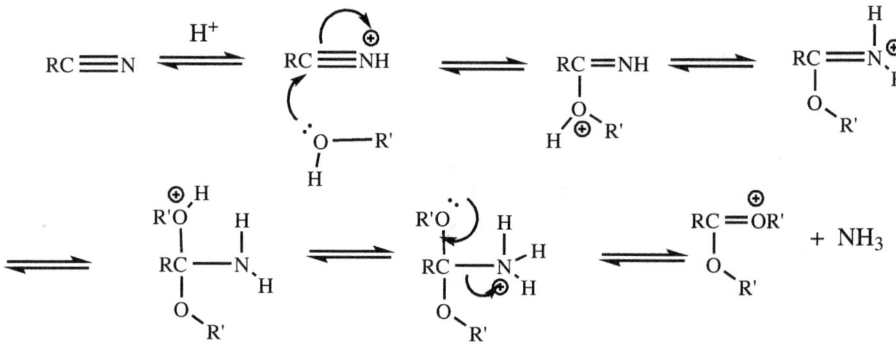

Reaction stops here until water is added. Then the following can occur.

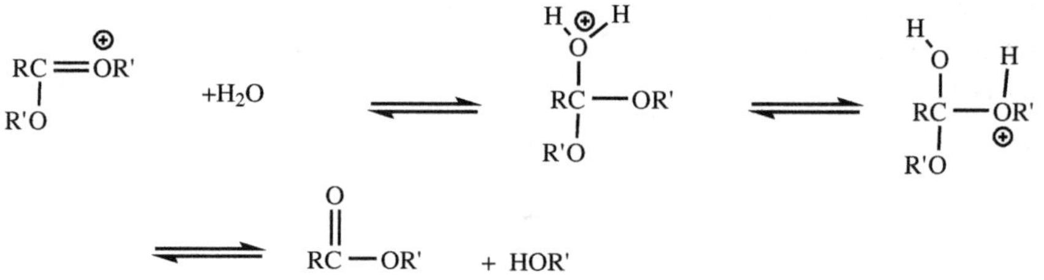

8-10.

$$CH_3CH_2CH_2OH \xrightarrow{CrO_3} CH_3CH_2\overset{O}{\overset{\|}{C}}OH \xrightarrow{CH_3OH/H^+} CH_3CH_2\overset{O}{\overset{\|}{C}}OCH_3$$

(a)

$$(CH_3)_3C\overset{O}{\overset{\|}{C}}-OH \xrightarrow{SOCl_2} (CH_3)_3C\overset{O}{\overset{\|}{C}}-Cl \xrightarrow{NH_3} (CH_3)_3C\overset{O}{\overset{\|}{C}}-NH_2$$

(b)

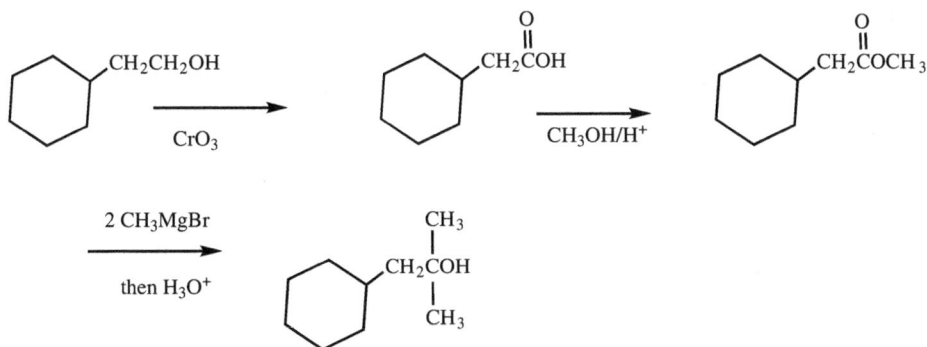

NB: Note that the Grignard reaction **CANNOT** be done on the acid. It must be converted to the ester first!

(c)

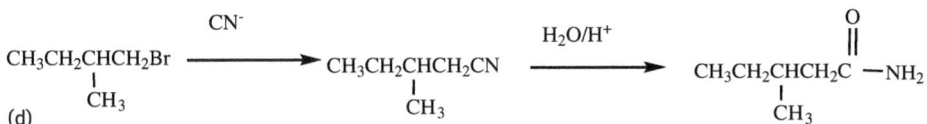

CN⁻

(d)

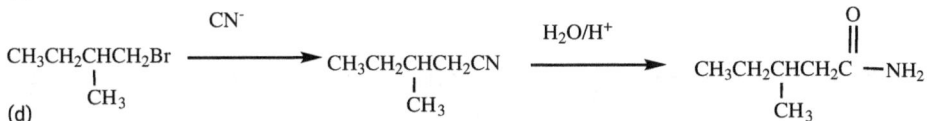

8-11.

(a)

HOCH₂CH₂CH₂COOH

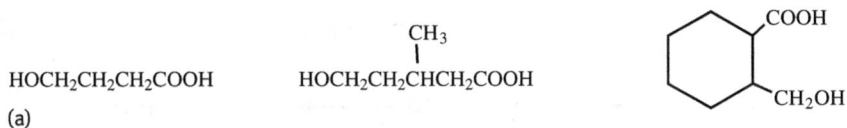

(b)

HOCH₂CH₂CH₂CH₂COOH

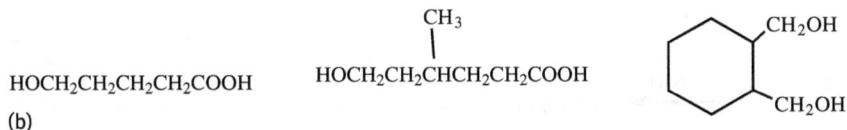

8-12.

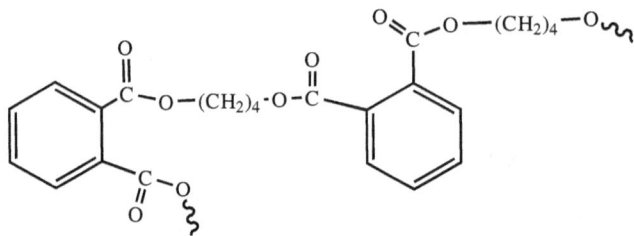

8-13.

(a) Test with moist litmus paper. The acid will turn it red whereas the ester will not.

(b) Treatment with $SOCl_2$ will give gas evolution with the acid but not with the acid chloride.

(c) see part (a)

8-14.

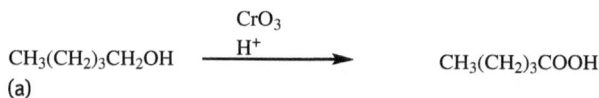

(a)

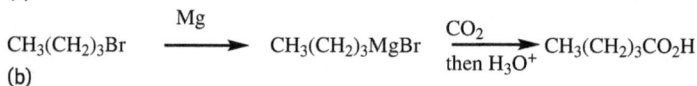

(b)

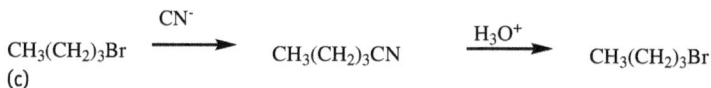

(c)

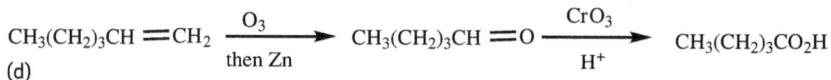

(d)

same as (c)

(e)

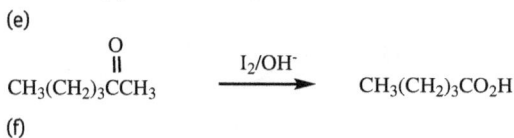

(f)

8-15.

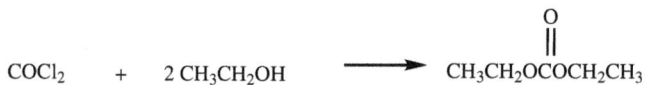

8-16.

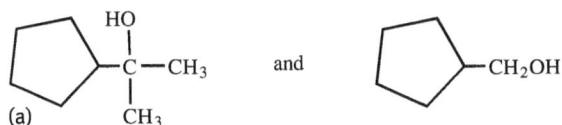

(a)

(b)

(c)

Chapter 9

9-1. The first, second sixth and seventh are aromatic.

9-2.
(a) $-101 - (4 \cdot -23) = -9 \, \text{kcal/mole}$.
(b) It is very small and thus cyclooctatetraene is not aromatic

9-3.

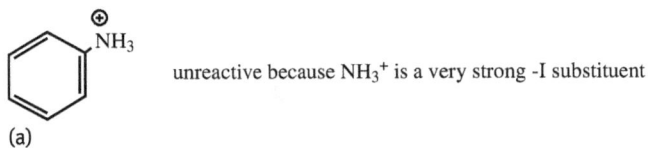

unreactive because $NH_3{}^+$ is a very strong -I substituent

(a)

(b) See Fig 10.3
(c) The meta isomer of nitroaniline would be expected since the $(-I)$ $-NH_3^+$ is meta-directing.

9-4.

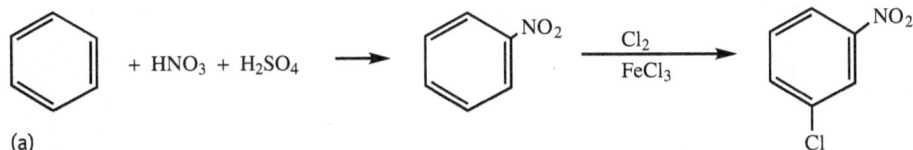

(a)

(b)

(c)

(d) from (a)

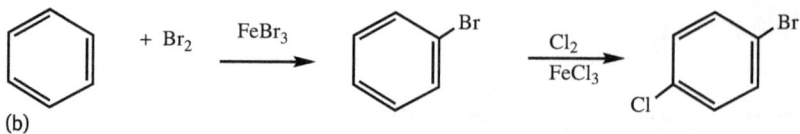

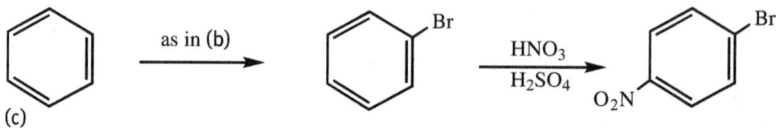

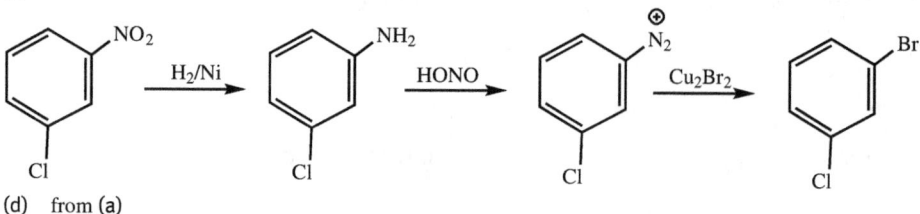

(e) Not possible with reactions studied.

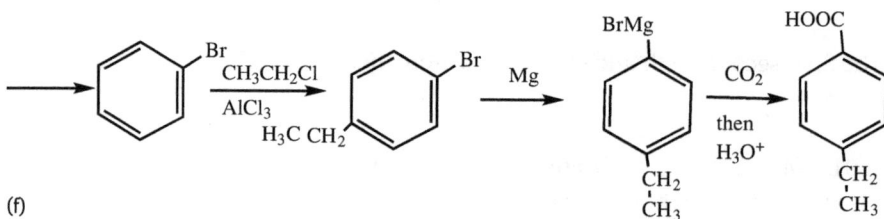

(f)

9-5.

(a)

(b)

(g)

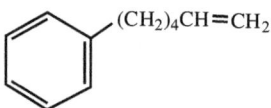

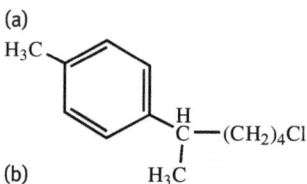

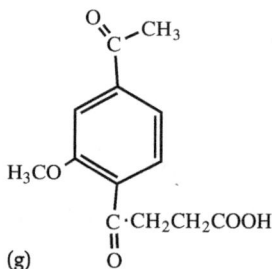

(h) HONO then H_2PO_3

(c)

(d)

(e)

AlCl₃ +

(i)

(j)

(k)

(f)

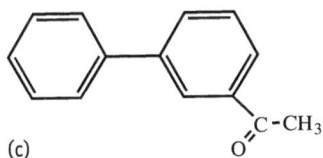

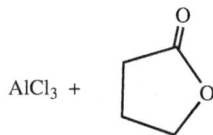

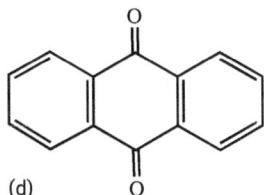

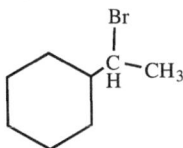

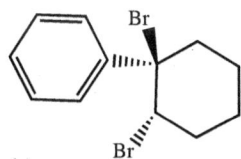

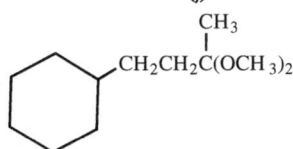

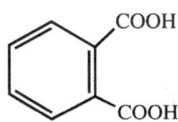

9-6. A must be either a methyl ketone or the related secondary alcohol. Phthalic acid has 8 carbons, of which 6 are part of a benzene ring. Therefore, A must have a benzene ring which has ortho substituents. Since A has 9 carbons, the carbon skeleton must look like I (below). Since there is only one oxygen atom, in order to get a methyl ketone, the oxygen must be located as in II. If the oxygen is present as a ketone, the molecular formula of A would be $C_9H_{10}O$. Therefore, A must be III.

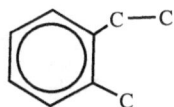

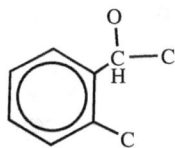

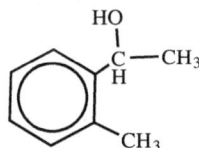

I

II

III

9-7.

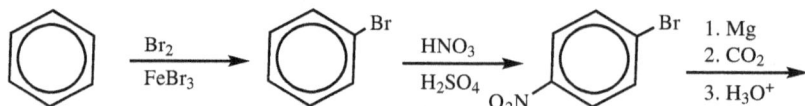

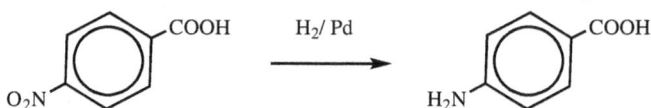

(a)

PABA

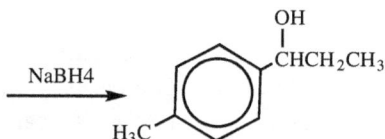

(b)

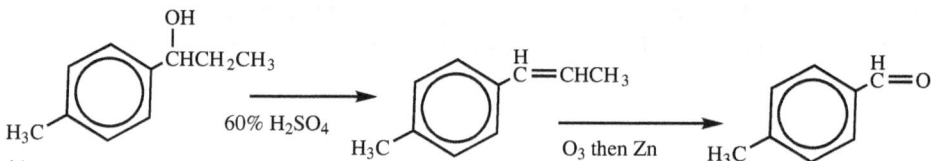

(c)

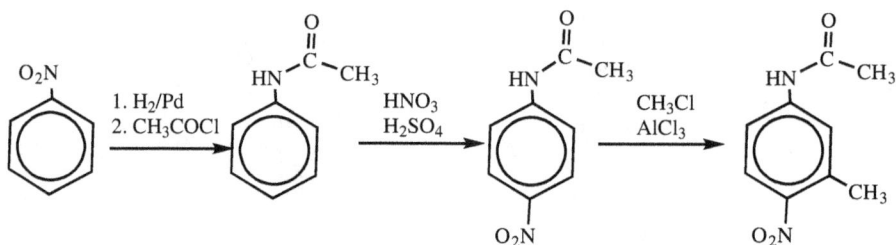

(d)

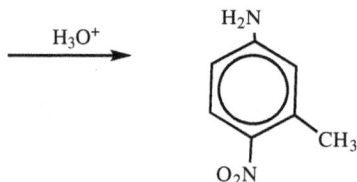

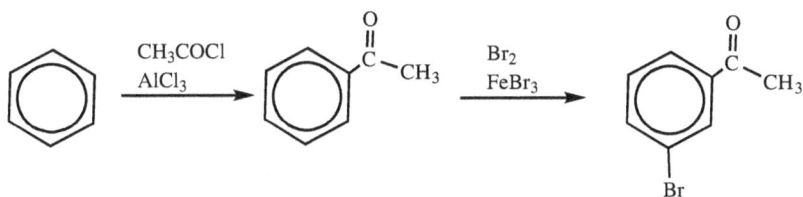

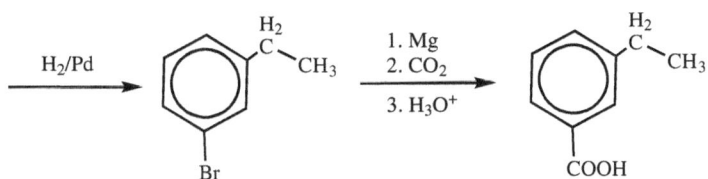

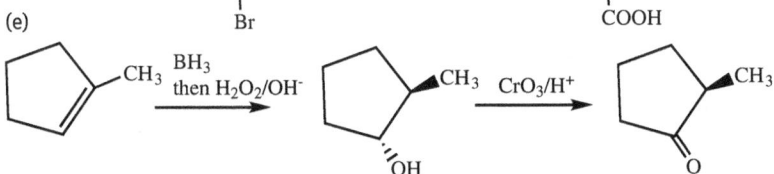

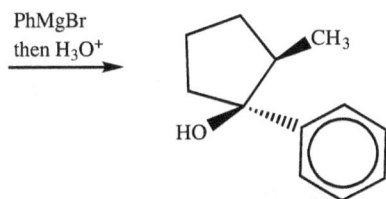

(e)

(f)

9-8. Resonance forms which preserve the six-membered ring with 6π electrons impart more stability to the resonance hybrid than do those in which this arrangement is absent. Thus, form **I** is a larger contributor to the resonance hybrid than **II** is.

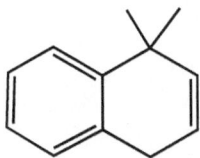

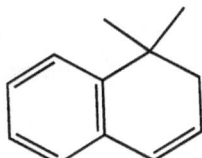

I   II

If the intermediate ions of a S_E2 reaction at the 1 and 2 positions of naphthalene are drawn, five structures can be drawn in each case but that from attack at the 1-position has two forms like **I** whereas that from attack at the 2-position has only one. Therefore, the intermediate in the latter case is less stable than the former and attack at the 1-position is favored kinetically.

9-9.

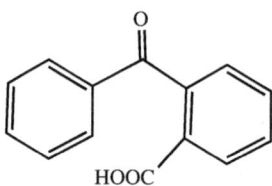

HOOC

A

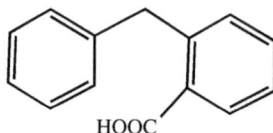

HOOC

B

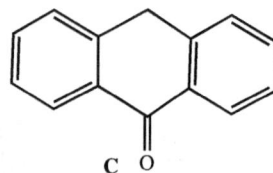

C O

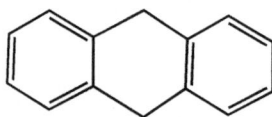

D

Note that the conversion of **D** to the final product is the reverse of hydrogenation.

9-10.

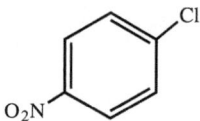

O₂N

(a)

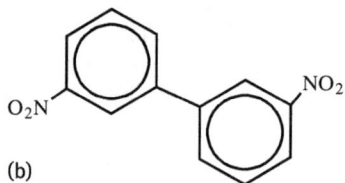

O₂N ... NO₂

(b)

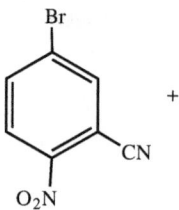

Br

O₂N ... CN

(c)

+

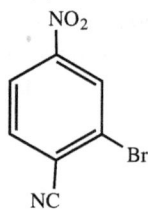

NO₂

NC ... Br

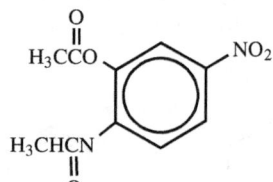

H₃CCO ... NO₂

H₃CHCN

(d)

9-11.

(a)

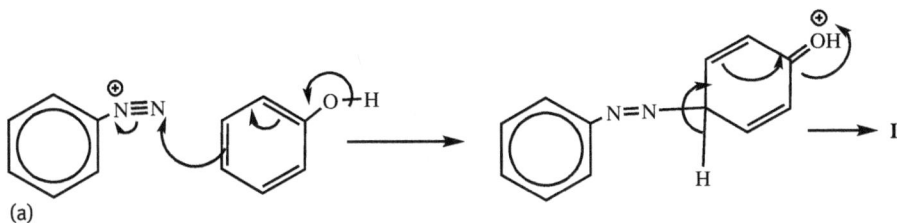

(b) The product is **II** which cannot form the phenol and is therefore not acidic.

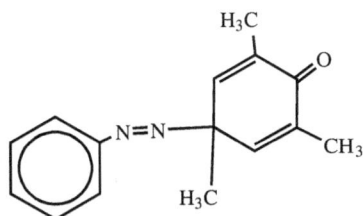

(b)   **II**

Chapter 10

10-1. Secondary, primary, tertiary, tertiary

10-2.
(a) pyridine > p-methoxyaniline > aniline > p-nitroaniline
(b) dimethylamine > methyl cyclohexylamine > N-methylaniline > N-methylac-etamide

10-3. Amines are basic and acids are acidic! Therefore, the reaction between them would be

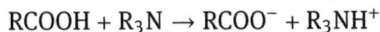

$$RCOOH + R_3N \rightarrow RCOO^- + R_3NH^+$$

10-4. The product is

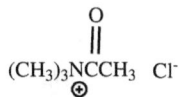

$(CH_3)_3NCCH_3 \ \ Cl^-$

Since the nitrogen is positively charged, it is a good leaving group. It is easily displaced by an alcohol.

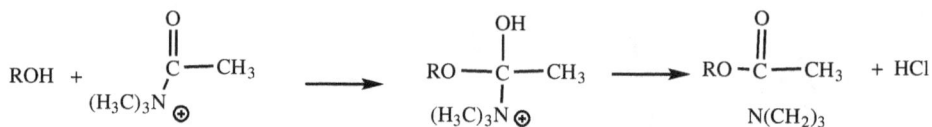

10-5. Pyrrole has 4π electrons in the carbon-carbon double bonds PLUS two π electrons on the N-atom. All are <u>coplanar</u> as shown below. Thus the number of electrons in cyclic conjugation is six and pyrrole is aromatic. Pyridine has the same electronic arrangement as benzene.

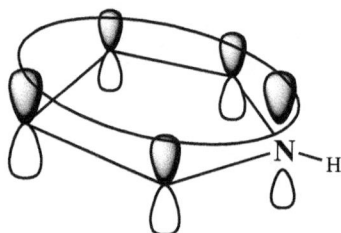

10-6.

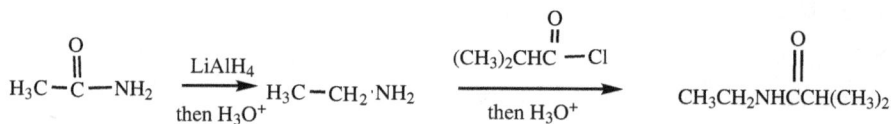

(a)

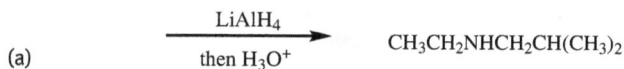

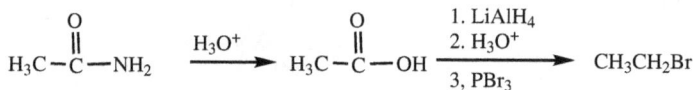

(b)

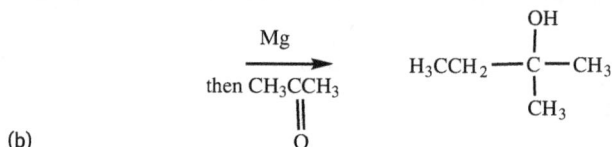

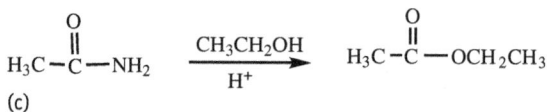

(c)

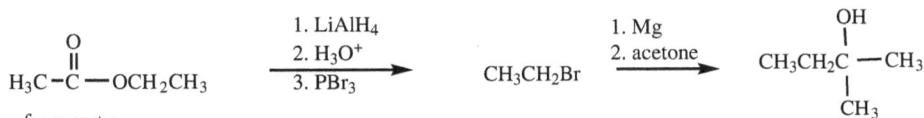

from part c

(d)

10-7. <u>Both</u> of the other two nitrogen atoms are adjacent to carbonyl groups and are therefore deactivated by resonance.

10-8. The four products shown are composed of two amines, one secondary ammonium salt and one quaternary ammonium salt. Salts are not soluble in ether, but are soluble in water. Extracting an aqueous mixture with ether thus gives a water solution of the salts and an ether solution of the amines. Dimethylamine is much more volatile than dibenzyl amine and these can be separated easily by distillation.

Treating the aqueous solution of the salts with base converts the secondary ammonium salt to dimethyl amine, but does not affect the quaternary salt. Extracting the basic solution with ether separates the amine (ether soluble) from the quat. salt (water soluble).

10-9.

$CH_3CH_2COO^- \; NH_4^+$

(a)

$CH_3CH_2CH_2\overset{\oplus}{N}(CH_3)_3 \quad \overset{\ominus}{Br}$

(b)

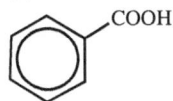

(c)

(d) H₃C

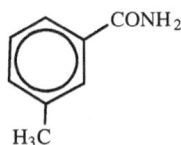

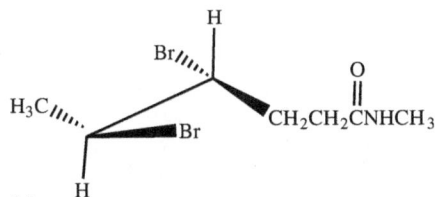

(e)

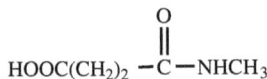

(f)

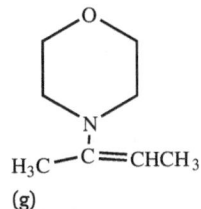

(g)

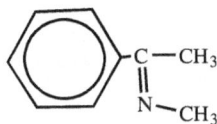

(h)

$(CH_3)_2CHCN$

(i)

10-10.

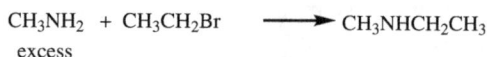

(a)

$CH_3(CH_2)_3COCl + NH_3 \longrightarrow CH_3(CH_2)_3CONH_2$

$CH_3(CH_2)_3C \equiv N + H_3O^+ \longrightarrow CH_3(CH_2)_3CONH_2$

(b)

(c)

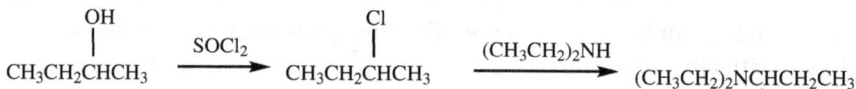

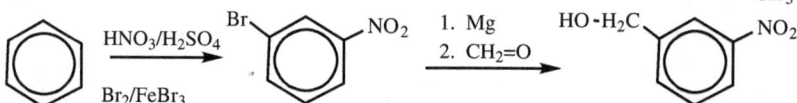

(d)

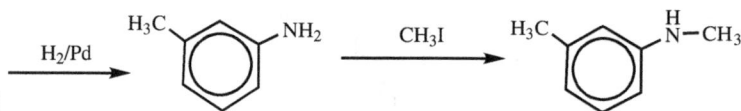

Chapter 11

11-1.
(a) isobutyl formate
(b) ethyleneglycol diacetate
(c) α methyl butyrolactone
(d) bromoacetonitrile
(e) chloropropionyl chloride
(f) butyrolactone

11-2. Because of the resonance between the carbonyl oxygen and the carbon-carbon double bond, there is a positive polarization on the carbon farthest from the carbonyl.

This deshields the attached proton and causes it to resonate farther downfield than the other hydrogen.

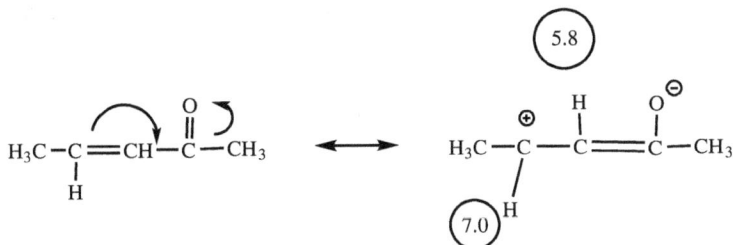

11-3.

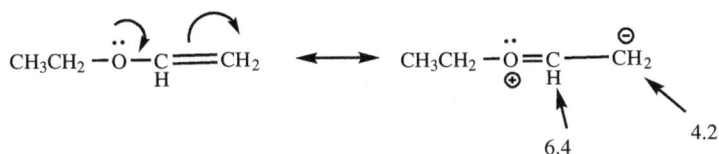

Here the resonance works in the opposite direction; it puts a **negative** charge on the end carbon.

Because of the stereochemistry around the double bond, the two hydrogens on the end carbon are different from each other; one is *cis* and the other is *trans* to the oxygen atom. As a result, they couple differently to the =CH group causing it to be two doublets (it is not a quartet because the peak ratios are not 1 : 3 : 3 : 1). In the same way, the two hydrogens of the =CH₂ are different and each couples to the =CH so there are two signals which are both doublets. The ethyl group triplet/quartet is normal.

11-4.

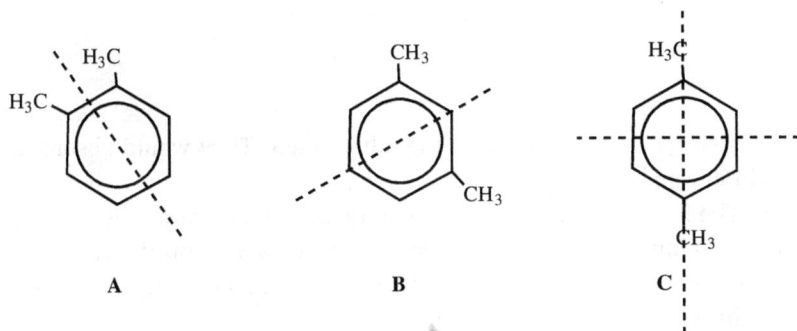

The three isomers of dimethyl benzene each have a different symmetry as shown below.

A will show 4 lines, **B** will show 5 lines, and **C** will show 3 lines. Therefore, the spectra, from top to bottom are the *ortho, para* and *meta* isomers, respectively.

11-5.

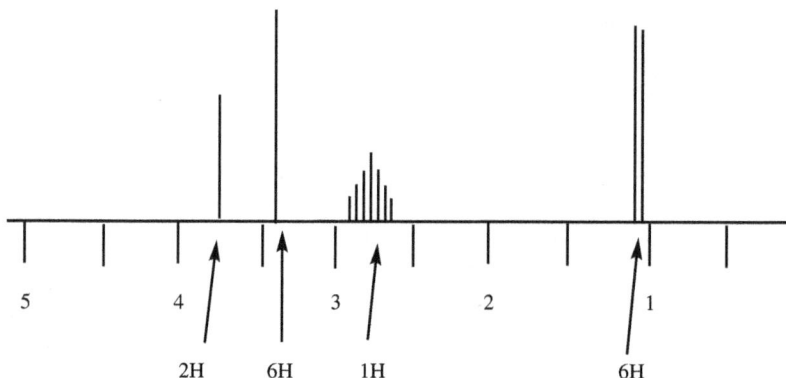

11-6. The following parts can be identified. There is a 3H singlet in the region characteristic of a methyl attached to a C=O. There is a CH_2 which is clearly attached to the O of an ester (IR data) and since it is a doublet, there is one adjacent H. There are also two identical methyl groups which appear as a doublet and the best way to accommodate this is a $CH(CH_3)_2$

Assembling these data, one arrives at the following structure.

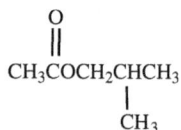

$$CH_3\overset{\displaystyle O}{\overset{\displaystyle \|}{C}}OCH_2CHCH_3$$
$$\underset{\displaystyle CH_3}{\big|}$$

11-7.

(a) Use IR. One carbonyl is conjugated and the other is not. They would absorb at 1710 cm^{-1} and 1680 cm^{-1}

(b) Use ^{13}C NMR. The first molecule has no symmetry and would show 7 peaks. The second one has a plane of symmetry and would show only 5 absorptions.

(c) Use ^1H NMR. The vinyl hydrogen I the left compound would be a singlet whereas that on the right compound would be a triplet.

11-8.
1. From the molecular formula, there must be one unit of unsaturation. The IR suggests a C=O and this is confirmed by the absorption in the ^{13}C spectrum at $\delta = 203$.
2. ^{13}C spectrum shows 4 peaks and therefore there is no symmetry in the molecule
    ^{1}H spectrum shows a 1 H quartet and a corresponding 3H doublet. Therefore, there must be a CH_3CH present. The chemical shift of the quartet suggests it is attached to an electronegative atom – Cl in this case.
3. The ^{1}H also shows a 3H singlet with a chemical shift ($\delta = 2.33$) suggestive of a methyl ketone ($CH_3C=O$).
    The ONLY possibility is **3-CHLORO-2-BUTANONE**

11-9.
1. From the molecular formula there must be one unit of unsaturation and this must be in the form of a carbonyl according to the IR. Since the IR also shows a band at 3400 cm-1 for an hydroxyl (OH) group and there are only two oxygen atoms in the molecule, the C=O must be an aldehyde or ketone
2. The ^{13}C NMR shows 4 carbons and so there are no elements of symmetry to worry about in the molecule, i.e., all 4 carbons are different. The absorption at $\delta = 210$ confirms the presence of an aldehyde or ketone.
3. The ^{1}H NMR integrations must be multiplied by 2 to get the correct number of H's. (Remember that the curves are only in the RATIO of types of H present). Therefore, there is a 2H singlet at 4.26 a typical CH3CH2- (triplet, quartet) pattern and a 1H broad signal.
4. Assembling what we know, there must be the following fragments:

$$C=O \quad CH_3 \quad CH_2X \quad YCH_2Z \quad OH$$

where X, Y and Z cannot have any H's that will couple included in them. One of X,Y and Z will be the OH group (recall that hydrogens on O do not couple since they are undergoing rapid exchange). The chemical shift of the CH_2 suggests that it is attached to both the OH and the C=O.
The only structure that fits all the data is **1-HYDROXY-2-BUTANONE**

11-10.
1. From the formula, there is one until of unsaturation and the IR suggests a carbonyl which cannot be an ester unless it is conjugated – which the units of unsaturation do not allow. There is no OH absorption in the IR, so the other two oxygen atoms must be ether types.
2. There are 6 carbons but the ^{13}C NMR shows only 5 signals, one of which ($\delta = 205$) confirms the presence of a ketone or aldehyde. There is also one other carbon which is significantly deshielded ($\delta = 101$), suggesting a carbon with significant electronegative effects.

3. The ^{1}H NMR shows a one H signal at $\delta = 4.8$ which is a triplet and a corresponding doublet ($\delta = 2.75$) which is due to two hydrogens. The presence of a fragment

$$XCH_2CHYZ$$

where X, Y and Z have no coupling hydrogens is confirmed.

4. There is a 6H singlet at $\delta = 3.36$. The only way this signal is possible is from the presence of two identical CH_3 groups, which accounts of the presence of only 5 ^{13}C signals. The shift of this singlet is suggestive of OCH_3 groups. Another 3H singlet at $\delta = 2.3$, suggestive of a methyl adjacent to a C=O completes the picture.

5. We now have the following pieces.

$$XCH_2CHYZ \quad C=O \quad 2OCH_3 \quad CH_3C=O$$

Assembling this, we realize that the elements X, Y and Z must be the C=O and the two OCH_3 groups which leads to the complete structure **4,4-DIMETHOXY-2-BUTANONE**

$$\underset{||}{\overset{O}{\;}}$$
$$CH_3CCH_2CH(OCH_3)_2$$

11-11.

1. The formula indicates only on unit of unsaturation and the IR indicates the presence of an ester – which uses this unit as well as both available oxygen atoms.

2. The ^{13}C NMR shows four different carbons (no symmetry allowed in the molecule), one of which is indicative of an ester C=O ($\delta = 167$) which confirms the IR assignment. There is one other C which is somewhat deshielded ($\delta = 62$), and this is probably the carbon attached to the oxygen of the ester.

3. The ^{1}H NMR shows a typical ethyl triplet quartet and a two-hydrogen singlet (no adjacent H's). Remember that we have a Br atom to accommodate as well.

4. The only possible structure that fits all the data is **ETHYL 2-BROMOACETATE**

11-12.

1. The formula requires one unit of unsaturation (review how the presence of nitrogen affects this calculation). The IR suggests that the oxygen is present as a ketone and not an amide which would show the carbonyl below 1700 cm^{-1}.

2. The ^{13}C NMR shows 7 absorptions, but there are 9 carbons in the formula, so there must be some element of symmetry present. The presence of a ketone carbonyl is also confirmed ($\delta = 209$).

3. The ^{1}H NMR is more complex and requires some analysis. There is a 6H triplet at $\delta = 1$ and the corresponding quartet is seen at $\delta = 2.5$ (more easily seen in the expanded scale in the insert). This indicates the presence of TWO equivalent ethyl groups with NO additional coupling.

4. There is a 3H singlet at the (now familiar) shift indicative of a methyl ketone.
5. There is a 2H signal consisting of 5 lines at δ = 1.7. One cannot have 4 H's on one carbon adjacent, so this must be a case of a CH_2 flanked on both sides by CH_2's which are essentially equivalent.
6. The expansion of the δ = 2.3 – 2.6 region shows, in addition to the quartet mentioned above, two triplets that are overlapping. These are the ones labeled 743.5, 736.5, 792.2 and 725.9, 718.9, 711.5, respectively. Each of these can be assigned to a 2-proton integral. The only way these signals and the signal at δ = 1.7 can be accommodated is to have three contiguous CH2's where the center one is coupled equally (same coupling constant) to the one on each side. This will lead to the observed 5-line pattern at 1.7 even though, chemically, the two flanking methylene groups are not the same.

$$XCH_2CH_2CH_2Y$$

$$a \quad b \quad c$$

Putting this all together, we have the following fragments:

$$2CH_3CH_2 \quad XCH_2CH_2CH_2Y \quad CH_3C{=}O \quad N$$

which accounts for all the atoms and therefore the X and Y must be the N and C=O leading to the structure

$$\overset{\displaystyle O}{\overset{\displaystyle \|}{CH_3CCH_2CH_2CH_2N}}\overset{\displaystyle \diagup CH_2CH_3}{\underset{\displaystyle \diagdown CH_2CH_3}{}}$$

11-13.
1. The formula requires 5 units of unsaturation – one of which is an ester according to the IR spectrum. To get that many units of unsaturation into a molecule with 10 C's, there almost HAS to be an aromatic ring in the structure (gives 4 units).
2. The ^{13}C NMR shows the ester C=O, four carbons in the region of δ = 120 – 150 which is where aromatic carbons are found. There must be six carbons in the aromatic system and so there is some symmetry here. Four signals from 6 aromatic carbons can be generated in several ways: monosubstituted, meta-disubstituted, etc. In addition, there are 3 'aliphatic' carbons present, one of which looks like the oxygen-attached carbon of the ester (δ = 65).
3. The ^1H NMR shows 5 H's in the region of δ = 7 which tells us that there are 5 aromatic protons and therefore the ring is monosubstituted. There is a 3H singlet at a shift suggestive of a $CH_3C{=}O$ and two coupled CH_2 units, one of which is much more deshielded than the other.

4. The correct structure is

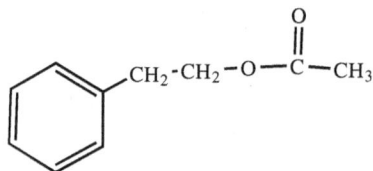

11-14.

1. The formula demands that there are 5 units of unsaturation and the IR indicates the presence of both a non-conjugated ester and an OH group.
2. The ^{13}C NMR says that there are two sp^3 carbons, a C=O and 4 signals for aromatic carbons, and since the ^1H NMR shows only 4 aromatic this must mean that the aromatic ring is para-disubstituted.
3. The ^1H NMR also shows a methyl singlet at a shift indicative of being attached to an O, so the presence of a methyl ester is indicated. The remaining signal is a 2H singlet and its chemical shift is almost as far downfield as the CH_3 attached to oxygen. However, this is not possible so it must be between two deshielding groups and have no adjacent H's.
4. The only possible structure is

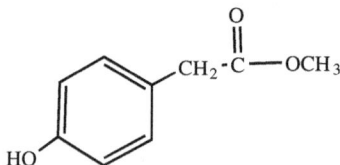

11-15. The correct structure is

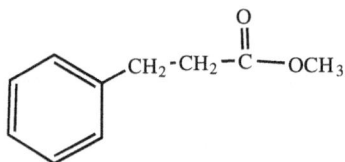

11-16. The correct structure is

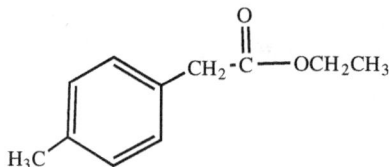

Chapter 12

12-1.

(a) $(CH_3)_3CO^-$

(b) H^+ via protonation of the OH and a carbocation intermediate

(c) H^+

12-2.

(a)

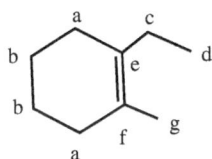

4 absorptions in 15 to 25 ppm range for a, b, c, d, and g
2 absorptions around 120 ppm for e and f

(b)

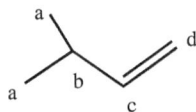

2 absorptions in the 15 - 30 ppm range for a and b
1 absorption ~110 for d and 1 at ~ 120 for c

(c)

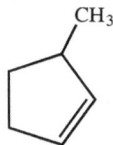

4 absorptions in 10-30 range
2 absorptions in 125 range

(d) There are two conformations that allow an antiperiplanar arrangement of the Br and a hydrogen atom. One has the R group and the Ph in a synclinal arrangement and the other in an anticlinal arrangement. The latter is preferred sterically and E2 elimination gives the *trans* product.

12-3.

E2 elimination with bulky base gives less substituted alkene

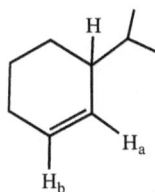

(a)

Ha = doublet of doublets

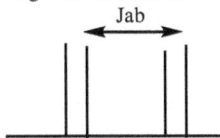

Hb = triplet of doublets

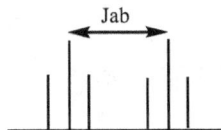

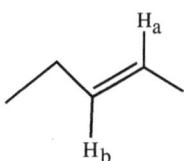

Ha = doublet of quartets

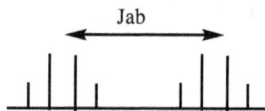

Jab

Hb = doublet of triplets (see part a)

(b)

12-4.

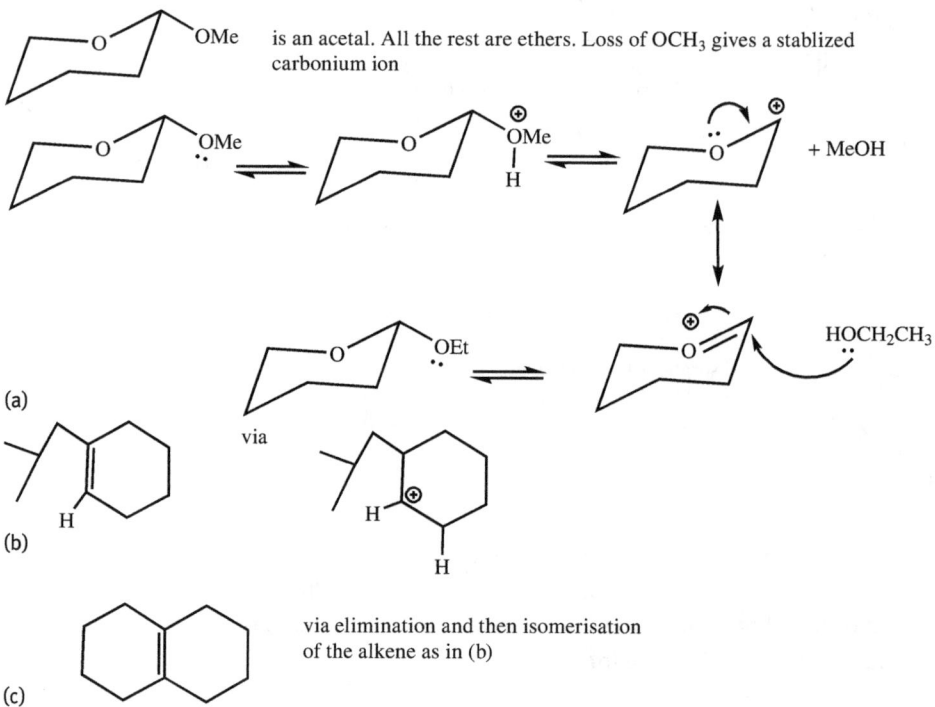

OMe   is an acetal. All the rest are ethers. Loss of OCH₃ gives a stablized carbonium ion

+ MeOH

HOCH₂CH₃

(a)

via

(b)

via elimination and then isomerisation of the alkene as in (b)

(c)

12-5.

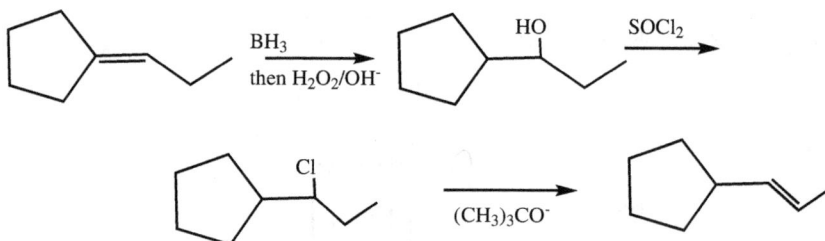

BH₃
then H₂O₂/OH⁻

HO

SOCl₂

Cl

(CH₃)₃CO⁻

12-6.

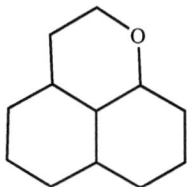

The attached OH participates in the addition to the
intermediate carbocation instead of the water

Part II: **The Naming of Organic Compounds**

14 Introduction

When two people wish to communicate, they must be able to speak the same language. One person may be completely lucid in expressing his or her thoughts, but if these thoughts are presented in a language which the listener does not understand, then no communication takes place. Chemistry has a special language of its own and a very important part of this is called nomenclature. It is the language that allows the chemist to specify unambiguously exactly to which chemical structure he or she is referring. Without this ability, communication of information about any given compound would require that the structure be drawn, a situation that makes verbal exchanges of information difficult.

In the early days of organic chemistry, when there were very few known compounds and few practicing chemists, communication was not difficult. At that time, two facts influenced the choice of many of the names for newly isolated organic materials. Most of these were isolated from plant or animal sources and it made good sense to assign them names that reflected their source. Such names of course had little to do with the structure of the compound which was, in most cases, completely unknown. Also, the large majority of the organic chemistry carried out in the late 19th century was done by German-speaking chemists and the choice of names reflected this heritage. For example, the German word for acid is "Säure" and the acid that is responsible for the foul odor of rancid butter was given the name "butter acid" or "Buttersäure," which became butyric acid when translated into English. These names are often referred to as "trivial names." As the number of known compounds increased, this method of naming compounds rapidly became unworkable. Also, much greater understanding of the structure of organic compounds was reached and it became clear that a system must be developed that would allow the structure of the compound – i.e., which atoms are joined and in what manner – to be readily determined from the name. The International Union of Pure and Applied Chemistry (IUPAC), a body established to guide the progress of chemistry, laid down a set of "Rules" which, with very few modifications or additions, are still the basis of standardized names used today. The IUPAC rules are not, in fact, one set, but rather two. There are "substitutive" names and "functional class" names. With very few exceptions, the former are the standard and these are the ones that will be the focus of this little text.

It must be noted that chemists, like all other forms of humanity, are not unanimous in their acceptance of any set of rules. Therefore, one will still see and hear "trivial names" used (acetone, butyl alcohol, acetic acid, etc.). These names survive and will probably always be a part of the practicing organic chemist's lexicon. Other systems have also been developed. One of the most important of these is that used by Chemical Abstracts. Chemical Abstracts is a compendium of information on all known chemicals and chemical information and is consulted very frequently by chemists. The Chemical Abstracts system of nomenclature is designed to facilitate *indexing* of names

https://doi.org/10.1515/9783110778311-015

so that information can be easily found. It differs from the standard IUPAC system in some small, but important ways. However, if you understand the IUPAC system, you will have no difficulty adapting to the Chemical Abstracts system as required in the future.

For an introductory course in organic chemistry, you will be expected to master the basics of the IUPAC system only. For that reason, the following text will concentrate *almost* exclusively on this system. A very few trivial names will be mentioned at some points because, in the future, you will undoubtedly have to become familiar with these.

How to Use This Text

This book is designed to supplement the text for the first course in organic chemistry. The material will not only teach the nomenclature system, but also reinforce such important topics as stereochemistry and isomerism. To be fully effective, the book must be used in the manner intended; that is each question should be answered in the given sequence and if an incorrect answer is given, the reason for the error should be *understood* before further questions are attempted.

Finally, many students may arrive in this course having received some prior instruction in organic nomenclature, frequently in high school. It is my experience that this exposure is usually well done, but can be incomplete and some errors, particularly in numbering systems, are common. It is to the students' advantage to work the material in the suggested sequence even if it seems extraordinarily simple at first glance.

Additional insights into nomenclature can be found in any introductory textbook on organic chemistry. In addition, there are several, more comprehensive, texts devoted entirely to nomenclature. However, they contain material that is not appropriate to this course and this booklet is an attempt to provide a focused view of the essentials of organic nomenclature.

15 Basic Principles

The Guinness Book of Records states that the world's longest word is the name of a
chemical. It happens to be a protein that contains many amino acid residues, each
of which contributes a three letter code to the word. To the layman, this "word" is
both mystifying and incomprehensible. To the chemist, it is both informative and hu-
morous. The latter is true because the addition of one more amino acid to the protein
would generate an even longer "word." In truth, as you will see, the concept of the
longest word is meaningless since one can always add one more piece to a molecule
and so increase the size of its name.

The situation is akin to the concept stressed in calculus where a whole is divided
into an infinite number of pieces. No matter how many pieces *you* divide it into, *I* can
divide it into more. As you will have learned, organic compounds are composed primarily of the elements carbon and hydrogen. Since carbon atoms form the backbone
of these molecules and it is always possible to add one more carbon to a molecule, an
infinite number of organic compounds is possible. How can an orderly system of naming be designed to cope with this array of molecules? Before we examine that question,
let us consider one fact which simplifies the problem greatly.

The Functional Group Concept

If there are an infinite number of possible organic compounds and if each of these
could undergo only one reaction, there would obviously be an infinite number of
possible reactions. In fact, the situation is worse than this since, on average, each
molecule can undergo several different reactions. If one was required to remember
or even record all these, the task clearly would be impossible. The salvation for this
situation lies in the fact that, of all the atoms making up a molecule, it is possible
in many cases to select a very few at which reactions will occur. These combinations
of atoms recur in many compounds and they behave to a large extent in a manner
that is completely independent of the rest of the molecule. To a first approximation,
it doesn't matter what is attached to these reactive sites, the ensuing reaction will be
the same.

For example, the group -OH attached to a carbon will react in the same fashion
regardless of whether the molecule contains 1, 4, 25, or 59 carbon atoms. Such reac-

https://doi.org/10.1515/9783110778311-016

tive groups are called *functional groups*. The study of the reactions of organic compounds is simplified enormously by this fact. Instead of needing to understand and know the reactions of an infinite number of compounds, the problem simplifies to knowing those of the common functional groups which are about 20 in number. Just as the study of the reactions of molecules is organized around the functional groups, so the *nomenclature* reflects this and simplifies the task. This book is organized by functional group. Each chapter or section is concerned with one functional group.

Definitions

We must start with some definitions.
1. A *hydrocarbon* is a compound that contains *only* the elements Carbon and Hydrogen.
2. A *saturated hydrocarbon* is a molecule that contains only carbon and hydrogen and in which all the carbon atoms are sp^3-hybridized. It therefore can not possess any double or triple bonds.
3. An *unsaturated* hydrocarbon is a molecule that contains only carbon and hydrogen, but in which there are one or more multiple (double or triple) carbon–carbon bonds.
4. *Isomers* are molecules with the same molecular formula but different structural formula – i.e., the same atoms in the same ratios are present in each compound but they are joined differently. There are several different kinds of isomers and we will meet many of these during this course. Each kind of isomer introduces its own nomenclature difficulties and solutions.
5. *Homologues* are molecules that differ from each other only by the addition of a CH_2 unit.
6. Organic compounds can be divided into two general classes: compounds that do not contain rings (*aliphatic*)and those that do (*cyclic*). Each of these can further be subdivided into compounds that are *hydrocarbons* and those that are not. The latter compounds contain *heteroatoms* (i.e., atoms that are different from carbon and hydrogen). This can be represented as follows. Some of these terms will not be examined until later in this text.

You should be aware that the various hybridized forms of carbon each has its own geometric arrangement of the attached atoms. The most common form – sp^3 hybrid – is tetrahedral. The most important feature of this geometric shape is that the corners of the tetrahedron – i.e., the four atoms attached to a sp^3-hybridized carbon atom – are all equidistant from each other. The angle between these bonds is about 109° – an important number to remember. It becomes very time-consuming to write the three-dimensional structure out completely and the practice of most chemists is to ignore

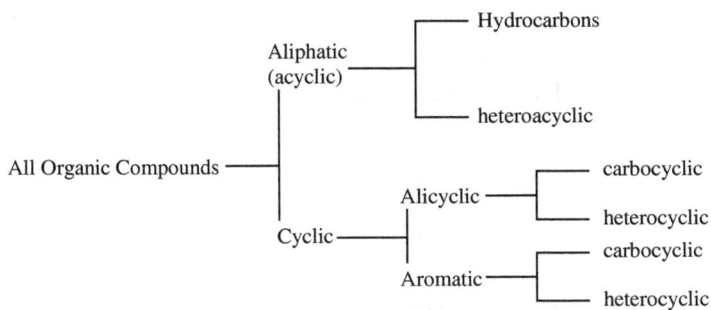

Fig. 15.1

this important feature unless it is necessary. Thus, the simplest organic molecule is usually written as either

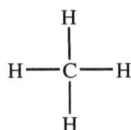

Fig. 15.2

or, more commonly as CH_4. The danger in this simplification is that the beginner will forget that the molecule is really three-dimensional. Why is this important? Consider the following two structures.

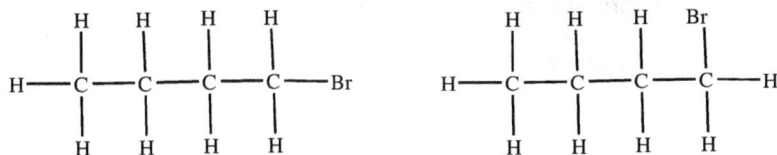

Fig. 15.3

It might appear that these are different, but in fact they represent the same molecule. Building a model of the molecules will make this more obvious.

> The purchase and use of a good set of molecular models is *highly recommended*. It will get you used to visualizing molecules in three dimensions, which is an important skill in many facets of organic chemistry.

Drawing molecules in the manner shown above (expanded formula) is useful because it specifies very exactly which atoms are joined. However, it is very time- and space-consuming. Chemists prefer to draw *condensed formula*. These require less time and space, but require the student to be aware of certain limitations.

The molecules shown above can be partially condensed by omitting all the C–H bonds. They would then be written as

$$CH_3-CH_2-CH_2-CH_2-Br$$

A second condensation omits all the C–C bonds as well and the condensed formula would look like

$$CH_3CH_2CH_2CH_2-Br$$

Note that in these cases, it might appear that a carbon in the interior of the chain is bonded to hydrogen atoms on the adjacent carbon. This is misleading but unavoidable.

Q15-1. Write the condensed structural formula for

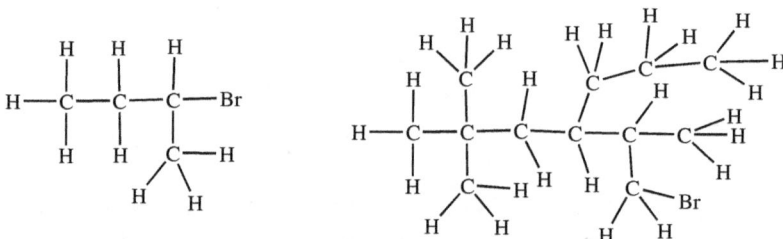

Q15-2. Write the expanded formula for

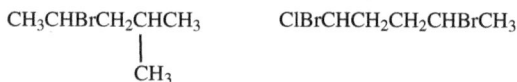

CH₃CHBrCH₂CHCH₃
        |
        CH₃

ClBrCHCH₂CH₂CHBrCH₃

The Basics: Alkanes

The starting point for the introduction of organic nomenclature, is *saturated aliphatic hydrocarbons*, compounds whose IUPAC classification is called *alkanes*. The name of any organic molecule can be divided into three parts: the prefix, the root, and the suffix. Each of these parts has a specific task.

Root specifies the number of carbons in the longest contiguous chain of carbon atoms in the molecule.

Suffix specifies what *kind* of a molecule the compound is. The functional group(s) that are contained in the molecule are listed here.

Prefix The part of the word that describes all the other points necessary to make the name complete and unambiguous.

$$[prefix(es)] + [root] + [suffix(es)] = name$$

The starting point for the study of the naming of organic molecules is always the class called *alkanes*. The suffix *ane* indicates that all the atoms present must be either carbon or hydrogen and that all the carbon atoms are sp^3-hybridized. Using alkanes as an example, the methodology of the construction and interpretation of the name of molecules will now be introduced.

The root words that describe the number of carbon atoms in the longest contiguous chain must be learned. Those above ten (dec) are used less frequently but still form a part of the required body of knowledge.

No. of carbons	Root word	No. of carbons	Root word
1	meth	7	hept
2	eth	8	oct
3	prop	9	non
4	but	10	dec
5	pent	11	undec
6	hex	12	dodec

The combination of the root "meth" or "hex" or "oct" with the suffix "ane" gives methane, hexane, or octane, which are the complete and unambiguous IUPAC names for the molecules whose condensed formula are CH_4, $CH_3CH_2CH_2CH_2CH_2CH_3$, and $CH_3CH_2CH_2CH_2CH_2CH_2CH_2CH_3$, respectively.

What happens with a structure like?

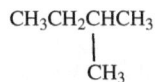

$CH_3CH_2CHCH_3$
     |
    CH_3

Fig. 15.4

Only four of the five carbons can be linked in one contiguous chain. Therefore, the basis of the name [root + suffix] will be *butane*. However, the molecule whose complete name is butane has a total of only four carbons. How do we indicate the presence of the additional CH_3 group? The solution lies in the *prefix*.

Q15-3. The root of the names of the hydrocarbons with the carbon chains shown are _____, _____, and _____ respectively

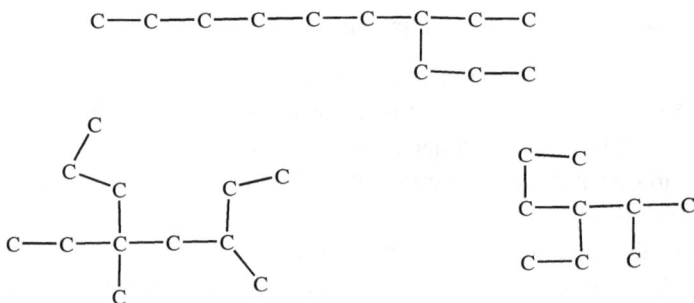

C—C—C—C—C—C—C—C—C
                  |
                  C—C—C

Any group that is not a part of the main chain (root) or functional group (suffix) is called a substituent because it replaces an H atom. It is indicated by using a prefix. Some common heteroatom (a heteroatom is an atom other than carbon or hydrogen) substituents and their names are:

−Cl	chloro
−Br	bromo
−I	iodo

Note that the chemical symbol for chlorine is Cl *not* CL and the symbol for bromine is Br, *not* BR! Also note that *There is an "H" in the word Chloro.* As people literate in science, you must learn to use the correct symbols and spellings of words!

If, as is the case illustrated above, the substituent is one that could be derived by removal of one H from a hydrocarbon chain [e.g., CH_4–H ⟶ CH_3–] the substituent name is derived from the same root, but using the suffix "yl" ("meth" (1 carbon) + "yl" = *methyl*) to indicate that it is not the complete chain.

In the same way a *substituent* with the structure CH_3CH_2– or $CH_3CH_2CH_2$– would have the name "ethyl" or "propyl," respectively.

The generalized name for a substituent derived from an alkane is *alkyl*. Thus, methyl (one carbon), ethyl (two carbons), butyl (four carbons), and octyl (eight carbons) are all *alkyl groups*.

Alkanes that have all carbon atoms in one continuous chain are often called "normal alkanes" whereas those that have alkyl substituents are termed "branched alkanes." The position of the alkyl substituent is called the "branch point."

Q15-4. Answer the following questions about the structures shown below.

(a) Which are normal alkanes?

(b) Which are not alkanes?

(c) Which are unsaturated hydrocarbons?

(d) Which are alicyclic hydrocarbons?

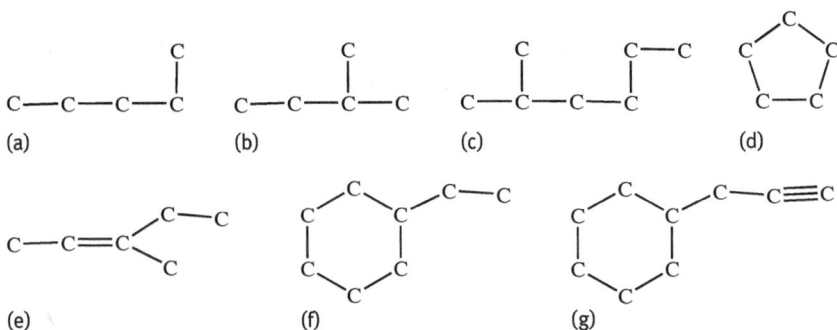

Consider the difference between

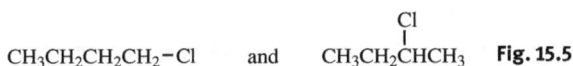

$$CH_3CH_2CH_2CH_2-Cl \quad \text{and} \quad CH_3CH_2\overset{\displaystyle Cl}{\underset{\displaystyle |}{C}}HCH_3 \quad \textbf{Fig. 15.5}$$

In both cases we have replaced one hydrogen of butane with a chlorine atom. The molecular formula of both of these is C_4H_9Cl but they clearly have different structures. They are perfect examples of *isomers*. If the nomenclature system is to be unambiguous, we must be able to distinguish between these. How to do this is the subject of the next paragraphs.

Q15-5. Identify the longest chain in the following compounds and assign the correct IUPAC names,

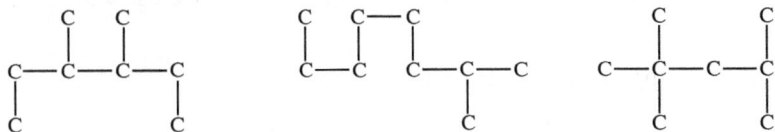

The main chain is numbered and the number of the carbon to which a substituent is attached is incorporated in the name. (The number is often referred to as a locant.) The direction the numbering takes is very important. The chain must be numbered to give the first encountered substituent the lowest number. If substituents are so placed that numbering from either end gives the same number to the first encountered substituent, the direction that gives the lowest number to the first difference is used. For example:

$$CH_3CH_2CH_2\underset{\underset{Cl}{|}}{C}HCH_3$$ **Fig. 15.6**

If the 5-carbon chain is numbered left to right, the chlorine atom is on C4, but if the numbering is done from right to left, the chlorine atom is on C2. The latter is the proper way. The correct IUPAC name is 2-chloropentane. Note that there is no space between the substituent name and the root.

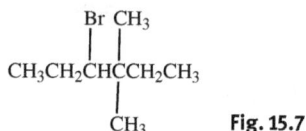

$$CH_3CH_2\underset{\underset{CH_3}{|}}{C}H\underset{\underset{}{|}}{C}CH_2CH_3$$ **Fig. 15.7**

Numbering this molecule left to right gives the locants 3,4,4 but right to left gives 3,3,4. The first number is the same so the second is used to determine the correct direction. The correct name is 4-bromo-3,3-dimethylhexane. Note that when two identical substituents are present on the chain, the prefixes di- (two), tri- (three), tetra (four) are used. There are two methyl groups on this molecule and so the name contains the fragment dimethyl.

The correct method of naming a compound also requires that the substituents be listed in alphabetical order. For this the prefixes di, tri, tetra, etc. are *not* considered part of the name for alphabetizing purposes; i.e., ethyl would be put before dimethyl.

In the very rare case where all the numbers are the same, but the substituents are different, the substituent coming first in an alphabetized list is given the lowest number. *The punctuation in the IUPAC system is very important.* Numbers are always separated from numbers by commas and numbers from words by hyphens. Look at the name of the last compound (4-bromo-3,3-dimethylhexane) more closely to verify this.

Q15-6. Give the correct IUPAC names for the following structures

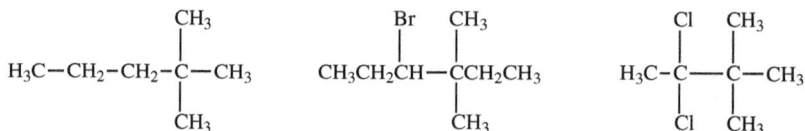

A common misconception is that the *sum* of the locants should be minimized. This is *not* the case. For example, consider the skeleton below.

The proper name for this is 2,7,8-trimethyldecane, not 3,4,9-trimethyldecane, even though the sum of locants in the latter is the smaller of the two.

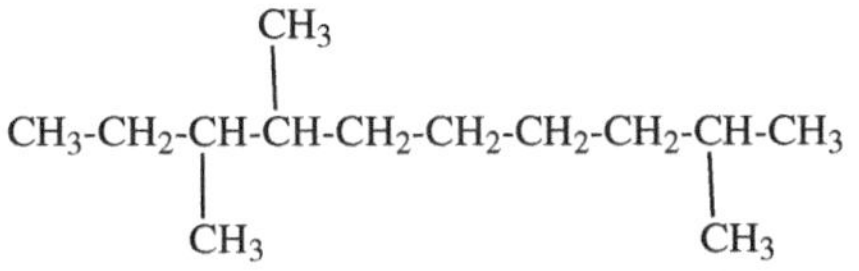

$$\text{CH}_3$$
$$|$$
$$\text{CH}_3\text{-CH}_2\text{-CH-CH-CH}_2\text{-CH}_2\text{-CH}_2\text{-CH}_2\text{-CH-CH}_3$$
$$|\qquad\qquad\qquad\qquad\qquad |$$
$$\text{CH}_3\qquad\qquad\qquad\qquad \text{CH}_3$$

Fig. 15.8

Q15-7. Draw the condensed structural formula for:
(a) 2-chloro-2-methylpentane
(b) 2,5-dibromo-2,3-dimethylhexane
(c) 2-chloro-4-ethyl-5,5-dimethylheptane

If there are two chains of equal length, then the one which has the *most substituents* is chosen as the main chain. For example, consider the skeleton

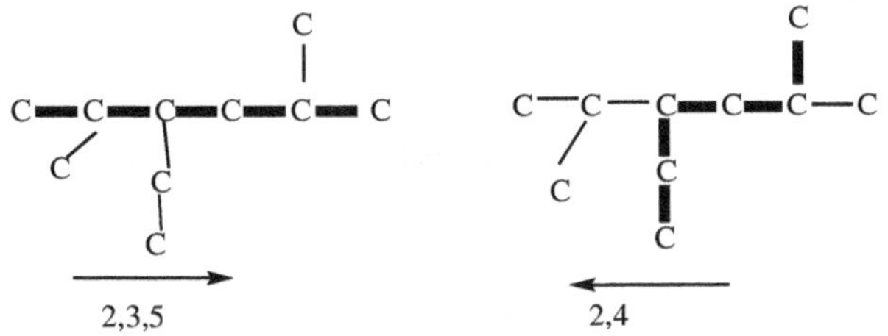

Fig. 15.9

There are two different chains, both of six carbon atoms which might be chosen as the basis for the name. One has three substituents numbered 2,3,5 and the other has two substituents numbered 2,4. The chain with three substituents is the correct choice.

Frequently, the substituents on a main chain themselves have branches. For example,

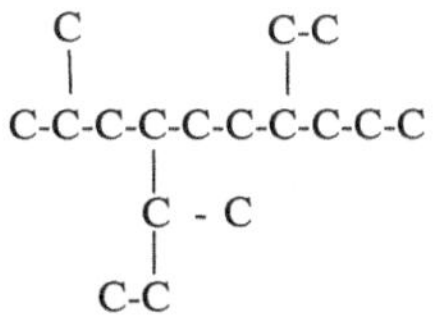

Fig. 15.10

The main chain has ten carbons and it must be numbered left-to-right, so the locants are 2,5,8. The substituent on C-5 is a propyl group, itself substituted with a methyl group. The general method for coping with this situation is to find the longest chain in the substituent (3 C's), number its chain with #1 as the carbon attached to the main chain, and then indicate the substituents on it in the usual way but putting these in square brackets. The name of the compound above would be 8-ethyl-2-methyl-5-[*1-methylpropyl*]decane. The type in italic face is the name of the complex substituent. In the specific case of the substituent -CH(CH3)2, the term *isopropyl* has been retained. It is synonymous with 1-methylethyl.

The Index of Hydrogen Deficiency (IHD)

A quick check of all the complete molecular formula for alkanes (saturated hydrocarbons) will confirm that for any compound with n carbon atoms, there are $2n+2$ hydrogen atoms. Halogen substituents replace one hydrogen atom each and therefore must be added to the hydrogen count. Oxygen atoms have no effect on the equation. This little mathematical expression will be of considerable use in predicting structures of organic molecules based on their elemental formula.

Q15-8. Correct the following IUPAC names:
(a) 4-chloro-5-methyhexane
(b) 2-methyl-4-chloropentane
(c) 3,3,6-trimethylheptane
(d) 2-bromo-5-chloro-6-isopropyloctane

Q15-9. The molecular formula of an acyclic alkane of unknown structure had 15 hydrogen atoms, one chlorine atom, and two bromine atoms. How many carbons did it have?

Summarizing Problems

15-1. Provide acceptable IUPAC names for the following

(a)

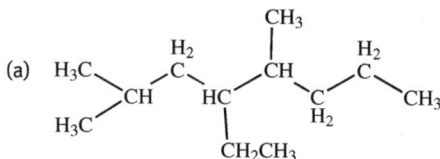

(b)

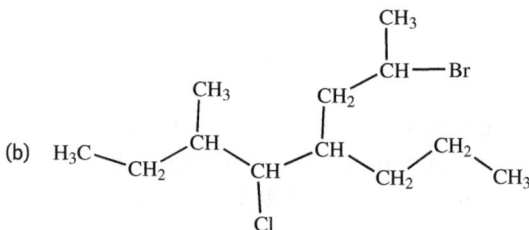

(c)

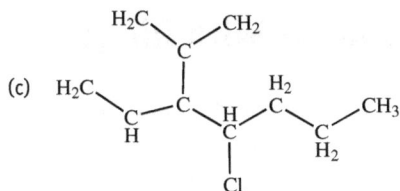

(d)

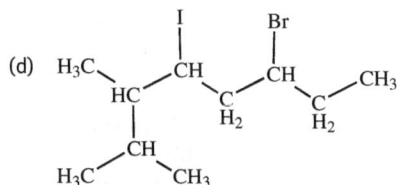

(e)

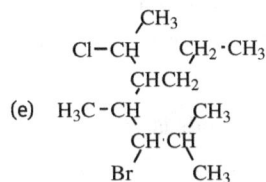

15-2. Give an expanded structural formula for each of the following names.
(a) 5-bromo-2,3-dichloro-5-methylheptane
(b) 3-ethyl-2,3,6-trimethyloctane
(c) 3-bromo-4-choro-2,5-dimehylhexane
(d) 5,5-dibromo-3-iodo-2-methyldodecane

15-3. The following names are almost correct according to the IUPAC system. Identify the error in each and provide a correct name.
(a) 3,6-dimethylheptane
(b) 2,5-dichloro-4-methylhexane
(c) 3-bromo-3-chloro-4,6-dimethylheptane
(d) 2-chloro-3-propylhexane
(e) 6-bromo-2-iodo-4,4-dimethylheptane

16 Alkenes, Alkynes, and Compounds With Rings

Alkenes are compounds that have at least one carbon–carbon double bond (C=C). The older name for this class of compounds is *olefin*. The IUPAC ending for the names is *ene*. This is the first class of compounds we will see that contains a functional group (the C=C). This is indicated by the change in the name suffix from *ane* to *ene*. If you know and understand the nomenclature methodology for alkanes, the change to alkenes is easy. The same rules regarding selection of the main chain and handling of substituents apply with two extra restrictions.

The chain chosen as the main chain *must include the double bond*, even if this means that it is not the absolutely longest chain in the molecule. In addition, the numbering must be done in such a way as to give the *functional group* (the double bond), and not the substituent, *the smallest number*. We will find that these restrictions apply to all organic molecules with the exception of alkanes where there is no functional group.

Q16-1. The root word of the IUPAC names of the following compounds is

(a)_____   (b)_____   (c)_____

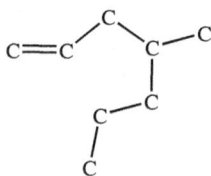

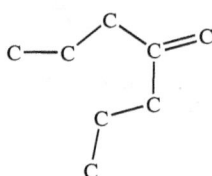

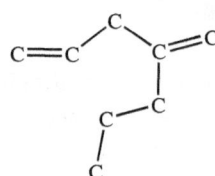

(a)   (b)   (c)

The number indicating the position of the double bond can appear in one of two positions in the name. For example, C–C=C–C–C can be written 2-pentene or pent-2-ene. Either is acceptable.

Compounds that have more than one double bond are dienes (2 C=C), trienes (3 C=C), etc. The naming and numbering systems are covered by the rule given above. The main chain must contain as many of the double bonds as possible and the numbering must give precedence to the functional groups. To make the names of these compounds easier to pronounce, the root word describing the chain length is combined with the word "diene" or "triene", etc., using the letter "a" as a bridge. Thus, names like 1,3-octadiene or 1,3,5-hexatriene are formed.

https://doi.org/10.1515/9783110778311-017

Q16-2. Provide IUPAC names for the following compounds

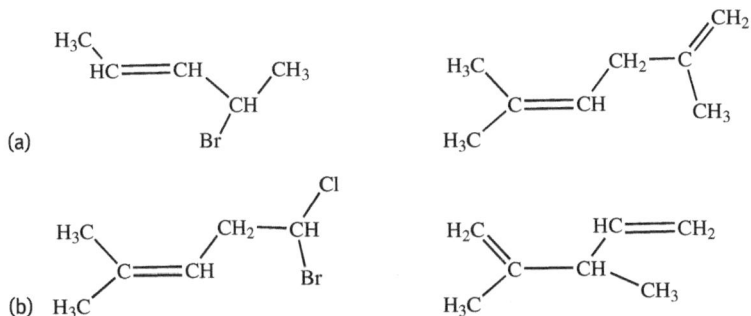

(a)

(b)

If chains containing double bonds are substituents, the usual methods apply. Thus, the name of $CH_2=CH-$ as a substituent is ethenyl. Propene ($CH_2=CHCH_3$) can give rise to three possible substituents depending on where the hydrogen is removed from the chain. These are $-CH=CHCH_3$ (1-propenyl), $CH_2=CHCH_2-$ (2-propenyl), and $CH_2=CCH_3$ (1-methyl-1-ethenyl). The point of attachment of the substituent chain to the main chain is always given the number 1. You should be able to see these derivations.

Q16-3. Provide complete IUPAC names for the following compounds

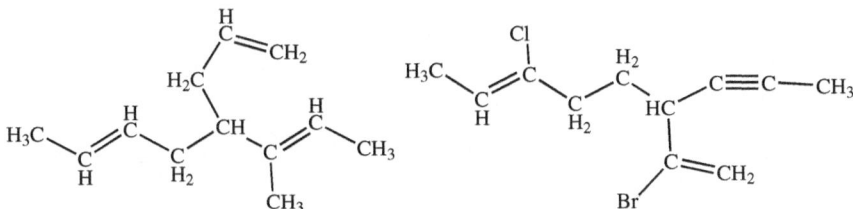

The Index of Hydrogen Deficiency II

As was outlined in the previous chapter on alkanes, the ratio of carbon to hydrogen in any alkane is given by the expression C_nH_{2n+2}. In order to create one double bond in a chain and maintain the valency of carbon at four, two hydrogens must be removed from adjacent carbons. This will then make the general formula C_nH_{2n}. This simple fact allows the calculation of the number of double bonds or their equivalents (the importance of this phrase will be seen shortly) in any molecule just from the molecular formula. Calculate the number of hydrogens that would be needed to create an alkane from the given number of carbon atoms. Subtract from this the number of hydrogens actually present. Divide this difference by two and the result is the number of double bonds or their equivalents present in the molecule. This is the Index of Hydrogen Deficiency (IHD). Some texts refer to this as SODAR (*Sum Of Double bonds And Rings*) but the concept is the same.

Assigning Stereochemistry

The presence of a second bond in a double bond prevents rotation around that bond. Also, the *shape* of the sp^2 carbons is planar. The combination of these facts leads to the result that there are frequently two different ways to assemble molecules containing double bonds. These two ways have the same molecular formula and different structural formula, so they fit the definition of isomers. However, they are different to the type of isomer we saw in the last chapter. In those cases, the two isomers actually had different atoms joined. For example, 2-methylbutane and pentane have the fifth carbon atom joined to different places on the original 4-carbon chain of butane. In the present case, this is not true. All atoms are joined to exactly the same atoms in both isomers, but the *three-dimensional arrangement in space* is different. Such isomers are referred to as *stereoisomers*.

> *Stereoisomers*
> Isomers that differ only with respect to how atoms are arranged in space and *not* with respect to which atoms are joined.

As you will find in the coming pages, there are several different *kinds* of stereoisomers, all of which fit this definition. The kind of stereoisomer being introduced here is called a *geometric isomer*.

> *Geometric isomer*
> Stereoisomers due to restricted rotation about double bonds or rings.

If our nomenclature system is to work, it must be unambiguous and we must have a way of distinguishing between these forms. As is frequently true, there are two systems in use for handling this situation. Both are in constant use, and therefore, you must be able to use either. One will be introduced here and the consideration of the other will be postponed until the next chapter.

The older system uses the terms *cis* and *trans* to describe the shape of a double bond in the following way. Consider shape of the double bond *in the main chain*.

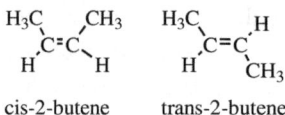

cis-2-butene   trans-2-butene

Fig. 16.1

If the substituents on the sp² carbons of the double bond are on the *same side*, the molecule is referred to as *cis*, whereas if they are on opposite sides, they are referred to as being *trans*. In the drawings below, the main chain is shown in **bold** lines. Stereoisomers of this type are called *geometric isomers*.

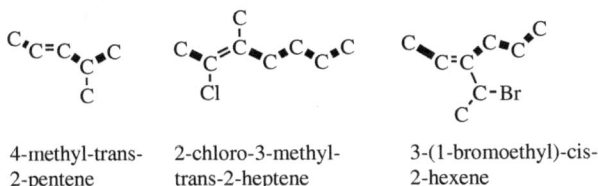

4-methyl-trans-   2-chloro-3-methyl-   3-(1-bromoethyl)-cis-
2-pentene   trans-2-heptene   2-hexene

Fig. 16.2

If there is more than one double bond, each requires its own stereochemical descriptor. Thus, the full name of the following compounds are

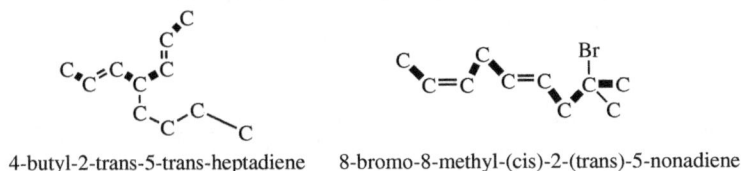

4-butyl-2-trans-5-trans-heptadiene   8-bromo-8-methyl-(cis)-2-(trans)-5-nonadiene

Fig. 16.3

Q16-4. Provide stereochemical descriptors where necessary of the compounds in questions 15-2 and 15-3.

Alkynes

Alkyne is the IUPAC term for the class of compounds that contain a carbon–carbon triple bond. As expected from the class name, the suffix for the name is *yne*. The non-systematic name for this class is *acetylene*. The carbons are joined by one σ bond and two π bonds and must be sp-hybridized. Exactly the same nomenclature rules apply to alkynes as were developed for alkenes. However, because sp-hybridized carbons are linear, there is no question of stereochemistry and no terms like *cis* and *trans* are required.

When the only functional group present in the molecule is a triple bond, the rules for naming are the same as for alkenes. However, now that we have more than one possible functional group, the question arises as to what to do when a molecule contains both a double and a triple bond. This question will become more important as the number of functional groups we can use increases. In such cases, there is a hierarchy of functional groups – i.e., *one will be "more important"* than the rest and it will be the one to control the numbering system. In the current case, if one chain cannot be found which contains *both* the C=C and the C≡C, then the main chain is the one containing the C=C. If a main chain which contains both the C=C and the C≡C is present, the lowest number is given to whichever of the groups appears first (the usual situation). If this does not produce a difference, the C=C is given precedence over the C≡C. Using these rules, the compounds shown below have the indicated names.

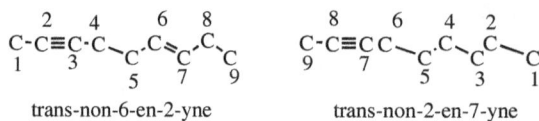

trans-non-6-en-2-yne   trans-non-2-en-7-yne

Fig. 16.4

Q16-5. Give IUPAC names including stereochemical descriptors (use the *cis, trans* method) for the following

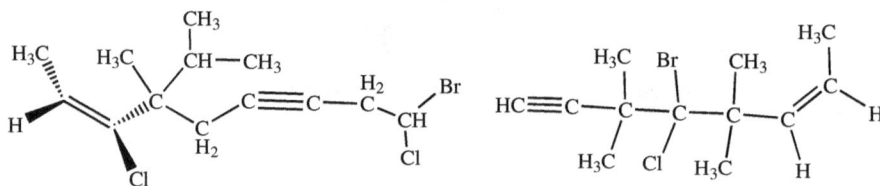

Index of Hydrogen deficiency (III)

It should not be hard to see that the formula of a saturated alkane (C_nH_{2n+2}) must lose *four* hydrogens to form one triple bond. The general formula of an alkyne is therefore C_nH_{2n-2}.

Cycloalkanes

The formation of a C=C bond from an alkane requires the removal of two hydrogen atoms from *adjacent* carbons. If two hydrogens are removed from *nonadjacent* carbons and the bonds used to form a C–C bond, the result is a compound that contains a *ring*. The molecule is still an *alkane* (remember the definition), but the formula looks like an alkene (C_nH_{2n}). This is the "double bond equivalent" referred to earlier. (In fact, in many ways, it is very convenient and helpful to think of a double bond as a two-membered ring). The naming of such compounds is almost the same as that for alkanes, but the fact that the main chain is in the form of a ring is indicated by the prefix "cyclo" in front of the root. Thus, cyclopropane, cyclobutane, cyclopentane, cyclohexane, etc., are the IUPAC names of the molecules shown below.

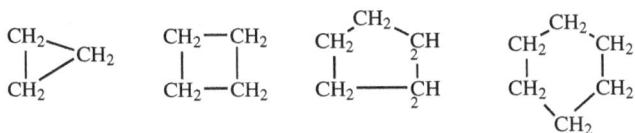

Fig. 16.5

It is usual to go another step in the condensation of these structural formula by omitting all the atoms as well. Such "stick structures" are quick and easy to write but the meaning of each part of the drawing must be well understood. Each corner of the drawing represents a carbon atom *with as many hydrogen atoms as is required to bring the valency of the carbon to four*. The stick drawings of the four molecules shown above would be as shown below and each corner represents one carbon and two hydrogens.

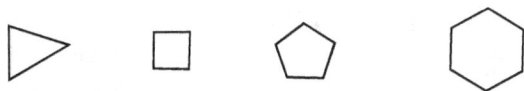

Fig. 16.6

As noted previously, the natural bond angle for sp^3-hybridized carbon is about 109°. Since the bond angles of an equilateral triangle are 60°, there is significant *strain* in cyclopropane. This raises the energy of this molecule which will affect its *reactivity* significantly. The bond angles in cyclobutane, cyclopentane, and cyclohexane might appear to be 90, 108, and 120°, respectively. However, remember that sp^3 carbons are not flat, but tetrahedral in shape. The cycloalkanes larger than cyclopropane are three-dimensional in shape which allows their bond angles to get closer to the desired 109° value. In fact, cyclohexane can achieve the perfect 109° quite readily. Three-dimensional representations of these three cycloalkanes are shown below and will be the subject of some discussion in Part One of this text.

Fig. 16.7

How are substituted cycloalkanes named? Since no "end" of the chain is present, position #1 can be defined anywhere. If only one substituent is present, it is understood to be on carbon #1. (The locant is frequently omitted.) If there is more than one substituent, choose #1 to give the lowest locant to the first difference. If no difference can be found, number so as to give the lowest locant to the first substituent in an alphabetized list of substituents. These rules are essentially the same as for alkanes. Some examples follow.

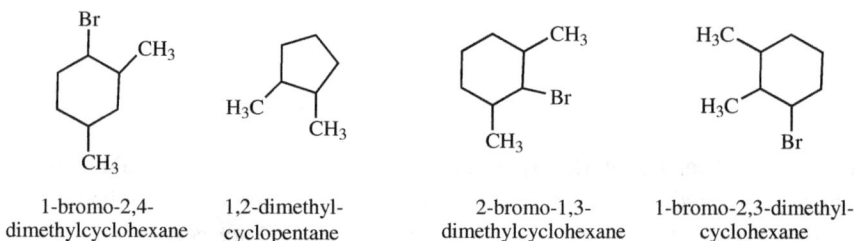

| 1-bromo-2,4-dimethylcyclohexane | 1,2-dimethyl-cyclopentane | 2-bromo-1,3-dimethylcyclohexane | 1-bromo-2,3-dimethyl-cyclohexane |

Fig. 16.8

Remember that in alkene nomenclature, the position of the double bond takes precedence in numbering the chain. For cycloalk*enes* (rings containing a C=C), the lowest possible position for the C=C is #1. Therefore, one of the sp² carbons must be numbered 1 and the other must be numbered #2. Which is which is determined by the position of other substituents if they are present. The numbering *must* begin at one end of the double bond and proceed through it. Consider the following examples.

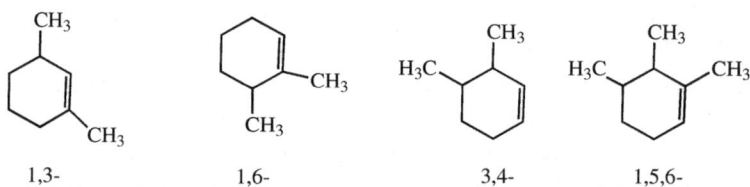

| 1,3- | 1,6- | 3,4- | 1,5,6- |

Fig. 16.9

In each case, if the sp² carbon is substituted, it is #1 and the numbering proceeds from there. If the sp² carbons are *not* substituted, numbering is started at the end of the double bond which gives 1 and 2 to the double bond carbons *and* the lowest number to the first encountered substituent – i.e., the usual rule. If more than one double bond is present in a ring, both double bonds must receive the lowest numbers and only then the substituents are considered.

Remember the analogy between double bonds and rings. Just as it is possible to have two stereoisomers of a double bond because rotation is not allowed, the same is true of substituted cycloalkanes. Rotation about any one bond of the ring is not possible without rotation around them all. Therefore, *geometric isomers* are possible. When you are introduced to the three-dimensional shape of cyclic systems (cyclohexane in particular), you will find that substituents on sp³ carbons can be either above or below the plane of the ring. Therefore, in disubstituted cycloalkanes, if the substituents are *not* on the same carbon atom, it is possible to have *cis* (substituents on the same side) or *trans* (substituents on opposite sides) stereoisomers. At this time, the notation using wedges to signify bonds above the plane of the ring and hashes to signify bonds below the plane of the ring will start to be used.

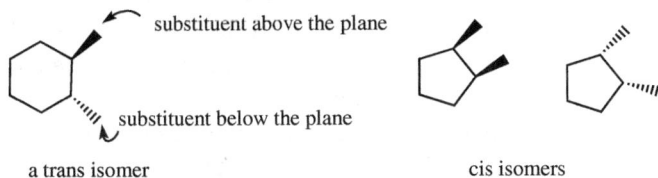

a trans isomer   cis isomers

Fig. 16.10

Therefore, the three possible stereoisomers of the third molecule shown before (2-bromo-1,3-dimethylcyclohexane) can be represented as

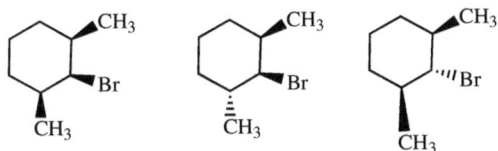

Fig. 16.11

Note that bonds attached to sp² carbons are in the plane of the ring, not above or below it. This is a consequence of the planar nature of sp²-hybridized carbon atoms. Referring to the structures on the previous page, there are no *cis* or *trans* isomers of the first two compounds since the bond to the sp² carbon is in the plane of the ring. However, there are two possible *cis* or *trans* isomers for the last two shown. These are shown below.

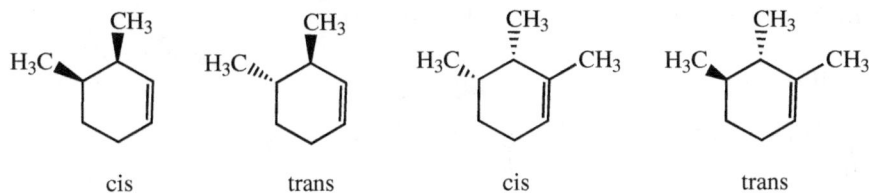

cis   trans   cis   trans

Fig. 16.12

At this point, we will make no distinction between the *cis* isomers that have both sub-stituents above the plane or below the plane of the ring, or between the *trans* isomers that have different substituents above or below the plane. This subject will be intro-duced later.

At this point, it must be noted that there are a few substituents that have names which may not appear obvious, but which you should know because they are so com-mon. Benzene is a molecule with the molecular formula C_6H_6. On the surface. it looks like cyclohexane with three double bonds. However, as you will find out, this is a gross simplification. For our purposes at present, recognize that all six carbons in benzene are sp^2-hybridized and that the whole molecule is *flat*. When one hydrogen is removed from benzene and the rest of the molecule is used as a substituent, the name given to that substituent is *phenyl*. (Later you will hear of a *molecule* called phenol. Do *not* con-fuse the two. One is a substituent (i.e., it lacks a hydrogen), and the other is a complete molecule.) When the whole substituent is a phenyl group bonded to a CH_2, the com-plete substituent is known as a *benzyl* group. It might appear as if the word benzyl should refer to the simple benzene ring, but for historical reasons, this is not the case.

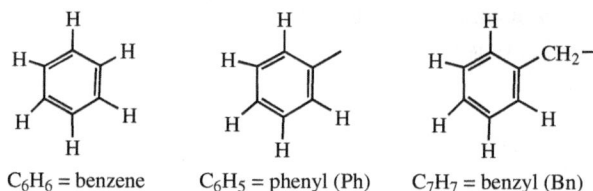

C_6H_6 = benzene C_6H_5 = phenyl (Ph) C_7H_7 = benzyl (Bn)

Fig. 16.13

Note the abbreviations Ph for phenyl and Bn for benzyl. You will encounter these very frequently in the literature of organic chemistry.

Summarizing Problems

16-1. Provide IUPAC names for the following compounds. Use the *cis, trans* system to indicate stereochemistry where this is required. (If stereochemistry is not explicitly shown in the drawing, it is not required in the answer).

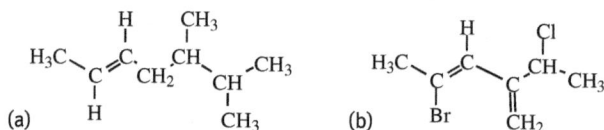

(a)

(b)

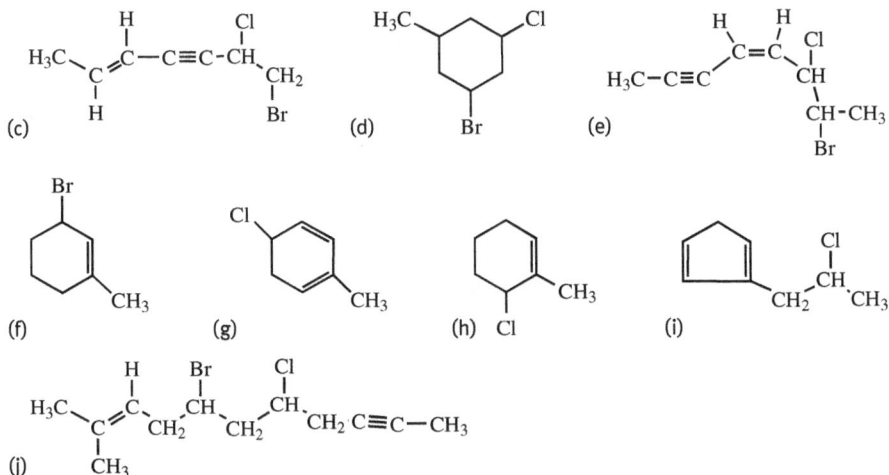

(c)

(d)

(e)

(f)

(g)

(h)

(i)

(j)

16-2. Calculate the molecular formula and the Index of Hydrogen Deficiency for each of the ten compounds shown in question 1.

16-3. Draw a structural formula for each of the following IUPAC names:
(a) 2,3-dimethyl-1,3-cyclohexadiene.
(b) *cis*-5-bromo-2-chloro-3-ethylhex-2-ene
(c) *trans*-4-methyloct-2-en-5-yne
(d) 2-butyl-3,5-dichloro-1-hexene
(e) 3-(1-chloroethyl)cyclopentene

16-4. The following names are almost correct. Identify the error in each and provide a correct IUPAC name.
(a) 2-chloro-3-methylcyclohexene
(b) *trans*-4-bromo-2-methyl-2-hexene
(c) *trans*-hept-3-en-5-yne
(d) 4-ethenyl-2-heptene

16-5. Draw structures that show stereochemistry for the following compounds.
(a) *cis*-1-bromo-3-methylcyclopentane
(b) *cis, trans*-5-methyl-2,5-octadiene

16-6. How many positional and geometric isomers are there of a cyclobutane ring substituted with one chlorine atom and one methyl group?

17 More Stereochemistry – The CIP System

Since the beginning organic chemistry, nomenclature systems have changed and evolved to accommodate advances in knowledge and understanding of reactions and structure. The time of changeover from one system to another is not short. It is very common to have more than one [sometimes several!] systems in use simultaneously. This is particularly true of stereochemical nomenclature. As noted in Chapter 2, the old way of denoting the stereochemistry of alkenes is by the use of the words *cis* and *trans*. When that system was introduced, it was noted that another system which is more generally applicable to alkenes and also is very useful in other, more complex systems would be introduced later. The time has come for this to happen.

The chronology of the introduction of this new system is important. It was originally designed to cope with a completely different stereochemical problem and was subsequently expanded to apply to the stereochemistry of alkenes. Nevertheless, the subject will be introduced to you in reverse order.

The Cahn–Ingold–Prelog System (CIP)

The principle ideas behind this method of describing stereochemistry of alkenes are extremely simple. It consists of three steps.
1. Divide the alkene in two by a line perpendicular to the double bond and through its center.
2. On the basis of the Sequence Rules, assign priorities (importance) to the two substituents attached to each sp^2-hybridized carbon.
3. Put the molecule back together and determine if the highest priority group on both sp^2 carbons are on the *same* side or *opposite* sides of the molecule. If the former is true, the stereochemical descriptor Z is used, but if the latter is the case, the stereochemical descriptor E is used.

To illustrate, consider the following.

The molecule 1-bromo-1-chloro-propene can be divided into two halves as shown [Step #1]. The priorities are assigned on the basis of the Sequence Rules as shown [Rule #2]. The two highest priority groups,

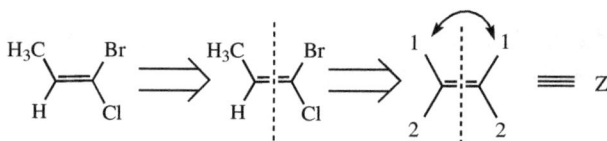

Fig. 17.1

https://doi.org/10.1515/9783110778311-018

one on each end of the double bond are found to be on the *same* side of the molecule and thus the stereochemical descriptor for the stereoisomer shown is Z [Rule #3].

The only thing missing is the way to determine which group or atom has highest priority. This is done on the basis of rules called *Sequence Rules*. They are called this for a reason – they must be applied in sequence. If the first rule gives an unambiguous answer, you must stop at that point. If it does not, then you proceed to use the second rule, etc. These rules are as follows:

1. The atoms directly attached to the sp²-hybridized carbons are compared and the one with the higher atomic *number* takes precedence.
2. If these atoms are the same, proceed down each chain, atom by atom, until a difference is encountered. At the point of difference, the atom with higher atomic number takes precedence.
3. If a branch point in the chain is encountered, take the path that leads to the highest priority group in the fewest number of steps.

For examples, consider the following molecular fragments.

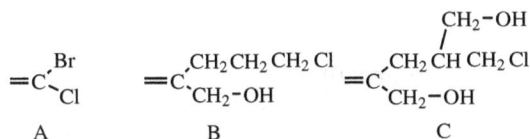

Fig. 17.2

In A, the atomic number of bromine (35) is greater than that of chlorine (17). (Note that you don't need to *know* these numbers – only their relative size – which is simple if you know that bromine is *below* chlorine in the periodic table of elements.) Therefore, Br has priority #1.

In B, the atoms directly attached to the sp² carbon are both carbons, so no difference is present. The next atom in the top chain is again carbon but in the lower chain it is oxygen which has a higher atomic number (8) than carbon (6). Therefore, the lower chain is of higher priority.

The same is true for structure C – the lower chain has higher priority.

There is a very useful notation method which simplifies the consideration of more complex examples of this type of comparison. Each carbon in a chain under consideration is noted as C(X,Y,Z), where X,Y,Z are the atoms attached to the carbon *in decreasing order of their atomic numbers*. Using this system, the designations of the first carbons in the chains of structure C above would be C(C,H,H) [upper chain] and C(O,H,H) in the lower. The difference in them is obvious (O>C) and leads to the same result as above.

Q17-1. Arrange the following atom or groups of atoms in order of increasing priority

Cl–   CH₃–   CH₃CH₂–   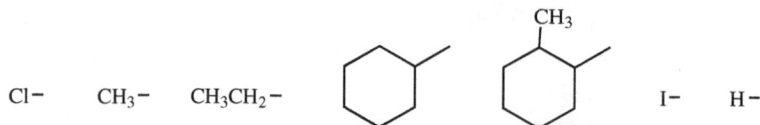   I–   H–

As a final example, consider the following structure.

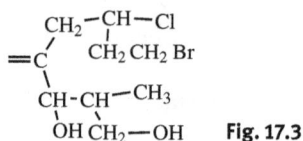

Fig. 17.3

Designating the carbons in the way suggested above gives the following result.

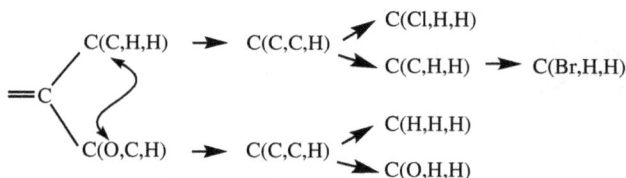

Fig. 17.4

The curly arrow indicates the point of first difference and the lower chain has the higher priority.

Q17-2. Assign the E or Z descriptor to the following molecules

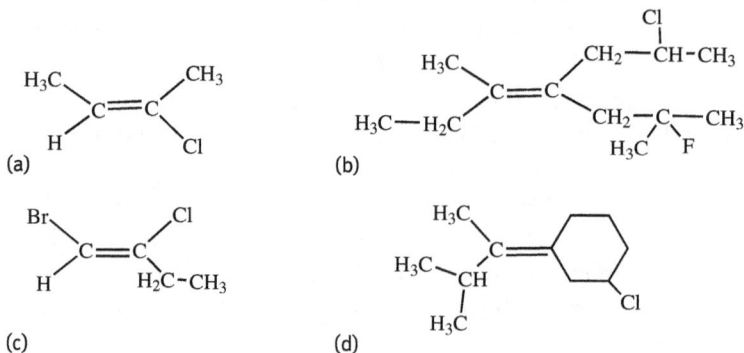

(a)

(b)

(c)

(d)

Q17-3. Assign full IUPAC names to the molecules shown in Q16-2 (a), (b), and (c)
Note what happens with a minor modification of the structure C.

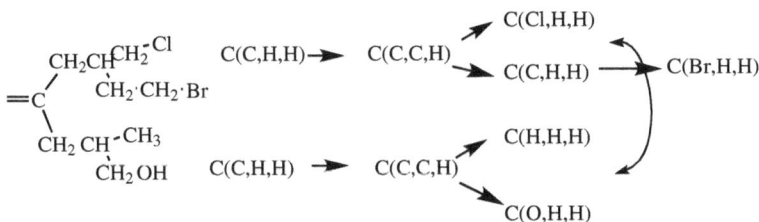

CH₂CH–CH₂·Cl   C(C,H,H)→   C(C,C,H)   C(Cl,H,H)
  CH₂·CH₂·Br   C(C,H,H) ——→ C(Br,H,H)
=C
  CH₂ CH–CH₃   C(H,H,H)
     CH₂OH   C(C,H,H) → C(C,C,H)
                                                 C(O,H,H)

Fig. 17.5

In this case, the lower CH₂OH group has been moved one carbon to the right. There is now no difference in the chains until after a branch point is passed. Rule #3 says take the path *in each chain* that leads to the first difference and then take the path to the higher one. This produces a C(Cl, H, H) in the upper one and a C(O,H,H) in the lower one. Comparison of these leads to the conclusion that the upper chain now has the higher priority.

Q17-4. Draw the structures that correspond to the following IUPAC names:
-   (E) 3-bromo-4-methyl-2-hexene
-   (2Z,4Z) 2-bromo-4-chloro-5-methyl-2,4-octadiene
-   (E) 7-chloro-4-chloromethyl-3,6,6-trimethyl-3-heptene

There are several questions in this chapter that test your skill in this area. Also, the methodology will be very useful when other kinds of stereochemistry are encountered.

The CIP system was actually designed to cope with a completely different stereo-chemical problem and was subsequently expanded to include the use you have just learned. The original intent was to unambiguously describe stereochemistry around single tetrahedral (sp³) carbon atoms. The concept of mirror images, superimposabil-ity, enantiomers, and diastereomers is beyond the scope of this text, but you will cer-tainly encounter it in lectures and in Part One of this text. However, the use of the CIP rules to *describe* such stereochemistry is a part of the nomenclature system and will be outlined next. It will be assumed that you have learned about and mastered the manipulation of a Fischer projection and the operations for converting three-dimensional drawings and Newman projections into Fischer projections.

A tetrahedral carbon attached to *four different* groups (note these are *not* neces-sarily single atoms) can exist in two possible arrangements that are mirror images of each other. One is left handed and the other is right handed. The system for defining the *absolute configuration* of one of these is as follows. The molecule is oriented such that the *lowest* priority substituent is behind the plane of the paper. If the order of priorities of the *remaining three groups* describes a right-handed (clockwise) arc, the carbon is designated as R (Rectus, Latin for right), whereas if a left-handed (counterclockwise) arc is described, the carbon is designated as S (Sinister, Latin for left).

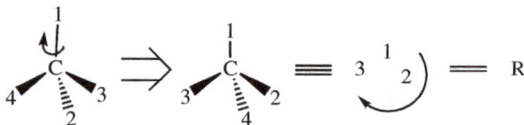

Fig. 17.6

You should practice using these ideas in the problems at the end of this chapter.

Two further additions to the Sequence Rules are necessary to cope with some fairly common situations.

4. When a multiply bonded atom is encountered, for purposes of determining the priority, that atom is expanded such that an atom with n bonds to the next atom of atomic number y is replaced by one with single bonds to n atoms of atomic number y.

Consider the following situations:

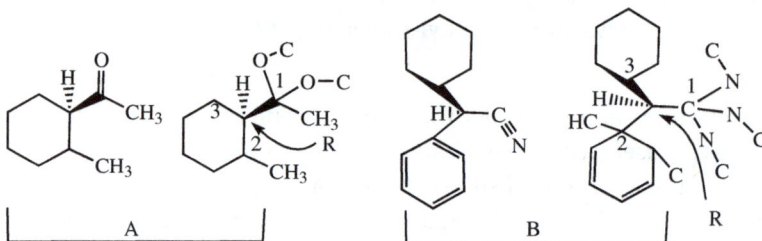

Fig. 17.7

In A, the carbon–oxygen double bond is "replaced" by two C–O single bonds, i.e., the C=O can be designated as C(O,O,C). This leads to the priority sequence shown. Similarly in B, the C≡N can be "replaced" with a carbon attached to three nitrogen atoms and the double bond of the phenyl substituent is "replaced" by a carbon singly bonded to two carbons. The designations of the three carbons attached to the chiral center are C(N,N,N), C(C,C,C), and C(C,C,H), which leads to the descriptor R as shown.

5. If such an expansion leads to a situation where a *real* carbon has the same substituents as an imaginary one, the real one takes precedence.

For example:

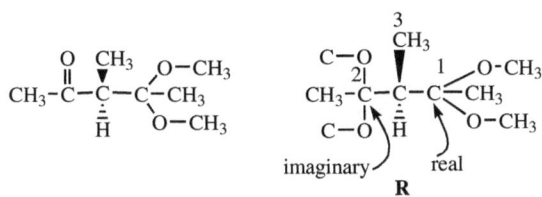

Fig. 17.8

Q17-5. Assign the priorities and the correct stereochemical descriptor to the central 🛈
carbon atom in the following structures

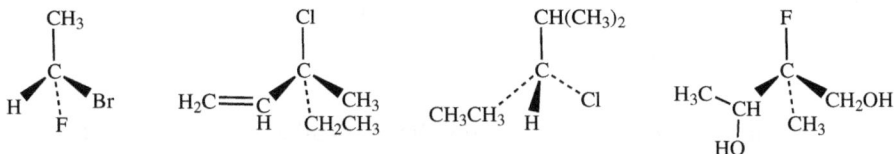

Using the same ideas as the system shown above, two centers C2 and C4, would both
be designated as C(O,O,C). However, one of these is imaginary and one is real, so the
real one has the higher priority of these two.

Summarizing Problems

17-1. Answer question #1 from the end of Chapter 16, but use the E/Z system of designating stereochemistry.

17-2. Use the E/Z nomenclature system to designate the stereochemistry of the double
bonds in the three compounds shown below. Provide complete IUPAC name for (a).

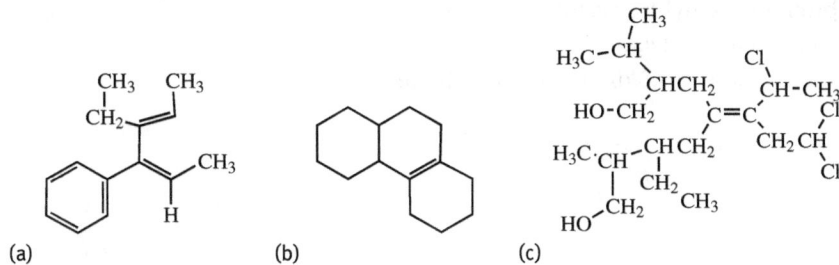

(a)   (b)   (c)

17-3. For the following compounds, provide complete stereochemical descriptors (E, Z, R, or S) for all centers and/or bonds that require them.

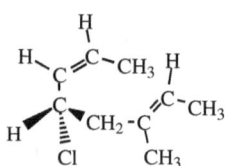

(a)

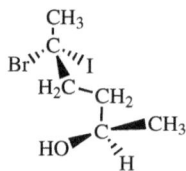

(b)

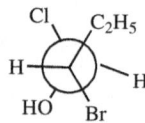

(c)

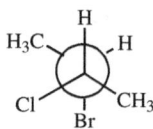

(d)

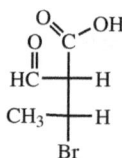

(e)

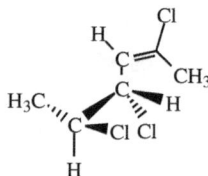

(f)

17-4. For the following compounds, provide complete IUPAC *names* including all required stereochemical descriptors (E, Z, R, or S).

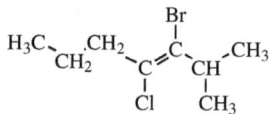

(a)

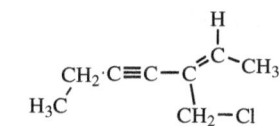

(b)

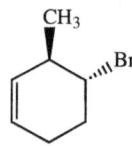

(c)

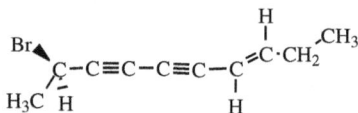

(d)

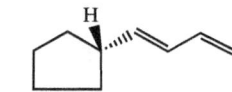

(e)

17-5. Draw the structure for each of the following compounds. Use a drawing that shows the indicated stereochemistry.
(a) *cis* 3-bromo-7-methyl-1,4-cycloheptadiene
(b) (2E,4E,6Z) 7-methyl-2,4,6-nonatriene
(c) (4R,7R,6E) 7-bromo-4,5-diethyldec-5-en-1-yne

18 Functional Groups

As you have learned, both the chemistry (reactions) and the systematic nomenclature of organic compounds are organized around groups of atoms called functional groups. Up to this point, all the types of compounds treated have been hydrocarbons (alkanes, alkenes, or alkynes) and derivatives of these where a hydrogen atom has been replaced by a substituent (e.g., chlorine). Now it is time to expand the scope and consider other common functional groups. Each has its own peculiarities, but in particular, we must learn the order of precedence of the functional groups – when more than one is contained in a single molecule, which of them controls the numbering system. The following list shows the major functional groups in order of precedence (highest to lowest):

- Acids
- Esters
- Acyl halides
- Amides
- Nitriles
- Aldehydes
- Ketones
- Alcohols
- Hydrocarbons

We will consider these in approximately reverse order.

Alcohols and Ethers

Alcohols are compounds formed by replacing one of the hydrogen atoms of water with a carbon atom. They contain the functional group OH. When this is the highest-ranking functional group in the molecule, the IUPAC system uses the ending "ol" to replace the final "e" of the name of the alkane from which it was formed. The general name for an alcohol in the IUPAC system is an *alkanol*. Other functional groups (e.g., double or triple bonds) can be indicated as suffixes, but the OH group is dominant in that it controls the numbering of the entire molecule. The following examples illustrate these points.

$CH_3CH_2CH_2CH_2OH$ **Fig. 18.1:** 1-butanol (4 C's, no unsaturation, alcohol)

CH_3CHCH_3
  |
  OH   **Fig. 18.2:** 2-propanol

https://doi.org/10.1515/9783110778311-019

$$CH_2=CH-\overset{\displaystyle OH}{\underset{\displaystyle |}{CH}}-CH_3$$ **Fig. 18.3:** 3-buten-2-ol

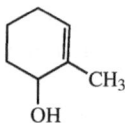

Fig. 18.4: 2-methylcyclohex-2-en-1-ol *or* 2-methyl-2-cyclohexen-1-ol

Note that in the last two examples, the numbering system has been changed from that seen when substituted alkenes (i.e., where there were no functional groups other than a double bond) were considered. This illustrates the hierarchy or relative order of importance of functional groups. The numbering system gives the lowest possible numbers to both functional groups, but the OH group ranks first in importance.

Q18-1. The IUPAC names (including stereochemical descriptors) for the three compounds shown below are _____ , _____ and _____ .

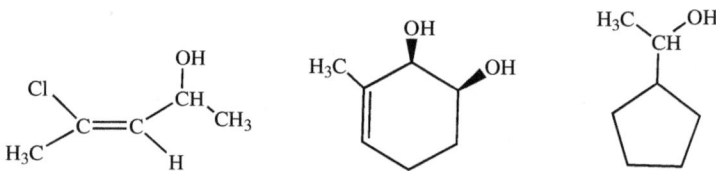

Alcohols, like haloalkanes, can be classified according to how many hydrogen atoms are attached to the carbon bearing the OH group. This carbon is called the *carbinol carbon*. If there are 2H's, the alcohol is *primary*, if there is only one hydrogen atom, the alcohol is *secondary*, and if there are no hydrogens on the carbinol carbon, the alcohol is *tertiary*.

Q18-2. Draw the structures that correspond to the following names:
(a) (E) 4,5-dibromo-3-penten-1-ol
(b) 5-ethyl-3,6-heptadien-2-ol
(c) 3-methyl-5-vinylhept-5-en-2-ol

Q18-3.
(a) Draw the structural formula for the secondary alcohol whose molecular formula is C_4H_9OH.
(b) Draw the structural formulas for the secondary alcohols with molecular formula $C_5H_{11}OH$. (There is more than one. Draw them all.)

If more than one OH group is present, the molecule is a diol, triol or tetrol (2, 3 or 4 OH groups). The final "e" is not dropped from the name of the parent alkane in these cases to aid in pronunciation.

Q18-4. Give the IUPAC names for the diols with the structures shown below:

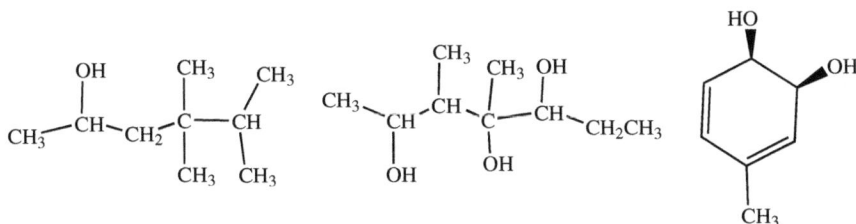

An older name for diols is *glycol*. Thus, the IUPAC name for the substance commonly used as automotive antifreeze ($HOCH_2CH_2OH$) is 1,2-ethanediol, whereas the common (trivial) name is ethylene glycol.

If an OH group is a *substituent*, and not the controlling functional group, the *prefix* "hydroxy" is used. This can occur when more important (in the nomenclature sense) functional groups are present and/or when the structure is complex enough to make it impossible to include all the OH groups in the main chain. For example, consider the following case

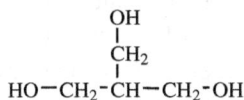

Fig. 18.5

The longest chain that contains functional groups is three and no matter how you number it, one of the OH groups is not on the main chain. The correct IUPAC name for this *triol* is 2-hydroxymethyl-1,3-propanediol.

Ethers

Ethers are compounds where *both* of the hydrogen atoms of water have been replaced by carbon atoms. There is no systematic ending for a substitutive name for ethers. The IUPAC system we are using designates the O–R as a *substituent* on the main chain and therefore it occurs in the name as a *prefix*. The term "alkoxy" is applied to the group –OR, where R is any alkyl group. Names like methoxy, ethoxy, propoxy, and 2-propoxy designate the groups CH_3O-, CH_3CH_2O-, $CH_3CH_2CH_2O-$, and $(CH_3)_2CHO-$. Ethers must have two alkyl chains attached to the oxygen atom, but only one of these can be the "main chain." As usual, it is selected as being the longest or the most heavily substituted if the lengths are the same. Also, the presence of functional groups in the chains must be considered when the main chain is being chosen. The chain that contains the highest-ranking functional group must control the name. Some examples are shown below.

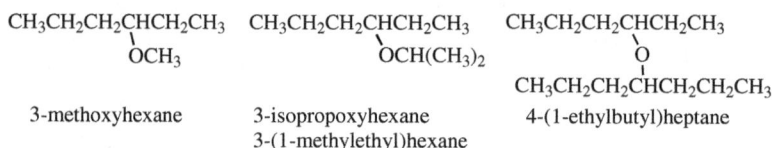

3-methoxyhexane   3-isopropoxyhexane   4-(1-ethylbutyl)heptane
                       3-(1-methylethyl)hexane

Fig. 18.6

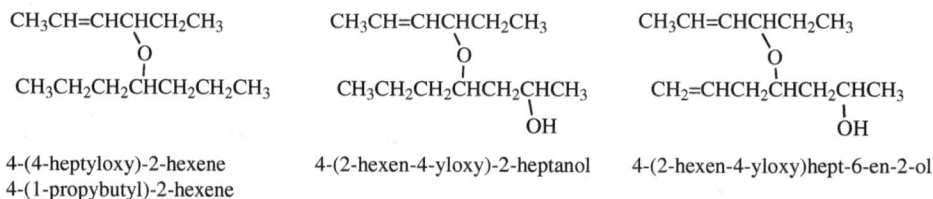

4-(4-heptyloxy)-2-hexene   4-(2-hexen-4-yloxy)-2-heptanol   4-(2-hexen-4-yloxy)hept-6-en-2-ol
4-(1-propybutyl)-2-hexene

Fig. 18.7

However, in the case of ethers, another IUPAC-approved type of name is most common – a so-called functional group name (see Introduction, Page 1 for a comment about this type of name). In these names, the *separate word* "ether" is preceded by the separate words (alphabetized) which describe the two alkyl groups attached to the oxygen atom.

Q18-5. Draw the structure that corresponds to the following names:
(a) *trans*-2-*cis*-4-cyclodecadien-1-ol
(b) *trans*-6-hydroxymethyl-2-methyldec-2-en-1,6-diol

Searching the later volumes of Chemical Abstracts will reveal that the nomenclature of ethers is treated in yet another way. Ethers are considered to be *hydrocarbons* where one of the methylene (CH_2) groups has been replaced by an oxygen atom. The *prefix* "oxa" denotes this substitution. For example,

 $CH_3CH_2CH_2CH_2CH_3$ = pentane

 $CH_3CH_2OCH_2CH_3$ = diethyl ether (functional group name – IUPAC approved)

 = ethoxyethane (substitutive name – IUPAC approved)

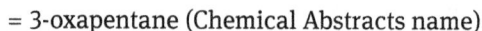 = 3-oxapentane (Chemical Abstracts name)

In cases where any functional group other than a double or triple bond is present, the substitutive name is used and the ether is treated as a substituent.

Aldehydes and Ketones

The next level of oxidation of oxygen-containing compounds are those containing the C=O (carbonyl) group. When this group is attached only to carbon or hydrogen, a set of two new functional groups is created.

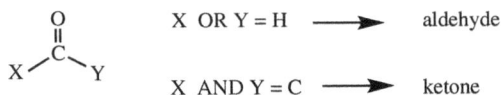

X OR Y = H ⟶ aldehyde

X AND Y = C ⟶ ketone

Fig. 18.8

The aldehyde group *must* come at the end of a chain and is frequently written as - CHO. Don't confuse this with an alcohol. If it was an alcohol, the hydrogen atom would be bonded to the oxygen and the carbon would have only two bonds! For the same reason, the notation COH should be avoided as this really looks as if the oxygen and the hydrogen atoms are joined. Ketones cannot come at the end of a chain and therefore the same ambiguity does not arise. The *generic* names for aldehydes and ketones are "alkanal" and "alkanone."

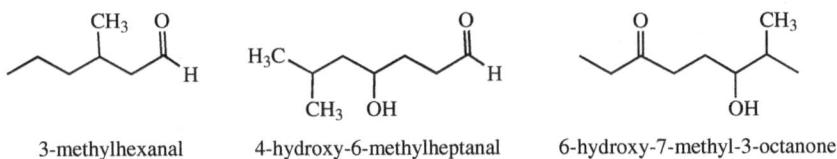

3-methylhexanal   4-hydroxy-6-methylheptanal   6-hydroxy-7-methyl-3-octanone

Fig. 18.9

Aldehydes & Ketones: Substitutive Names

Ketones and *acyclic* aldehydes are named by replacing the final "e" of the parent alkane with either "al" (for aldehyde) or "one" (rhymes with clone) (for ketones). (The distinction between "al" and "ol" is important and your handwriting must distinguish between them.)

According to the hierarchical list given at the beginning of this chapter, the position of aldehydes and ketones will determine the numbering system used for all molecules with functional groups we have seen so far. Since the aldehyde group must come at the end of a chain, it must be #1 and the numbering proceeds from that point. The position of ketone carbonyl groups must be specified and, in the usual way, is assigned the lowest possible number.

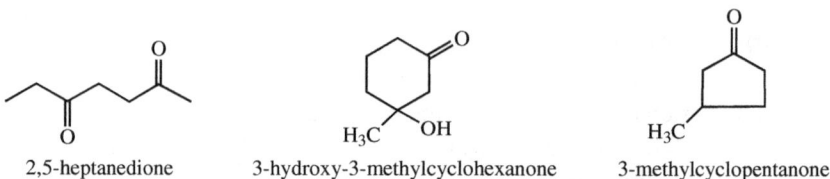

2,5-heptanedione   3-hydroxy-3-methylcyclohexanone   3-methylcyclopentanone

Fig. 18.10

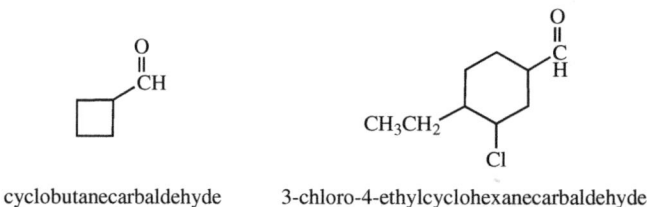

cyclobutanecarbaldehyde   3-chloro-4-ethylcyclohexanecarbaldehyde

Fig. 18.11

If the aldehyde group is attached to a ring, the compound is named by adding the word fragment "carbaldehyde" or "carboxaldehyde" (used by Chemical Abstracts) to the name of the ring compound. For example:

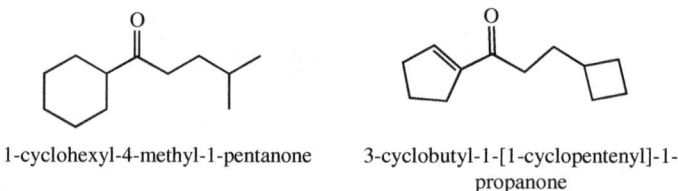

1-cyclohexyl-4-methyl-1-pentanone   3-cyclobutyl-1-[1-cyclopentenyl]-1-propanone

Fig. 18.12

Q18-6.
(a) Draw the structural formula (including stereochemistry) for 3-bromopentanal, 6-chloro-2-heptanone, Z 6-phenyl-5-heptenal.
(b) Give acceptable IUPAC names for the structures below.

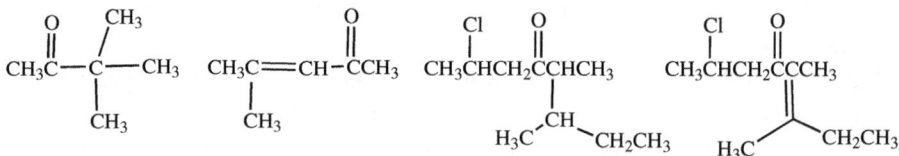

Because aldehydes and ketones are so closely related in both their structure and their reactions, there is a move towards combining their nomenclature. Chemical Abstracts, the publication that lists all chemical compounds known and where to locate their reactions and directions for their preparation, is moving towards naming both aldehydes and ketones using the "one" suffix. In this scenario, the aldehyde would be a "1-one." This is not yet in general use. However, for molecules like those shown below where the usual substitutive name is not obvious, the "1-one" nomenclature for aldehydes is taking hold.

For example:

Q18-7. Two substances produced by some species of ants have the structures shown below. Derive acceptable names for them.

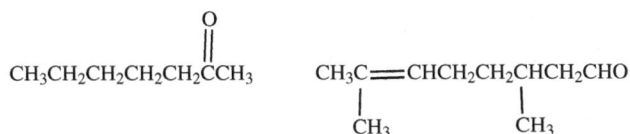

$$CH_3CH_2CH_2CH_2CH_2\overset{\overset{\displaystyle O}{\|}}{C}CH_3$$

$$CH_3C\!\!=\!\!\!=\!\!\!CHCH_2CH_2\underset{\underset{\displaystyle CH_3}{|}}{C}HCH_2CHO$$
$$\qquad\underset{\underset{\displaystyle CH_3}{|}}{}$$

Aldehydes and ketones are very common perfume and odor constituents. In addition, there are many of these types of molecules which are so common they are often best known by their trivial (nonsystematic) names. Four of these are:

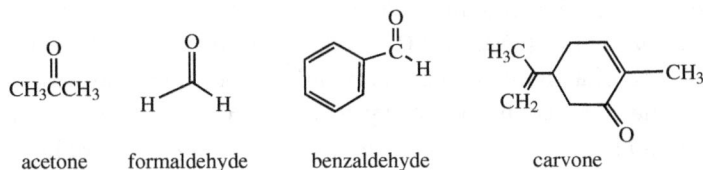

acetone   formaldehyde   benzaldehyde   carvone

Fig. 18.13

Acetone is a common solvent. Formaldehyde is a constituent of some plastics (urea-formaldehyde foam insulation for example) and is used in the preservation of biological specimens. It is known to have negative health implications for some people. Benzaldehyde has the odor of almonds.

Q18-8. Assign acceptable IUPAC names to the compounds with the trivial names acetone and formaldehyde.

Just as the carbonyl group "outranks" the OH group, there are functional groups that take precedence over aldehydes and ketones. When one of these is present in a molecule, the carbonyl group must be considered as a *substituent* and indicated in the *prefix* of the name. The prefix "oxo" indicates the presence of a doubly bonded oxygen atom. (Be very careful not to confuse "oxo" and "oxa." The first refers to a doubly bonded oxygen atom, whereas the latter refers to an oxygen atom singly bonded to two carbons. See the discussion of *ethers* for more on this.) For example:

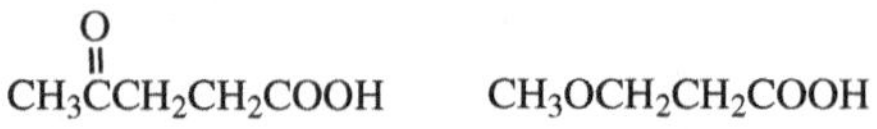

$$\underset{\text{4-oxopentanoic acid}}{CH_3\overset{\overset{\displaystyle O}{\|}}{C}CH_2CH_2COOH}$$ $$\underset{\substack{\text{4-oxapentanoic acid} \\ \text{(3-methoxypropanoic acid)}}}{CH_3OCH_2CH_2COOH}$$

Fig. 18.14

Aldehydes & Ketones: Functional Group Names

Ketones are sometimes named by multiple words ending in the word "ketone." So, for example, 2-butanone can be called methyl ethyl ketone and 3-methyl-2-butanone can be called methyl isopropyl ketone. However, this system is becoming obsolete and generally you should use the substitutive name.

Carboxylic Acids

The functional group which consists of a carbonyl group attached to an OH group is called a *carboxylic acid*. It has the structure –C(=O)OH, is called a carboxy or carboxyl group, and is often written as COOH or CO_2H. Like aldehydes, this functional group must come at the end of a chain (a chain-terminating group). The carboxylic acid group is the next-highest oxidation state of oxygen-containing molecules and it takes precedence in the numbering of chains over all the functional groups we have so far seen.

The IUPAC ending for substitutive names for carboxylic acids uses "oic acid" to replace the final "e" of the alkane name. Thus, names like butanoic acid, 2-methylpentanoic acid, and (Z) 3-hexenoic acid represent

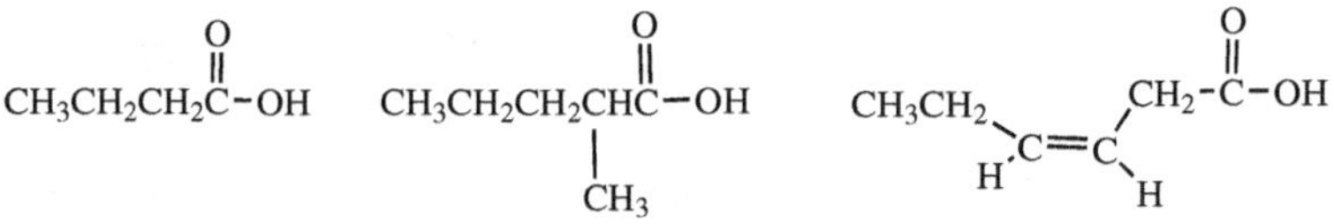

$$CH_3CH_2CH_2\overset{\overset{\displaystyle O}{\|}}{C}-OH \qquad CH_3CH_2CH_2\underset{\underset{\displaystyle CH_3}{|}}{CH}\overset{\overset{\displaystyle O}{\|}}{C}-OH \qquad \underset{H}{\overset{CH_3CH_2}{\diagdown}}C=C\underset{H}{\overset{CH_2-\overset{\overset{\displaystyle O}{\|}}{C}-OH}{\diagup}}$$

Fig. 18.15

Compounds that have two carboxylic acids in the main chain are named as "dioic acid" without dropping the final "e" of the alkane name (cf. alcohols and diols).

$$\text{HOOCCH}_2\text{CH}_2\text{COOH}$$

$$\overset{\displaystyle \text{CH}_3}{\underset{\displaystyle |}{\text{HOOCCH}_2\text{CH}_2\text{CHCH}_2\text{COOH}}}$$

butanedioic acid   3-methylhexanedioic acid   **Fig. 18.16**

If the carboxylic acid group is attached directly to a cyclic fragment, the nomenclature method changes in the same manner as previously outlined for aldehydes. The functional group COOH is denoted by attaching the words "carboxylic acid" to the name of the cyclic alkane. For example,

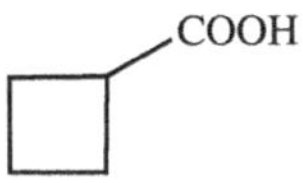

cyclobutanecarboxylic acid 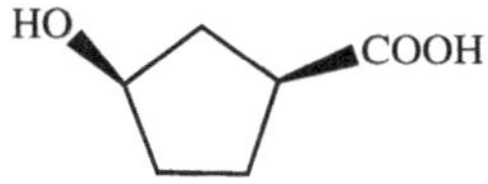   cis 3-hydroxy-cyclopentanecarboxylic acid 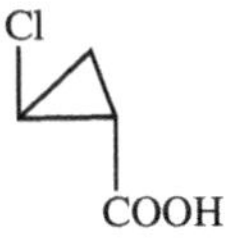   trans 2-chloro-cyclopropanecarboxylic acid

Fig. 18.17

Q18-9. Provide IUPAC names for the following compounds

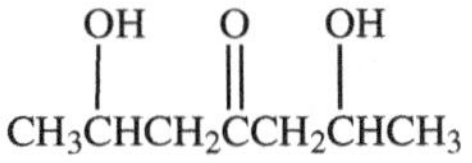

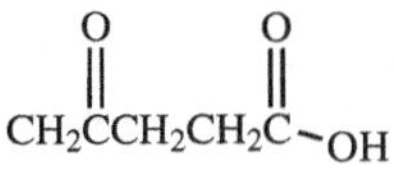

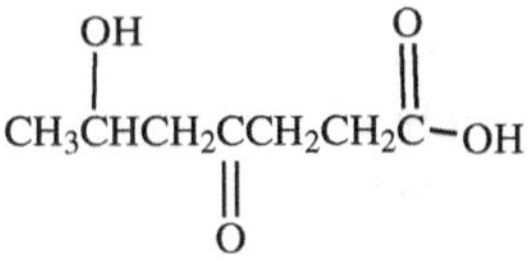

(a)
$$\underset{\text{CH}_3\text{CHCH}_2\text{CCH}_2\text{CHCH}_3}{\overset{\text{OH}\quad\text{O}\quad\text{OH}}{}}$$

$$\text{CH}_2\text{CCH}_2\text{CH}_2\text{C}-\text{OH} \quad (\text{O},\text{O})$$

$$\underset{}{\overset{\text{OH}}{\text{CH}_3\text{CHCH}_2\text{CCH}_2\text{CH}_2\text{C}-\text{OH}}}$$

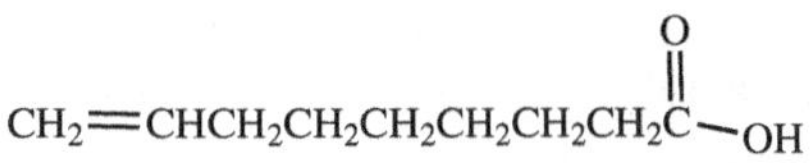

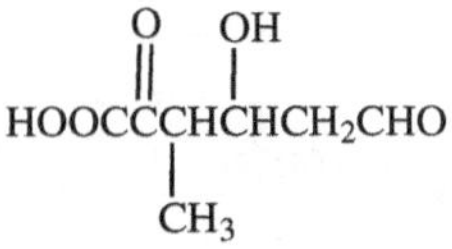

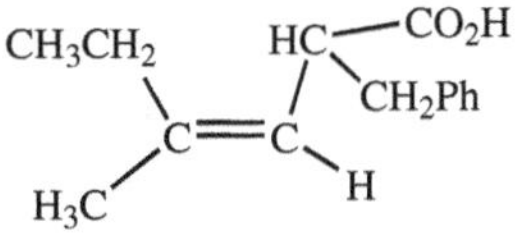

(b)
$$\text{CH}_2{=}\text{CHCH}_2\text{CH}_2\text{CH}_2\text{CH}_2\text{CH}_2\text{C}-\text{OH}$$

$$\underset{\text{CH}_3}{\overset{\text{O}\quad\text{OH}}{\text{HOOCCCHCHCH}_2\text{CHO}}}$$

$$\underset{\text{H}_3\text{C}}{\overset{\text{CH}_3\text{CH}_2}{}}\text{C}{=}\text{C}\underset{\text{H}}{\overset{\text{HC}-\text{CO}_2\text{H}}{\overset{}{\diagup}}\text{CH}_2\text{Ph}}$$

Q18-10. Draw the structures that correspond to the following names:
(a) 3,4-dichloro-5-hydroxy-2-heptanone
(b) 4-chloro-3-hydroxy-3,5-dimethyloctanoic acid
(c) 2-bromo-3,4,5-trihydroxypentanoic acid
(d) 3-methyl-4-oxo-heptanedioic acid
(e) 3-oxo-4-methylcyclohexanecarboxylic acid
(f) 3-carboxymethylpentanedioic acid

Esters and Other Acid Derivatives

Acids are one of a class of molecules with the general structure R–(C=O)X, where X = OH. The R–C=O fragment is known as an "acyl" group. (Contrast this with the term "alkyl group"). Other functional groups are generated by using other fragments for "X." These are often referred to as "acid derivatives." These include

X	Functional class name
OH	Carboxylic acid
OC	Ester
NH_2, NH–C or N–C_2	Amide
OC(=O)R	Anhydride
Cl	Acyl halide

In addition, the equivalent of a carboxylic acid using the element nitrogen rather than oxygen is the *nitrile* (R–C≡N). A nitrile can be formed using one nitrogen atom whereas the acids and esters must use two oxygen atoms.

Acid Anhydrides

Replacement of the acidic H atom in an acid with an acyl group leads to an *anhydride* (so called because it is formally derived from two molecules of acid with the elimination of a molecule of H_2O). Some anhydrides require two molecules of a monobasic acid (one which has only one acidic hydrogen atom), whereas molecules that contain two carboxylic acid functional groups in the same molecule can form an anhydride by forming a ring. The names of the latter type or those formed from two molecules of the *same* acid are constructed by replacing the word "acid" with the word "anhydride." These points are illustrated by the following examples.

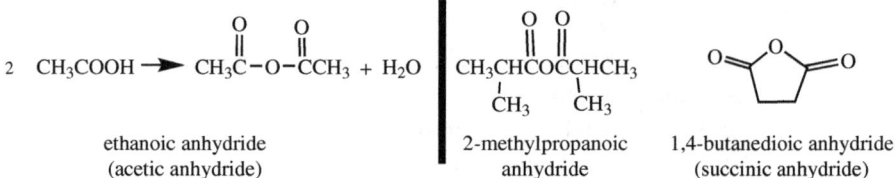

2 CH_3COOH ⟶ $CH_3\overset{O}{\overset{\|}{C}}-O-\overset{O}{\overset{\|}{C}}CH_3$ + H_2O

ethanoic anhydride
(acetic anhydride)

$CH_3CH\overset{O}{\overset{\|}{C}}O\overset{O}{\overset{\|}{C}}CHCH_3$
    | |
   CH_3 CH_3

2-methylpropanoic
anhydride

1,4-butanedioic anhydride
(succinic anhydride)

Fig. 18.18

Acid Halides (Acyl Halides)

When X = halide (almost always Cl), the molecular class is called an acid halide or, more properly an acyl halide. The nomenclature for acyl halides starts with the name of the corresponding acid. This is modified by dropping the "ic" ending of "oic" and replacing it with "yl" and adding the word "chloride," "bromide", etc. to describe the particular halide that is present. For example:

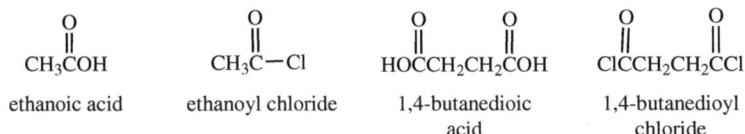

$$CH_3\overset{O}{\overset{\|}{C}}OH \qquad CH_3\overset{O}{\overset{\|}{C}}-Cl \qquad HO\overset{O}{\overset{\|}{C}}CH_2CH_2\overset{O}{\overset{\|}{C}}OH \qquad Cl\overset{O}{\overset{\|}{C}}CH_2CH_2\overset{O}{\overset{\|}{C}}Cl$$

ethanoic acid   ethanoyl chloride   1,4-butanedioic   1,4-butanedioyl
                                                 acid   chloride

Fig. 18.19

Q18-11. Provide an acceptable IUPAC name for the acids shown and also the acyl chloride and anhydride derived from them.

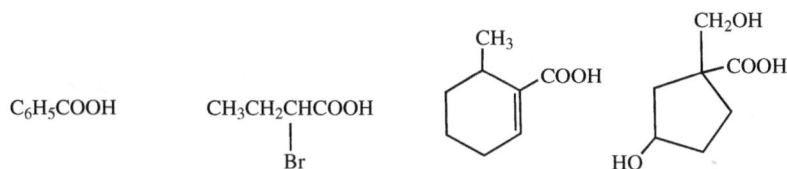

C_6H_5COOH $CH_3CH_2CHCOOH$
                            |
                            Br

Esters

Replacement of the acidic hydrogen atom in acids with a carbon atom generates the functional group known as an ester. (Note that there is no "H" in the word ester!) The nomenclature of esters is one that causes students of organic chemistry the most problems, but it is not difficult if the basic principles are known and followed. Consider the following structure.

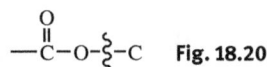

$$-\overset{O}{\overset{\|}{C}}-O-\overset{\xi}{\underset{}{}}-C \qquad \textbf{Fig. 18.20}$$

To generate the name of an ester, the ending on the name of the corresponding *acid* is changed from "oic acid" to "oate." This defines the left half of the molecule. The identity of the part of the molecule to the right of the wavy line is specified by using the corresponding *substituent* name which ends in "yl." This part of the ester name is placed FIRST and stands as a separate word.

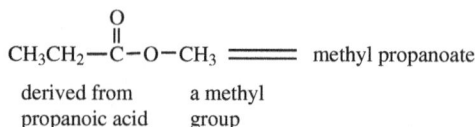

$$CH_3CH_2-\overset{\overset{\displaystyle O}{\|}}{C}-O-CH_3 \equiv\text{methyl propanoate}$$

derived from   a methyl
propanoic acid   group

Fig. 18.21

As you will have learned, the ester group is normally formed by reaction of an acid and an alcohol. The *actual bond that is formed* when an ester is made in this way is *not* the O–C bond to the alkyl group, but rather the bond between the C(O)– and the O–C (the acyl oxygen–carbon bond).

When more complex acids or alkyl groups are encountered, the rules do not vary but the names can *look* more complex.

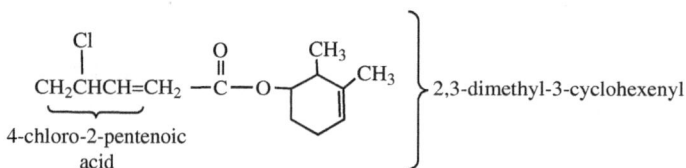

4-chloro-2-pentenoic acid   2,3-dimethyl-3-cyclohexenyl

Fig. 18.22

The full IUPAC name of this compound would be 2,3-dimethyl-3-cyclohexenyl 4-chloro-2-pentenoate.

Q18-12. Give IUPAC names for the following molecules

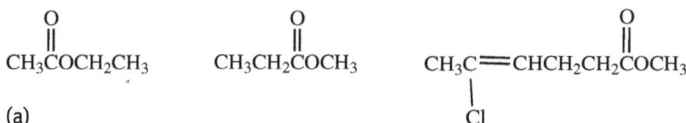

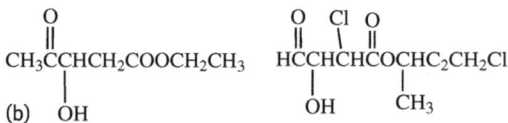

(a) $CH_3\overset{\overset{\displaystyle O}{\|}}{C}OCH_2CH_3$ $CH_3CH_2\overset{\overset{\displaystyle O}{\|}}{C}OCH_3$ $CH_3C\!\!=\!\!CHCH_2CH_2\overset{\overset{\displaystyle O}{\|}}{C}OCH_3$
$\underset{\displaystyle Cl}{|}$

(b) $CH_3\overset{\overset{\displaystyle O}{\|}}{C}CHCH_2COOCH_2CH_3$ $H\overset{\overset{\displaystyle O}{\|}}{C}CH\overset{\overset{\displaystyle Cl}{\backslash}}{C}H\overset{\overset{\displaystyle O}{\|}}{C}OCHC_2CH_2Cl$
    $\underset{\displaystyle OH}{|}$ $\underset{\displaystyle OH}{|}$ $\underset{\displaystyle CH_3}{|}$

Nitriles

Oxygen is always divalent. Therefore, a carbon atom that has one substituent and the rest of the valencies attached to oxygen requires *two* oxygen atoms – as in an ester or carboxylic acid. However, nitrogen can form *three* bonds (e.g., NH_3) and therefore only *one* nitrogen atom is needed to fulfill the role of the two oxygens. A molecule with the structure R–C≡N is called a *nitrile*. It is the nitrogen equivalent of a carboxylic acid.

However, note that the carbon atom of a nitrile is sp-hybridized, whereas that of an acid is sp^2-hybridized.

Naming nitriles is very straightforward and reflects the relationship with acids. The name of the corresponding acid is shortened by dropping the "ic" and adding the fragment "nitrile." Thus, the IUPAC name of the molecule $CH_3CH_2C\equiv N$ is propionitrile and that of $(CH_3)_2CHCH_2CH_2C\equiv N$ is 4-methylpentanonitrile.

Q18-13. Draw structures for the following names:
(a) cyclohexyl methanoate
(b) benzyl 2-methylpropanoate
(c) 2-chloroethyl 3-chloropropanoate
(d) 3-oxobutyl 5-methyl-5-hydroxyheptanoate

Summarizing Problems

18-1. Derive IUPAC names for the following compounds. (Note that line drawings are now being used. Remember what these mean. Each point represents a carbon with enough hydrogen atoms attached to bring the total valence of the carbon to *four*.)

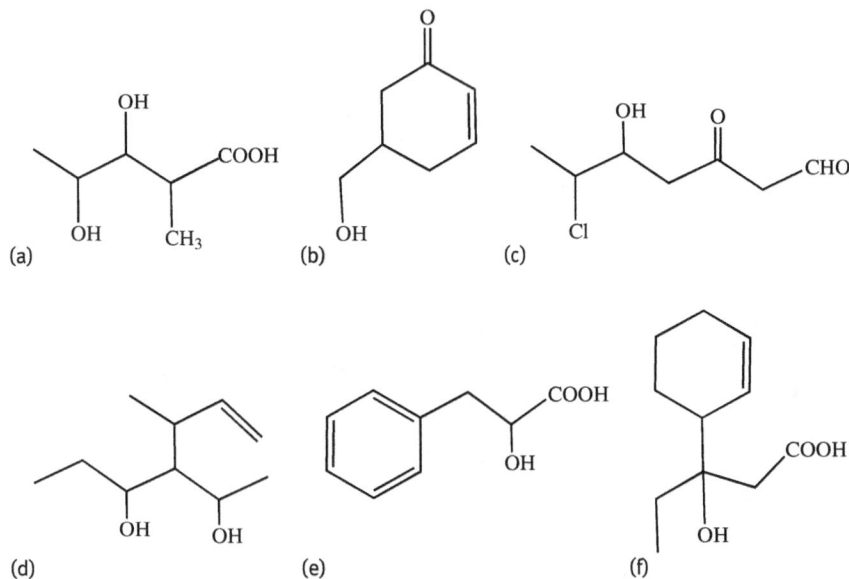

(a) (b) (c)

(d) (e) (f)

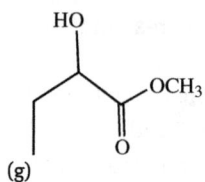
(g)

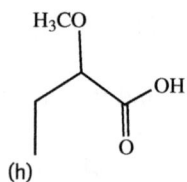
(h)

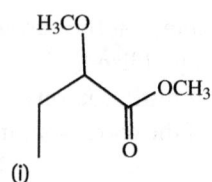
(i)

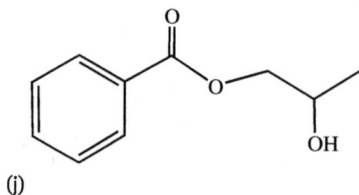
(j)

19 Aromatic Compounds

Introduction

The class of compounds referred to as "aromatic" is extremely large. Because of their stability, they were some of the earliest known and most investigated molecules. Much of the early work on aromatics was carried out long before the idea of a systematic nomenclature system evolved. Therefore, trivial (nonsystematic) names abound. Names such as toluene (methylbenzene), xylene (dimethylbenzene), aniline, and cresol can be cited as examples. Incorporation of one or more heteroatoms in the basic ring structure leads to literally hundreds of thousands of "heteroaromatic" compounds, each with its own characteristic problems of nomenclature. In this short chapter, only the simplest examples of aromatic compounds will be considered. Toward the end of the chapter, compounds called phenols and anilines will be introduced to give you an idea of how more complex systems are handled.

Hydrocarbons

Benzene (C_6H_6) is the parent aromatic hydrocarbon. Since all carbons and hydrogens in benzene are equivalent, it does not matter where on the ring a substituent is placed. The same compound will be generated so no number is needed in the name. Also, since all the carbons in benzene are sp^2-hybridized, there are no stereochemical details to worry about. The molecules are *flat*.

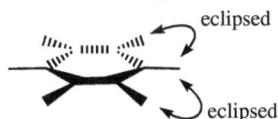

Fig. 19.1

Attaching an alkyl group to the benzene ring gives an *alkylbenzene* (e.g., methylbenzene, ethylbenzene, etc.). When two or more substituents are present on the same ring, their relationship must be specified. The best way to do this is using numbers in the same way that is used for cyclohexane. However, an older system is still very much in use and you should be aware of it. For *disubstituted* benzene rings the terms *ortho*, *meta*, and *para* (usually abbreviated *o*, *m*, or *p*) are used to designate the molecules whose substituents are separated by 0, 1, or 2 ring carbon atoms, respectively. Thus, *o*-disubstitued rings would be numbered 1,2-, *m*-disubstituted as 1,3-, and *p*-disubstituted as 1,4-. Some examples of this are shown below.

https://doi.org/10.1515/9783110778311-020

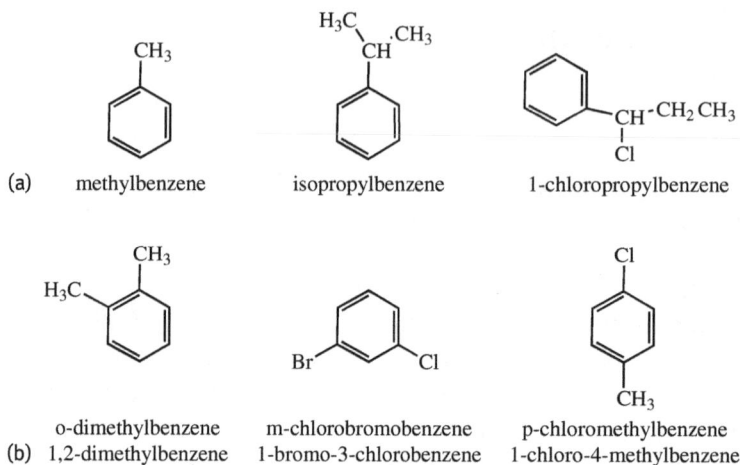

(a)   methylbenzene   isopropylbenzene   1-chloropropylbenzene

o-dimethylbenzene   m-chlorobromobenzene   p-chloromethylbenzene
(b) 1,2-dimethylbenzene   1-bromo-3-chlorobenzene   1-chloro-4-methylbenzene

Fig. 19.2

Note that the orientation of the drawing of the ring does not matter. It is the *relationship of the two substituents* that determines the o, m, or p prefix.

Benzene as a Substituent

If a benzene must be considered as a *substituent* on a chain, as is the case when other functional groups of greater precedence are present, the fragment C_6H_5 is called a *phenyl* group (cf. methyl, ethyl, etc.). The word "benzyl" is used for another purpose. As has already been mentioned (Chapter 2), the substituent $C_6H_5CH_2-$ is designated by this word. Some examples that illustrate these points are shown next.

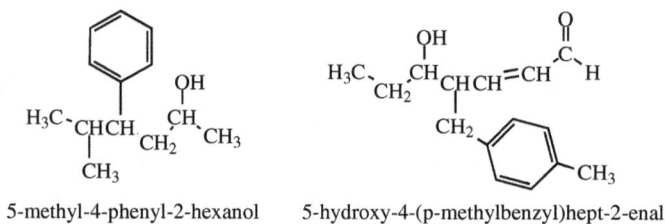

5-methyl-4-phenyl-2-hexanol   5-hydroxy-4-(p-methylbenzyl)hept-2-enal

Fig. 19.3

The use of the short form "Ph" for phenyl is very common and therefore the structure shown above left might be abbreviated as

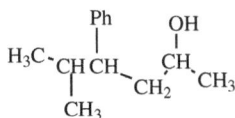

Ph   OH
 | |
H₃C‑CHCH   CH
    | ⟍CH₂⟍CH₃
   CH₃

Fig. 19.4

The foregoing structures also illustrate the point that the treatment of phenyl and benzyl groups as substituents is not different to that of (for example) methyl or butyl groups. The normal hierarchy of functional groups is maintained as is the numbering system.

Functionalized Benzene Rings

Consideration of situations where *functional groups* are attached *directly* to a benzene ring show that the IUPAC system becomes much more complex. Most of this material is beyond the scope of this little text. However, three situations are important enough to be treated.

When a benzene ring is attached to an aldehyde or carboxylic acid, the word fragment "benz" is used. This fragment is combined with "aldehyde" or "oic acid" to give the names of the two compounds shown below. Further modification of the name benzoic acid to give the names of esters, nitriles, etc., occurs in exactly the same manner as for other carboxylic acids.

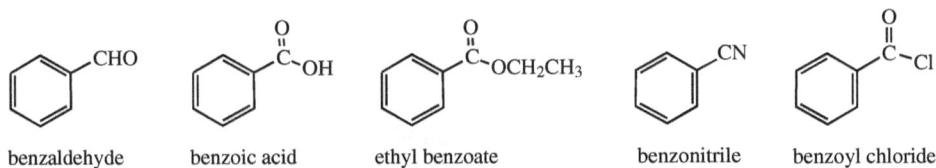

benzaldehyde   benzoic acid   ethyl benzoate   benzonitrile   benzoyl chloride

Fig. 19.5

(Note the difference between the word fragments "benzyl" [$C_6H_5CH_2$] and "benzoyl" [$C_6H_5C=O$]. It is a very important distinction!)

When a hydroxyl group is attached directly to the benzene ring, a class of compounds called *phenols* is generated. Also if the group attached directly is an amino group (NH_2), the class of compounds is called *anilines*.

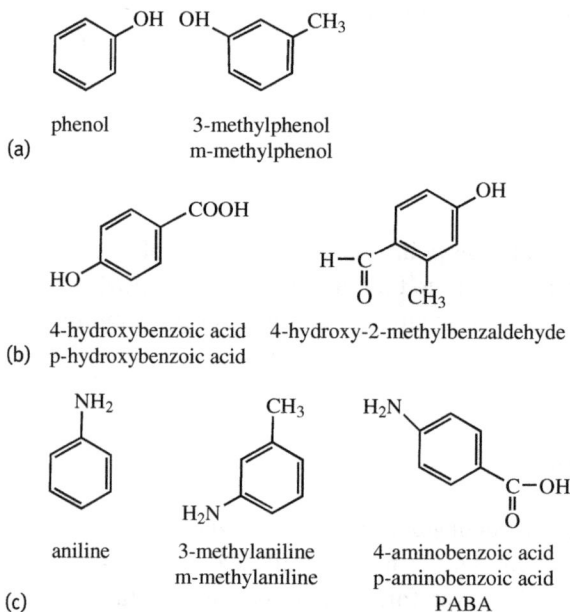

phenol

(a)

3-methylphenol
m-methylphenol

4-hydroxybenzoic acid   4-hydroxy-2-methylbenzaldehyde

(b) p-hydroxybenzoic acid

aniline

(c)

3-methylaniline
m-methylaniline

4-aminobenzoic acid
p-aminobenzoic acid
PABA

Fig. 19.6

Phenols (*don't confuse this word with phenyl*) are much more acidic than simple alcohols. This can be explained by resonance. See if you can develop the explanation. For the same reason, anilines are much weaker bases than simple amines and ammonia.

Summarizing Problems

19-1. Draw structures for the following compounds:
(a) *p*-methoxymethylbenzene
(b) *m*-bromomethylbenzene
(c) *o*-isopropylmethylbenzene
(d) *p*-methylphenol
(e) *o*-methylbenzoic acid
(f) *m*-bromobenzaldehyde
(g) *p*-methylbenzyl *o*-chlorobenzoate

19-2. Provide alternate names for each of the molecules given in question #1. *Do not* use the ortho, meta, and para system.

19-3. Show the difference between *m*-chloromethylbenzene and chloromethylben-zene. What is another name for the latter?

19-4. Give acceptable IUPAC names for each of the following structures. Use both the numbering and the *o*, *m*, *p*, system where this is possible.

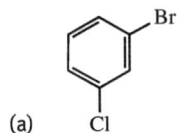

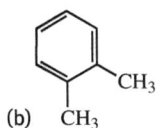

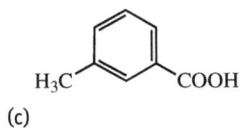

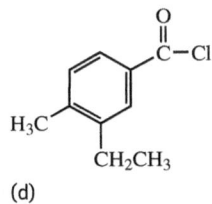

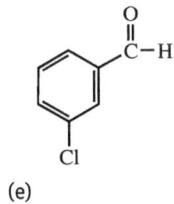

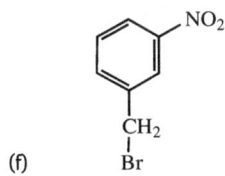

(a)

(b)

(c)

(d)

(e)

(f)

Appendix: Answers to Problems From Part II

Chapter 15

Problems From Text

Q15-1.

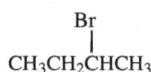

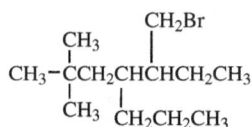

Q15-2.

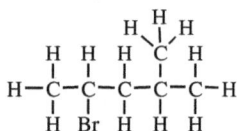

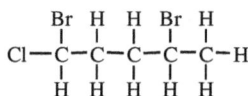

Q15-3. (a) dec   (b) oct   (c) hex

Q15-4.
(a) Normal alkanes: a
(b) Not alkanes: e,g
(c) Unsaturated hydrocarbons: e,g
(d) Alicyclic compounds d,f,g

Q15-5.
(a) Longest chain = 6: 3,4-dimethylhexane
(b) Longest chain = 8: 2-methyloctane
(c) Longest chain = 5: 2,2,4-trimethylpentane

Q15-6.
2,2-dimethylpentane
4-bromo-3,3-dimethylhexane
2,2-dichloro-3,3-dimethylbutane

https://doi.org/10.1515/9783110778311-021

Q15-7.

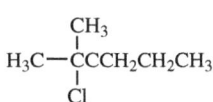 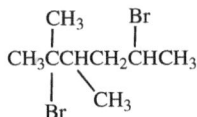 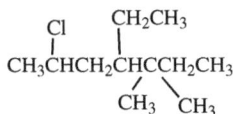

Q15-8. There are 18 hydrogen equivalents (15 H + 1 Cl + 2 Br). According to the formula C_nH_{2n+2}, there must be 8 carbons.

Q15-9.
(a) 3-chloro-2-methylhexane (numbered wrong way)
(b) 2-chloro-4-methylpentane (chlorine is lower in alphabet, therefore takes precedence when all else is equal)
(c) 2,5,5-trimethylheptane (numbered wrong way)
(d) 7-bromo-4-chloro-3-ethyl-2-methyloctane (use most substituted main chain)

Summarizing Problems

15-1.
(a) 4-ethyl-2,5-dimethyloctane
(b) 2-bromo-5-chloro-6-methyl-4-propyloctane (use most substituted chain as the root when more than one chain of same length)
(c) 4-chloro-3-ethyl-2-methylheptane
(d) 6-bromo-4-iodo-2,3-dimethyloctane
(e) 3-bromo-5-(1-chloroethyl)-2,4-dimethyloctane

15-2.

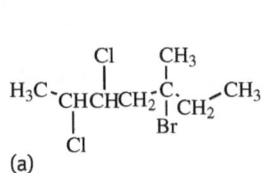 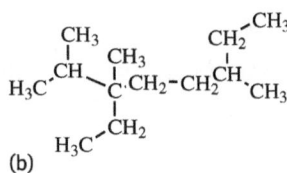

(a)   (b)

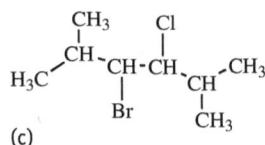 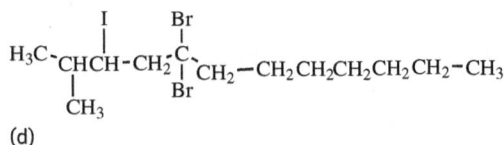

(c)   (d)

15-3.

(a) The proper name is 2,5-dimethylheptane. (The chain has been numbered the wrong way)

(b) The proper name is 2,5-dichloro-3-methylhexane. (The chain has been numbered the wrong way. Numbering in either direction gives #2 to the chlorines, so the decision has to be made by giving the lowest possible number to the *methyl* group)

(c) The proper name is 5-bromo-5-chloro-2,4-dimethylheptane. (The chain has been numbered the wrong way)

(d) The correct name is 4-(1-chloroethyl)octane. (The longest contiguous chain is *eight*)

(e) The correct name is 2-bromo-6-iodo-4,4-dimethylheptane. (When everything else is equal, the substituent first in an alphabetized list gets the lowest number)

Chapter 16

Problems From Text

Q16-1.

(a) hept

(b) pent

(c) pent

Q16-2.

(a) 4-bromo-2-pentene

(b) 2,5-dimethyl-1,4-hexadiene

(c) 5-bromo-5-chloro-2-methyl-2-pentene

(d) 2,3-dimethyl-1,4-pentadiene

Q16-3.

(a) 3-methyl-4-(2-propen-1-yl)-2,6-octadiene (use most substituted chain as the main chain)

(b) 6-(1-bromoethenyl)-3-chloronon-2-en-7-yne

Q16-4.

2(a) *trans*; 2(b) neither bond requires descriptor; 2(c) none; 2(d) none

3(a) *trans, trans* 3-methyl-4-(2-propen-1-yl)-1,2-octadiene

3(b) (E) 6-(1-bromoethenyl)-3-chloronon-2-en-7-yne

Q16-5.

cis 9-bromo-3,9-dichloro-4-isopropyl-4-methylnon-2-en-6-yne

cis 4-bromo-4-chloro-3,3,5,5-tetramethyloct-6-en-1-yne

Summarizing Problems

16-1.

(a) *trans* 5,6-dimethyl-2-heptene	C=C, not substituents, take precedence in numbering
(b) *trans* 4-bromo-2-(1-chloroethyl)-1,3-pentadiene	C=C's take precedence
(b) *trans* 4-bromo-2-(1-chloroethyl)-1,3-pentadiene	
(c) *trans* 7-bromo-6-chloro-2-hepten-4-yne	
(d) 1-bromo-3-chloro-5-methylcyclohexane	
(e) *cis* 7-bromo-6-chlorooct-4-en-2-yne	compare with (c).
(f) 3-bromo-1-methylcyclohexene	number through the double bonds and give first substituent lowest possible number
(g) 5-chloro-2-methyl-1,3-cyclohexadiene	
(h) 6-chloro-1-methylcyclohexene	
(i) 2-(2-chloropropyl)-1,3-cyclopentadiene	"1,3" is not required
(j) 5-bromo-7-chloro-2-methylundec-2-en-9-yne	

16-2.

Formula	IHD
(a) C_9H_{18}	1
(b) $C_7H_{10}BrCl$	2
(c) C_7H_8BrCl	3
(d) $C_7H_{12}BrCl$	1
(e) $C_8H_{10}BrCl$	3
(f) $C_7H_{11}Br$	2
(g) C_7H_9Cl	3
(h) $C_7H_{11}Cl$	2
(i) $C_8H_{11}Cl$	3
(j) $C_{12}H_{18}ClBr$	3

16-3.

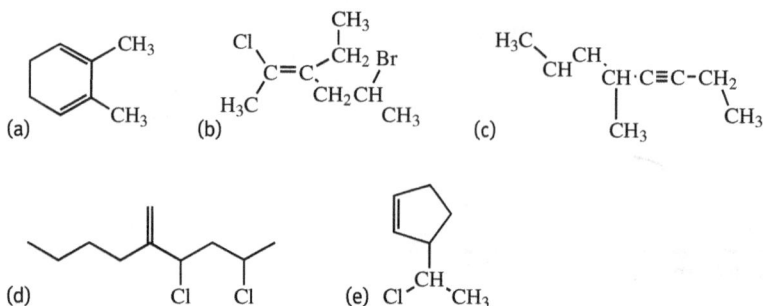

(a)

(b)

(c)

(d)

(e)

16-4.
(a) 1-chloro-6-methylcyclohexene (Numbered wrong way. Substituted end of C=C must get lowest number and other end of C=C gets #2)
(b) 4-bromo-2-methyl-2-hexene (No *trans* isomer is possible because both substituents on one end are the same)
(c) *trans* hept-4-en-2-yne (Numbered in wrong direction. The C=C takes precedence only when same locants are found)
(d) 3-propyl-1,4-hexadiene (The main chain must contain *both* functional groups if this is possible. Note that "vinyl" is the name of the *substituent* with the structure $CH_2=CH-$)

16-5.

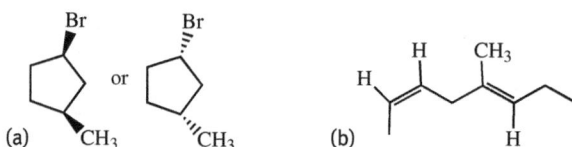

(a)

or

(b)

16-6. There are five isomers of these types. They are:
1-chloro-1-methylcyclobutane
cis 1-chloro-2-methylcyclobutane
cis 1-chloro-3-methylcyclobutane
trans 1-chloro-2-methylcyclobutane
*trans*1-chloro-3-methylcyclobutane

Chapter 17

Problems From Text

Q17-1. The sequence should be:
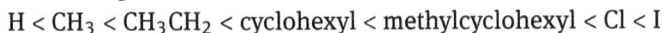
H < CH$_3$ < CH$_3$CH$_2$ < cyclohexyl < methylcyclohexyl < Cl < I

Q17-2. (a) E; (b) E; (c) Z; (d) Z

Q17-3.
(a) E 2-chloro-2-butene
(b) 4-(2-chloropropyl)-6-fluoro-3,6-dimethyl-3-heptene (use most substituted main chain)
(c) Z 1-bromo-2-chloro-1-butene

Q17-4.

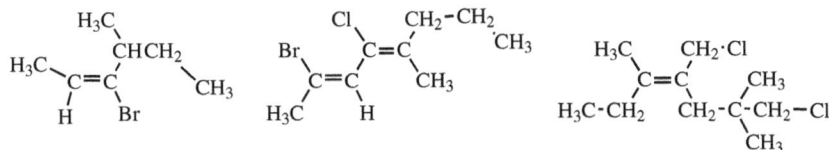

Q17-5.
(a) Priorities Br > F > CH$_3$ > H; configuration = S
(b) Priorities Cl > C= > CH$_2$CH$_3$ > CH$_3$; configuration = R
(c) Priorities Cl > CH(CH$_3$)$_2$ > CH$_2$CH$_3$ > H; configuration = R
(d) Priorities F > CH(OH) > CH$_2$(OH) > CH$_3$; configuration = S

Summarizing Problems

17-1.
(a) (E) 5,6-dimethyl-2-heptene
(b) (Z) 4-bromo-2-(1-chloroethyl)-1,3-pentadiene)
(c) (E) 7-bromo-6-chloro-2-hepten-4-yne
(e) (Z) 7-bromo-6-chlorooct-4-en-2-yne

17-2.
(a) (2E,4E) 3-ethyl-4-phenyl-2,4-hexadiene
(b) The double bond is Z
(c) The double bond is Z

17-3.
(a) The complete name is (2E, 5S, 6Z) 5-chloro-3-methyl-2,6-octadiene.
(b) The complete name is (2R,5R) 5-bromo-5-iodo-2-hexanol.
(c) The stereochemical descriptors are both S.
(d) The stereochemical descriptors are S for front carbon and S for the back one.
(e) The stereochemical descriptors are both R. (Note that the carbon bonded to Br is highest priority!)
(f) The double bond is E and the chiral centers are both S.

17-4.
(a) (E) 3-bromo-4-chloro-2-methylhept-3-ene
(b) (Z) 3-chloromethylhept-2-en-4-yne
(c) (3R,4R) 4-bromo-3-methylcyclohexene (note if the stereochemical drawing represents only *relative* configuration, the descriptor would be *trans* or E. The priorities around C3 are C–Br > C => C–CH₃ > H)
(d) (E, 9R) 9-bromodec-3-en-5,7-diyne
(e) (E) 1-cyclopentyl-1,3-butadiene

17-5.

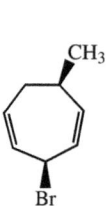

a)

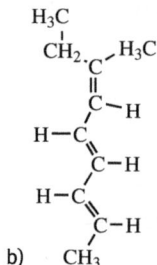

b)

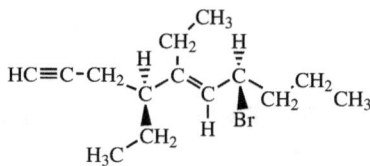

c)

Chapter 18

Problems From Text

Q18-1.
Z 4-chloro-3-pentene-2-ol
cis 3-methyl-3-cyclohexene-1,2-diol
1-cyclopentylethanol

Q18-2.

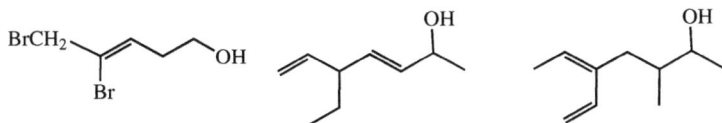

Q18-3.

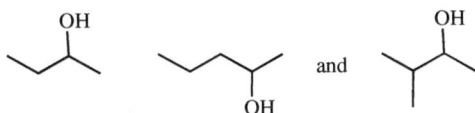

and

Q18-4.
4,5-dimethylhexane-2,4-diol
3,4-dimethylheptane-2,4,5-triol
4-methylcyclohexa-3,5-diene-1,2-diol

Q18-5.

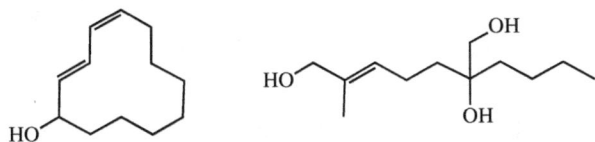

Q18-6.
(a)

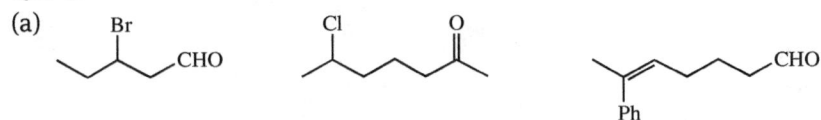

(b) 3,3-dimethyl-2-butanone
    4-methyl-3-penten-2-one
    2-chloro-5,6-dimethyl-4-octanone
    (E) 2-chloro-5,6-dimethyloct-5-en-4-one

Q18-7.
2-heptanone
3,7-dimethyl-6-octenal

Q18-8.
2-propanone
methanal (or methanone)

Q18-9.
2,6-dihydroxy-4-heptanone
4-oxopentanoic acid
6-hydroxy-4-oxoheptanoic acid
8-nonenoic acid
4-hydroxy-3-methyl-2,6-dioxohexanoic acid
Z 2-benzyl-4-methylhexanoic acid

Q18-10.

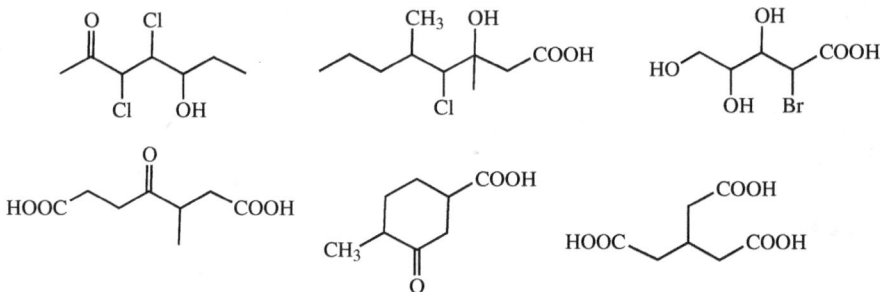

Q18-11.
benzoic acid
2-bromobutanoic acid
6-methyl-1-cyclohexenecarboxylic acid
3-hydroxy-1-hydroxymethylcyclopentanecarboxylic acid

Q18-12.
ethyl ethanoate
methyl propanoate
methyl 5-chloro-4-hexenoate
ethyl 3-hydroxy-4-oxopentanoate
4-chlorobut-2-yl 2-chloro-3-hydroxy-4-oxobutanoate

Q18-13.

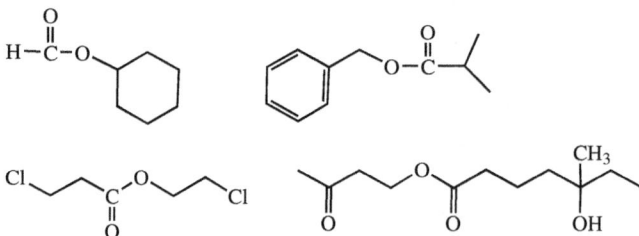

Summarizing Problems

18-1.
(a) 3,4-dihydroxy-2-methylpentanoic acid
(b) 5-hydroxymethyl-2-cyclohexenone
(c) 6-chloro-5-hydroxy-3-oxoheptanal [or 6-chloro-5-hydroxy-1,3-heptanedione]
(d) 3-(3-buten-2-yl)hexan-2,4-diol [or 3-(1-methyl-2-propen-1-yl]hexan-2,4-diol
(e) 2-hydroxy-3-phenylpropanoic acid
(f) 3-(2-cyclohexen-1-yl)-3-hydroxypentanoic acid
(g) methyl 2-hydroxybutanoate
(h) 2-methoxybutanoic acid
(i) methyl 2-methoxybutanoate
(j) 2-hydroxypropyl benzoate

Chapter 19

Summarizing Problems

19-1.

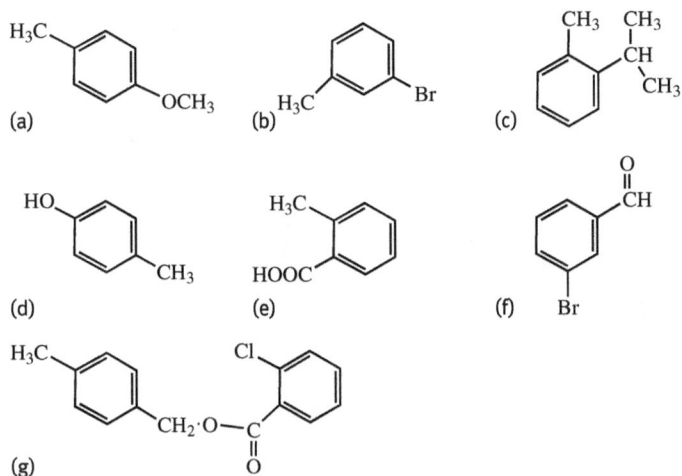

(a) (b) (c)

(d) (e) (f)

(g)

19-2.
(a) 1-methoxy-4-methylbenzene
(b) 1-bromo-3-methylbenzene
(c) 2-isopropyl-1-methylbenzene

(d) 4-methylphenol
(e) 2-methylbenzoic acid
(f) 3-bromobenzaldehyde
(g) 4-methylbenzyl 2-chlorobenzoate

(*Note*: In all the above cases, the number "1" could have been omitted.)

19-3.

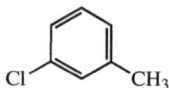

m-chloromethylbenzene   chloromethylbenzene

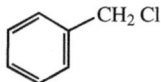

An alternate name for chloromethylbenzene is benzyl chloride.

19-4.
(a) *m*-bromochlorobenzene *or* 1-bromo-3-chlorobenzene
(b) 1,2-dimethylbenzene *or* o-dimethylbenzene (trivial name = *o*-xylene)
(c) *m*-methylbenzoic acid *or* 3-methylbenzoic acid
(d) 3-ethyl-4-methylbenzoyl chloride
(e) *m*-chlorobenzaldehyde *or* 3-chlorobenzaldehyde
(f) *m*-nitrobenzyl bromide *or* 3-nitrobenzyl bromide *or* 3-nitro-1-bromomethylbenzene

Index

https://doi.org/10.1515/9783110778311-022

www.ingramcontent.com/pod-product-compliance
Lightning Source LLC
Chambersburg PA
CBHW080702220326
41598CB00033B/5282